Computer-Integrated Manufacturing Handbook

THE McGRAW-HILL DESIGNING WITH SYSTEMS SERIES
Eric Teicholz *Consulting Editor*

Daryanani *Building Systems Design with Programmable Calculators* (1980)

Stitt *Systems Drafting: Creative Reprographics for Architects and Engineers* (1980)

Stitt *Systems Graphics: Breakthroughs in Drawing Production and Project Management for Architects, Designers, and Engineers* (1984)

Stitt *Designing Buildings That Work: The Architect's Problem Prevention Sourcebook* (1985)

Teicholz *CAD/CAM Handbook* (1985)

Gerlach *Transition to CADD for Architects and Engineers* (1987)

Radford and Stevens *CADD Made Easy* (1987)

Teicholz and Orr *Computer-Integrated Manufacturing Handbook* (1987)

Computer-Integrated Manufacturing Handbook

Eric Teicholz

Graphic Systems, Inc., Cambridge, Massachusetts

Joel N. Orr

Orr Associates, Great Falls, Virginia

McGraw-Hill Book Company

New York St. Louis San Francisco Auckland Bogotá
Hamburg Johannesburg London Madrid Mexico
Milan Montreal New Delhi Panama
Paris São Paulo Singapore
Sydney Tokyo Toronto

Library of Congress Cataloging-in-Publication Data

Teicholz, Eric.
Computer-integrated manufacturing handbook.

(The McGraw-Hill designing with systems series)
Bibliography: p.
1. Computer integrated manufacturing systems.
I. Orr, Joel N. II. Title. III. Series.
TX155.6.T45 1987 670.42'7 86-20825
ISBN 0-07-047774-4

1234567890 DOC/DOC 893210987

ISBN 0-07-047774-4

The editors for this book were Harold B. Crawford and Galen H. Fleck, the designer was Naomi Auerbach, and the production supervisor was Thomas G. Kowalczyk. It was set in Auriga by Techna Type. Printed and bound by R. R. Donnelley & Sons Company.

Contents

Contributors

DR. JOHN CANADA Industrial Extension Service, North Carolina State University, Raleigh, N.C. (*The Economics of CIM*)

ROBERT EDWARDS Industrial Extension Service, North Carolina State University, Raleigh, N.C. (*The Economics of CIM*)

DON EWALDZ Ingersoll Milling Machine Co., Rockford, Ill. (*Getting Started*)

DR. ROBERT E. FULTON School of Mechanical Engineering, Georgia Institute of Technology, Atlanta, Ga. (*An Approach to CAD/CAM Integration*)

GEORGE J. HESS Ingersoll Milling Machine Co., Rockford, Ill. (*Getting Started*)

DR. ALEX HOUTZEEL Weybridge, Surrey, England (*Group Technology*)

LEO ROTH KLEIN President, Manufacturing Control Systems, Inc., Cleveland, Ohio (*The Role of Materials Handling*)

DARRYL V. LANDVATER The Oliver Wight Companies, Essex Junction, Vt. (*Production Planning and Control*)

DR. LEONID LIPCHIN Manager, The Boston Consulting Group, Boston, Mass. (*Planning for a Competitive CIM Environment*)

MALCOLM L. MACFARLANE President, Market Navigators Inc., Cupertino, Calif. (*The Role of Quality in CIM*)

PETER A. MARKS Vice President, Marketing, Automation Technology Products, Campbell, Calif. (*Product Definition: The Role of CAD/CAM in CIM*)

HAL MATHER Hal Mather, Inc., Atlanta, Ga. (*Material Requirements Planning and Inventory Control*)

ANN MEISTER D. Appleton Co., Manhattan Beach, Calif. (*Numerical Control Systems*)

DR. M. EUGENE MERCHANT Metcut Research Assoc., Cincinnati, Ohio (*Future Trends and Developments*)

MICHAEL C. MILLS McDonnell Douglas Helicopter Co., Culver City, Calif. (*Obstacles to Computer-Integrated Manufacturing*)

RICHARD MORLEY Gould Inc., Amherst, N.H. (*CIM and Communications*)

WILLIAM MUIR Price Waterhouse, Chicago, Ill. (*Technology Management and Factory Automation*)

DR. JOEL ORR Orr Associates, Great Falls, Va. (*Process Planning*)

DAVID C. PENNING DCP Associates, Palo Alto, Calif. (*Considerations for Successful Implementation: Controls, Feedback, Benchmarking*)

ERIC TEICHOLZ President, Graphic Systems, Inc., Cambridge, Mass. (*CIM Industry Overview*)

DR. PETER M. WALSH Adaptive Automation, Inc., South Windsor, Conn. (*Robotics*)

Preface

CIM—computer-integrated manufacturing—can be achieved today. There are no technological obstacles to overcome before we can build computer-controlled factories that "run themselves." Automation professionals have little difficulty in designing fully automated facilities.

So why are there not more of them? Because the goal is easier to define than the process for attaining it. All manufacturers have an investment in plant and people, and it is not at all obvious what is to become of them under CIM.

There are many current efforts to refine system analysis methodologies for use in the factory. Before automation can be implemented, an organization must understand what it currently does, what it will do under automation, and the steps leading from one situation to the other.

Ignorance of how manufacturing organizations *really* work is the most troublesome obstacle to CIM. Our problem is that we don't know what we're doing! That is, workers know their jobs and managers know what they're supposed to manage—but the details of individual operations, and their interrelations, are hidden from company strategists by sheer numbers and by the pressure to produce.

On the other hand, although defining a "factory of the future" with current technologies is not difficult, nuts-and-bolts implementation must confront the issue of obsolescence. Since the change process will take time, how can the agents of change set a course that will avoid having to discard the new equipment before it is installed?

Implementation of CIM requires an appreciation of the way things are, a vision of how they are to be, and a clear plan for getting "from here to there" with minimal trauma to the manufacturing organization and its people.

The Gulliver that is CIM must be tied down by many small strings, including technologies, methodologies, and the experiences of others, anchored to the stakes of common sense. An overview, some case studies, and a discussion of obstacles to CIM open this handbook. Part 2 explores the technologies and their relations to one another. Methodologies for planning CIM are discussed in Part 3, and Part 4 deals with implementation issues by examples, narrative, and formulas.

The authors of this handbook are both practitioners and observers of CIM. Their points of view are not identical, but all of them have the authority of experience. There is no single best approach to CIM; you must work out your own path.

If you are contemplating CIM, read this book through from cover to cover. Then use it as a reference during the implementation process. It will give you a framework into which you can integrate CIM information, and it will provide you with inspiration and encouragement.

Eric Teicholz
Joel N. Orr

Computer-Integrated Manufacturing Handbook

PART 1

Introduction

CHAPTER 1

CIM Industry Overview

Eric Teicholz

Joel Orr

1.1 Introduction

"Computer-integrated manufacturing" (CIM) is the term used to describe the complete automation of the factory, with all processes functioning under computer control and only digital information tying them together. In CIM, the need for paper is eliminated, and so also are most human jobs. CIM is the ostensible evolutionary outcome of computer-aided design and drafting and computer-aided manufacturing (CADD/CAM).

Why is CIM desirable? Because it reduces the human component of manufacturing and thereby relieves the process of its most expensive and error-prone ingredient. But CIM is, for the most part, an unrealized dream. The application of computers to the activities that make up the manufacturing process occurred in bottom-up fashion; that is, the potential utility of automation was recognized at the working level of organizations long before it came to the attention of management. CADD/CAM was first applied to numerical control (NC) programming on the production side of the factory and to analysis on the engineering side. Later, it began to be used in detail drafting, and now it is being applied to conceptual design. The result has been "islands of automation" in which individual processes are automated without concern for compatibility with one another. The sought-after productivity has been deferred.

What decisions must management make to implement CIM? Automation in the factory is still very much a process of enhancing islands of automation, and few efforts have been made toward their integration. The situation will persist until management deals with the four greatest obstacles to integrated automation:

- The pressure of the pyramid
- The prerogatives of the priesthood
- The personality of the power tools
- The powerlessness of the person

1.2 The Pressure of the Pyramid

How does a factory work? Figure 1.1 is a simple schematic diagram of the entities and processes. Most CADD vendors and many manufacturing executives believe that this diagram is a reasonable abstraction of the factories with which they come in contact. It depicts well-defined areas of responsibility and authority with simple flows of information, goods, and services. The automation of such a factory would be fast, easy, and interesting—even enjoyable.

However, the diagram has a slight flaw. This minor imperfection does little to decrease the popularity of the viewpoint represented by the diagram, but its elucidation will yield important insights into our first CIM obstacle. The flaw is simply this: The diagram is meaningless because it bears no relation to reality. It is worth as much to an organizational analyst as Fig. 1.2 is worth to a physician.

In reality, the factory is a seething caldron of emotion, perspiration, nobility, foolishness, greed, sincerity, selfishness, idealism, vanity, and generosity. It is a far uglier sight than a beehive or an anthill, and it is far more difficult to comprehend. Its actual operation is almost impossible to diagram because it is shrouded in a fog raised by the heat of human activity.

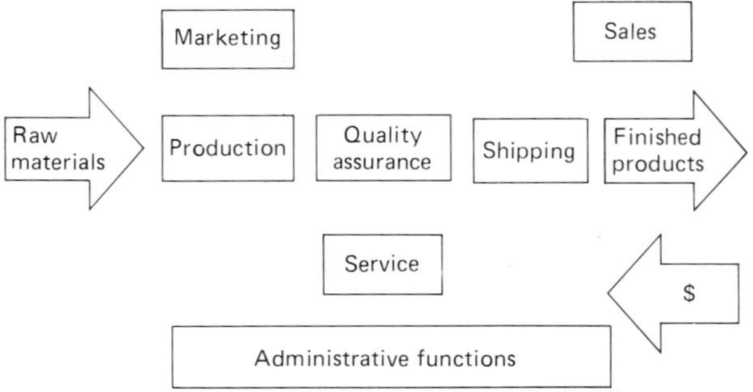

Figure 1.1 Factory functional block diagram.

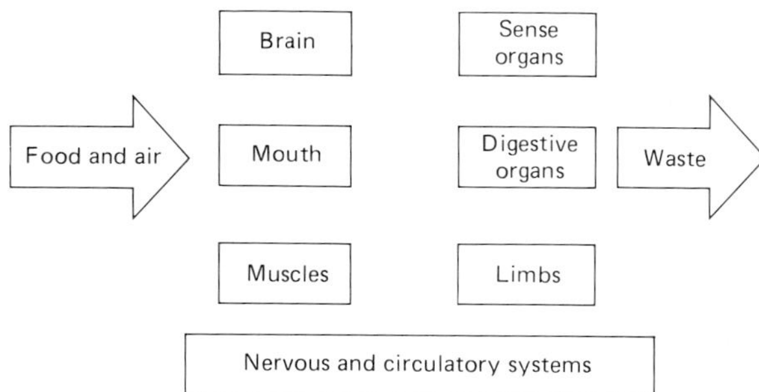

Figure 1.2 Human body functional block diagram.

How the factory got to be that way is easy to understand: It simply grew from its origins in an unplanned way. Each added element—a new machine, new management, a new product line—caused some turmoil while the organization adjusted to it and ultimately became part of a new equilibrium. Additions that did not "take" were eliminated, or they caused the organization to collapse.

Since CIM involves computers, the computer department is often in the vanguard of CIM implementation planning. Although computer professionals like to think of organizational growth as crystalline, with pure geometric accretions accumulating in well-ordered quanta, the propagation of human organizations is usually messy. Organizations are made up of people, each of whom is self-seeking and self-centered. In addition to their assigned responsibilities and tasks, they develop relationships and procedures to protect and further their interests.

The relationships and procedures are not documented; nor are they derived from the organization's explicit or implicit policies. But no modification of the organization can succeed without taking them into account, just as no surgery can be successfully performed without taking into account the complex networks involved in the functioning of the human body.

Almost all organization charts are hierarchical; there is one person at the top and there are many at the bottom of the hierarchy. That is the explicit aspect of the pyramid. The hidden side of it comprises the self-seeking behavior patterns of the individuals.

Natural principles also are at work. There is the principle of *inertia*—things tend to stay the way they are. People don't like change, and they will act to keep things from changing. At each level of the organizational pyramid, equilibrium is maintained. Most organizations start with some consonance between the goals of the organizations and the goals of the individual, but the pyramid always grows into rigidity. Maintaining equilibrium becomes more important than pursuing the goals of the organization. This is the *pressure of the pyramid*.

1.3 The Prerogatives of the Priesthood

Computers intimidate people who have not grown up with them. Initially costing millions of dollars and accessible only to specialists, these machines acquired mystical reputations. The metaphors of the computer as a deity, of the glass-enclosed, environmentally controlled computer room as a temple, and of long-haired social misfit programmers as priests are clichés.

But people best suited to working with computers are those who see situations as collections of black-and-white phenomena, yes-or-no decisions. The best programmers are known to be asocial, and they often are antisocial. When these professionals are called upon to create a model of the factory, the model usually winds up having many of the attributes of computer systems. Human needs and idiosyncrasies are generally left out.

Computers have thoroughly infested most companies, but workers are still mystified and frightened by them. The spread and use of computers are still planned and dictated largely by computer professionals, who have not learned to understand the needs of users. Technical issues overshadow functional considerations in many system design and equipment selection processes; in the final analysis, no one dares dispute the arguments of the technocrats. They are indisputable because of the prerogatives of the priesthood.

1.4 The Personality of the Power Tools

Achieving CIM requires harmonious interaction with computers at many levels. Most factory personnel will need intensive training in dealing with the particular systems they will encounter. Design engineers must use CADD systems, for example, and manufacturing engineers must have to do with both CADD and CAM systems. However, the computer systems in CIM installations will come from a variety of vendors. The ways in which users interact with them will be many and varied, and they will necessarily be inconsistent with one another. This raises a tremendous barrier to CIM. For people to use the systems, the personality of the power tools must be congenial and consistent.

1.5 The Powerlessness of the Person

People need to feel worthwhile. Put them in a situation in which they feel that they have no effect on the organization and they will react with rebellion or depression. In the pyramid of organizational hierarchy, only those near the top have much influence on the direction of the organization.

Organizations so segment tasks that individuals seldom have the opportunity to see anything through to completion. Products and ideas are thus orphaned and left to fend for themselves. At the same time, workers are bereaved of their brainchildren, and they feel sterile and frustrated. As the organization grows rigid with age, people in it feel more and more impotent. Many are impelled to leave. Those who remain in the ossifying structure are concerned with issues of seniority, turf, and pensions—not productivity.

1.6 Plucking Productivity from the Jaws of Organization

"More output with fewer resources" is a common definition of productivity improvement. It is often applied to the declared goals of organizations. But translating it into action steps for individuals is more than an exercise in management by objectives (MBO); it requires the full cooperation and participation of the individual in the translation process. Without cooperation, the goal cannot be achieved; without participation, there can be no cooperation. The individual must have a sense of ownership to adopt the goal.

The four obstacles described above must be overcome if the participation of the individual is to be obtained. First, the pressure of the pyramid must be relaxed or avoided. In *Intrapreneuring*, Gifford Pinchot III discusses the benefits of working within organizations with entrepreneurial techniques. "When any CEO calls for innovation, very little happens. This is not because of a lack of good ideas but because of the difficulties your people have in implementing them. If you are not hearing good ideas, it is because they are blocked or sanitized before they reach you."

Pinchot recommends giving individuals access to the corporate vision so that they can know how to direct their creativity. He then suggests that innovators be rewarded and given encouragement, authority, and resources to develop ideas about which they are passionate. CIM is radically different from what most manufacturers currently do. Without passionate advocates, it cannot become reality. Pyramid rule must be relaxed.

Second, the prerogatives of the priesthood must be abridged. The overriding considerations in the analysis and acquisition of computer-based systems must be functional, not technical. Adherence to computer standards must be considered only within the context of the application, not a management information system (MIS) at large. Computers must be looked upon as means to ends, not as ends unto themselves.

Third, the personality of the power tools—the "faces" systems present to users—must be congenial and consistent. We must reduce the amount of training and education that CIM will necessitate by working on the least expensive components: computers. People are expensive, and training is expensive; software is inexpensive by comparison. Today's CADD systems are largely uncongenial. Their personalities reflect those of their programmers, who have little in common with the system users. They are consequently hard to learn and hard to use.

In *Your Natural Gifts*, Margaret Broadley discusses some of the findings of the Johnson O'Connor Research Center—Human Engineering Laboratory. This organization has found, after several decades of study, that people have genetically determined aptitudes. These aptitudes can be developed, but not learned—and people with unexercised aptitudes are bound to feel frustrated. The cluster of aptitudes exhibited by successful engineers includes high structural visualization, good proportion appraisal, and good musical aptitudes. Surprisingly, however, engineers often have low capacities for inductive reasoning. Thus, they are particularly averse to complicated computing tools, which must be "figured out." The engineer's computing tools must be self-evident.

Since there is no generic engineering design methodology, operating procedures vary greatly from one vendor to another. In fact, operating procedures often vary within the confines of a single system: Solid-modeling modules are different in their operation from other parts of most CADD systems. Computer graphics comes to the rescue in two important ways: It provides a consistent definition of the product that endures throughout the design, production, and delivery process, and it offers a much broader information bandwidth for person-machine communications.

Fourth, we must empower the individual within the organization to take up the corporate vision. We must provide room for intrapreneurs to seek and achieve fulfillment of their personal needs in such a way that they will, in so doing, further the goals of the organization. We will not get to CIM otherwise.

1.7 Technology: Is It Ready for CIM?

In the minds of many manufacturing executives, the implementation of CIM awaits the development of proper technologies. But that is not true. Current CADD/CAM technology is equal to the task of fully automating the factory, but it is rarely applied to an entire manufacturing facility. Management has yet to be convinced of the validity of the automation vendors' productivity improvement promises. Computers are employed only in situations in which the short-term benefits are measurable and are likely to be realized. This has contributed to the isolation of the islands of automation.

Integration is still rare. Even the organizations that use CADD/CAM extensively employ hard-copy drawings. In most companies, departments using CADD/CAM *must* produce hard copy because other departments have no way to deal with digital information. The capabilities of CADD/CAM systems are thus constrained to fit within largely manual operations, and much attention is consequently given to issues, such as plot quality, that are irrelevant to CIM.

1.8 Does CIM Require Solids?

Implementation of CIM requires a product definition that is unique and consistent throughout the design and manufacturing process; it is computer graphics that makes a practical implementation of this dictum possible. At the heart of this definition is the *geometry* of the product—its shape. The world has three spatial dimensions, so models must be three-dimensional.

Three-dimensional CADD models can take three forms:

- Wireframe, which includes only points in space and the lines connecting them. Objects are represented by their edges.
- Surface, in which the edges of the wireframe are spanned by areas.
- Solid, in which the space enclosed by the surfaces also is defined.

Solids are the least abstract and most realistic of the three forms; they necessitate far more computing power for their creation and management. They attract designers because construction of complex models, especially those that lack symmetry, is usually much faster with solids than with surfaces. Wireframe representations of complex objects are very difficult to "understand" visually, because computer displays and plots seldom give an indication of depth. Seeing all the edges at once leads to perceptual confusion because of ambiguities.

A complex object can be decomposed into a much smaller number of solid components (called primitives) than surfaces or points and lines. This is evident when one considers that a solid primitive can be decomposed into surfaces, which can be broken down into points and lines. When there are fewer parts to manipulate, a model is more rapidly constructed.

Solid and surface models allow the generation of hidden-line and hidden-surface images, which are much more realistic than wireframe models. Some shapes can be represented by surfaces but not by solids. Thus, even designers who believe in the essential superiority of solid representations are forced to resort to surfaces for certain complex objects.

There are many commercial solid-modeling systems. However, almost all of them are poorly integrated with their respective CADD and CAM functions. Companies implementing CADD have difficulty seeing the significance of solids, because they expect to enhance existing islands, not develop CIM, in the near future. Solids, in conclusion, make CIM easier, but they are not essential to it.

1.9 Islands Are Growing

An interesting example of an evolving island is numerical control. The leading numerical controls now come from Japan; many of them actually contain powerful CADD systems, thanks to the plummeting costs of computer graphics hardware and software. The idea behind their operation is that the machine operator can simply enter the geometry of the part into the control, much as if it were being entered into a CADD system. In some cases, the initial toolpath is automatically generated by the control; in others, its development is facilitated by the control software.

The concept sounds wonderful, but it has a flaw—at least in the United States: Machine tool operators do not have the authority to make the design decisions required to take advantage of this arrangement. In fact, the growing popularity of CADD systems, with digital connections to the NC equipment, renders such controls superfluous. From the CADD side, NC has been a subject of many promises and poor performance. Few CADD system users who have toolpath generation capability in their systems use it for most of their parts; yet NC-related features often figure prominently in benchmark tests.

The main obstacle to the use of CADD-generated toolpaths is the wall between engineering and manufacturing. Toolpath creation is the right and obligation of manufacturing, not engineering, but seldom is the CADD system shared with manufacturing. The benefit of having the geometry in digital form is generally thrown away, and the usual drawings are passed to manufacturing. These are then used by part programmers to recreate the geometry on their NC programming systems, after which toolpaths are created. The only obstacle to integration of these processes is organizational, not technological.

Organizational obstacles to integration present themselves throughout the factory. It is easy to design a fully automated process, but it is difficult to design a series of steps leading from the present manual systems and islands of automation to CIM in an economical and humane sequence. For this reason, CIM cannot be sold as a packaged product or service; each factory requires entirely different considerations.

1.10 CIM-Related Products

As yet, there are no complete CIM products. The CADD/CAM market contains the following products:

- Mainframe- and mini-based CADD systems. The products that established Computervision, Calma, and others in this market are of this type—design and drafting systems built around large and medium-size computers such as the DEC VAX, the DG MV series, and the IBM 4300. Average price per seat: about $100,000.
- Distributed CADD systems based primarily on the products of Apollo. Average price per seat. about $80,000.

- Microcomputer CADD products, sold principally without hardware, such as Autocad. Average price per seat, including hardware: about $15,000.
- Graphical peripherals, such as digitizers and plotters.
- Analytical software packages, such as those used for finite-element analysis.
- NC programming systems, usually consisting of specialized text editors bundled with high-level languages, such as Compact II and APT, and appropriate output devices.
- Robots and robot-programming systems.
- CAE systems, which combine CADD and analysis software in configurations aimed at the individual engineer. These products are just beginning to emerge, with an average cost per seat of about $40,000.
- Networks and area computers. The computational hierarchy in the automated factory is logically centralized, though physically dispersed. As manufacturers move toward CIM, communication among the plethora of "smart" devices in the factory is becoming recognized as having critical importance. Specialized communication hardware and software to deal with these issues are being marketed. Inseparable from the communication issues are the new "traffic cops": the area and cell computers that serve as switchboard operators for the information flowing among the computer-controlled devices in the factory.
- Manufacturing resources planning (MRP) is not a new technology. It is universally recognized as essential to manufacturing productivity, but it is not yet widely implemented. The fact is, most manufacturing operations are not well systematized; consequently, it is very difficult to automate them. All MRP vendors are forced to undertake helping the customer systematize the flow of information within the factory as a prerequisite to the introduction of MRP; thus, services are sold along with software.
- The comments about MRP systems apply equally well to group technology and computer-aided process planning (GT/CAPP). It is commonly acknowledged that part classification and coding schemes (the heart of GT) increase productivity; although they are simple in principle, their implementation is organizationally complex. Computer-aided process planning leans heavily on GT.

1.11 CIM's Tomorrow

Fully automated factories are unlikely to become epidemic before the late 1990s. The social and organizational obstacles to such a radical change are proving more formidable than anyone expected. Technology is infiltrating engineering and manufacturing and bridges among the islands of automation now seem to be an inevitable intermediate step on the road to total integration.

This is a time of transition, and the successful products will be those that ease the transition. Automated drafting systems are a case in point: They do not seem to be a step toward CIM, but they provide measurable productivity improvement in the automation of a manual process. (Of course, by familiarizing users with computers, they do facilitate further automation. But that is only a side effect, not a measurable benefit.)

1.12 Implementing CIM

Management decision is the only prerequisite to CIM. The prescription for implementing CIM is simple:

1. Use this handbook, and other publications and educational forms, to gain a thorough understanding of CIM and its implications at large. Equip your staff and your management with a similar education.
2. Thoroughly analyze existing operations. Flowchart the explicit, business-oriented flow of information and materials. Describe in detailed narrative form the implicit interpersonal forces that are at work.
3. Design the fully automated operation, with both detailed flowcharts and extensive narratives. Involve workers at all levels in the design process.
4. Consider the description of existing operations as the initial state of a dynamic system and the design of the fully automated operation as the final state. Design a series of intermediate steps that will bring the organization from the initial state to the final state with minimum trauma to the organization and its people.
5. Set up an organizationwide implementation committee, and empower it to act.

This process is not simple. It is difficult and time-consuming. But there is no other.

CHAPTER

2

Case Studies

Malcolm Macfarlane

Joel Orr

2.1 Introduction

Computer-integrated manufacturing (CIM) is a goal, not a destination. A factory cannot purchase CIM off a shelf, or even locate two experts who can agree on what it is. Instead, a factory which has adopted the CIM goal of implementing a cohesive digital information base integrating manufacturing, design, and business functions begins by understanding its own operating processes and the trends and pressures influencing them. A series of coordinated, discrete acquisitions and implementations, chosen on a basis of most effective use of company resources, is the most efficient means of managing toward CIM.

Computer-aided design and drafting and computer-aided manufacturing (CADD/CAM) have raced ahead of the mechanical counterparts because of their well-defined design methodologies, and CIM for electronic products is following suit. In mechanical design, there is no clear boundary between design and drafting. At the same time, there is no simple way to measure design productivity. Because drafting productivity measurement is a simple matter, the purchase of most mechanical CADD/CAM systems is justified on the basis of drafting. The systems are given not to designers, but to drafters; as a result, the needs of designers are not expressed to the system manufacturers, but the needs of drafters are. Systems therefore improve in their drafting capabilities, but not in their design capabilities. But there is a slowly growing awareness of the limits of possible productivity gains derived from the mechanization of drafting. It is *design* that determines the ultimate quality and cost of manufacturing a product.

Most companies considering CADD/CAM focus on shortening the time required for design, and the automation of drafting is a simple, low-risk step toward that goal. In fact, drafting productivity improvement is virtually assured for drawings that undergo any number of changes in the course of their useful life. But drafting time and costs are generally only a small component of the overall manufacturing time and costs of a product. The greatest opportunities for increasing manufacturing productivity, reducing costs, cutting production time, and improving quality are in the design process.

For most products, design is a matter of trial and error. There are usually too many variables to admit a process based on a formula. Instead, initial designs are based on the experience of the designer. Older drawings are often used as a point of departure. When the design is complete, a prototype is constructed and tested; the prototype is then modified and tested again until its performance is, in some sense, satisfactory.

Both the prototype and the tests are made as realistic as possible. Together, building and testing prototypes are called design verification. They are usually the most time-consuming portion of engineering, and they are also the portion most susceptible to improvement by automation. Through computer graphics, CADD/CAM systems make it possible to create a digital prototype and digital testing sequences. For many products, they are faster, less costly, and more accurate than physical prototyping. Here are some examples:

- A major toy manufacturer has significantly reduced the time it takes to bring a new toy from concept into production. Conceptual designers use a CADD system that yields color-shaded images, which are submitted to marketing personnel for go/no go decisions. Previously, expensive and time-consuming mock-ups were constructed.
- Once the concept is approved, the engineering department can use the digital solid model as a starting point, rather than having to begin with sketches and drawings. By working entirely in digital form, the need for manual pattern making for plastic molds is obviated; molds are created directly from the digital database. The entire process now takes less than half the time of the manual process.
- Large airplanes at Boeing Commercial Airplane Company are now assembled entirely without shims, because of the precision of the digital modeling process. Modifications to commercial aircraft now take days rather than months. Changes can be designed, analyzed, and manufactured entirely without drawings; even the quality assurance process, with its coordinate measuring equipment, uses the digital database directly.
- A leading manufacturer of truck trailers won a major contract with a

trailer rental firm in a competition with tight deadlines through the use of CADD. Several designs were modeled in less time than a single design could have been produced manually. The completeness of the sets of drawings greatly impressed the rental company; after the contract was awarded, the same drawings were used for the production of the trailers, thereby demonstrating their validity.

These accounts testify that manufacturing productivity *can* be improved on a piecemeal basis—that it is, in fact, possible to automate portions of the process and integrate them afterwards. Although planning an entire integrated operation from scratch is far more attractive from a theoretical point of view, few companies can seriously entertain such a notion—they simply have too much invested in people, processes, and plant.

The greatest gains from automation appear when integrated information flow throughout the factory is viewed as the key objective, and when quality becomes an ingredient of the product rather than an afterthought. This case study examines a manufacturing operation which is well on its way toward the CIM goal, and pragmatic decisions about its next phase of implementation are being made. It does not center on hardware technology; instead, its operations are compared with corporate goals and its choices, in Pareto principle fashion, are ranged against them.

2.2 Electronics Manufacturing at Xerox

Xerox Systems Group Electronics Division Manufacturing, located in El Segundo, California, is well along its evolutionary path to CIM. During the five years from 1980 through 1985, Xerox manufacturing management gradually transformed its operations from disconnected activity areas to integrated workcenters coordinated through a cohesive base of data. By judging each supplier of CIM components on its own merits, Xerox has made a series of evolutionary implementation choices in its design, planning, manufacturing, and field support systems.*

An effective way to implement CIM is to choose one view of plant information and drive all CIM components toward that view. For example, a common view chosen is product design. Product characteristics form the fundamental data element, and other information sets are concentrated on contributing to that view. Because of the increasing complexity of the products to be built and the quality demands on those products, Electronics Division Manufacturing of Xerox chose the manufacturing process as its fundamental focus of information.

*The facts in this case study came from discussions held with Jorge L. Valledares, manager, plant operations, and Joseph M. Camarata, manager, manufacturing information systems.

The information view to which the other parts of the operation contribute is contained in a system called Equinox, an acronym for Electronic QUality INtegrated On-line for Xerox. Developed by the manufacturing people themselves, Equinox is the focal point for design, engineering, planning, and field support systems. Equinox provides routing control, data acquisition, and history data across the life cycle of all parts assembled in the factory. Design, revision level, components, assembly and test history, and field and repair history are all currently maintained in the system. Before describing the evolution of Equinox at Xerox, the plant is described as it appeared during the first half of 1985, in the middle of its CIM evolution.

2.3 Plant Description

The plant supplies Xerox operations worldwide with most of its assembled circuit boards for mechanical integration into printing and computer systems. Over 5000 boards per day are manufactured at this facility; they are drawn from over 300 types of boards current at any time. The overriding strategic demands on the plant are quality, cost, volume, and flexibility.

2.3.1 Incoming inspection

Inspection varies by part and depends on statistical quality records. Some parts, particularly the bare circuit boards, are not inspected, because the plant monitors the quality records of its suppliers in an aggressive supplier quality control program. Statistical histories are maintained for integrated circuits and other components. As a component develops an error-free history, the samples inspected become fewer and fewer until shipments become "skip lots" and are not inspected at all. If problems are encountered in skip lot components, 100 percent inspection is performed on the next two incoming lots. If no problems are found, the component regains its skip lot status. Otherwise, it must again earn its status through the established sampling process.

2.3.2 Raw materials inventory

Following incoming inspection, parts are put into automated inventory devices—stacker cranes with bins and carousels. The part numbers, counts, and locations are entered and maintained in the corporate inventory database, currently the Xerox computer services inventory software. Cycle count procedures periodically verify the accuracy of the inventories.

2.3.3 Kitting

The master schedule drives the kitting operation. In the marshaling area components converge into precise amounts required for the assembly run.

Once assembled, these binned kits are inspected for completeness—it being much cheaper to check for correctness than to start a run with an incomplete kit.

Following binning, the kit is sent to the preparation stations, where it is split by the technique used to assemble it. Components to be inserted by automatic-insertion equipment are prepared, inspected, and put into carts required by the equipment or into tubes or magazines. Boards are stamped with their unique part numbers. Bar-coded serial number labels are attached and input to the tracking system used to follow each board through its manufacturing and operational lifetime. Appropriate preparation is performed on any remaining hand-inserted components.

Bar codes play an important part in the implementation of the system. They ensure integrity of the test path, and they allow the association of the process with the quality assurance (QA) data on the product. Because they allow identification of each board manufactured with a unique number, the life cycle of each board can be tracked and preserved.

2.3.4 The assembly process

Of the 5000 boards manufactured every day, 20 percent are assembled, soldered, and tested on a single continuous assembly process (CAP) line built by Panasonic. This line is composed of a series of numerically controlled automatic-insertion devices connected by conveyor, each of which performs a specific insertion or inspection activity. Inspection and test results are entered into the database via the bar-code readers or direct electronic connection with the testers. The rest are manufactured on a combination of another CAP line, a surface-mounted technology (SMT) line, and stand-alone insertion and wave-soldering equipment. The company goal is to perform most of the assembly on the CAP and SMT equipment and thereby eliminating the largest remaining source of random product defects: those caused by human assembly errors. Following the initial installation of the first CAP line, rework in the plant was reduced by 55 percent.

The SMT line, manufactured by Zevatech AG, a Swiss company, provides the plant with the capability to manufacture boards populated with more than 1300 parts, 700 of which are surface-mounted devices (SMDs) on the back side of the board. The line is similar to the CAP line; again it consists of a series of numerically controlled machines connected by conveyor, each of which performs a different part of the SMT process. The continuous SMT line applies glue, places the components, and then cures the glue. After the glue is cured, the board is ready to be soldered. Rework is performed before the glue is cured, so there is inspection following the component placement operation. Measurements are made and records are kept on the accuracy of device placement. When the accuracy is below established limits, the machine is brought down and recalibrated.

Components which cannot be placed on boards by CAP or component-insertion equipment because of their odd shape are stuffed into the boards by hand. This human process produces the highest rate of random manufacturing error in the plant. Robots are being researched to replace humans in this activity to reduce the variability of quality of this process step.

At each separate step in the CAP, SMT, or human assembly process, where key data elements or validation control is desired, the bar code on the board is read. The assembly history is maintained once the board is complete for tracking into system assembly and field performance. Each of the 5000 boards is read 10 to 15 times during the assembly and test process. At the end of both CAP lines, boards are inspected by humans. A vision system which will perform this operation is in its experimental stages. The wave soldering machine on CAP line 1 is built by Sensbey of Japan. The measurements made within the machine are self-contained, and the history is local. The wave soldering machine of the second CAP line is fully computer-controlled. It was built by Electrovert, USA.

At the end of the CAP lines, following wave soldering inspection, an Intellidex robot inserts finished boards into a functional inspection device run by a Xerox 820 computer. Vacuum positions a board into the edge connector and bed-of-nails combination tester, and the trials are run. The robot moves the board to the pass and fail bins based on the results of the test. If the failure rate becomes large, the line is stopped and trouble shooting is done in real time, rather than wait until the end of the day. Boards not completed on the CAP lines are tested by using conventional testers. Following testing, passed boards are moved to finished goods inventory to be shipped. Failed boards are moved to rework and then inserted back into the process at the appropriate point. Boards which pass through rework three times are scrapped.

2.4 The Evolution of CIM at Electronics Division Manufacturing

During the 1970s, Xerox was beset with competitive pressures both foreign and domestic. While the Japanese were attacking the small-copier market, IBM and Kodak were threatening the high end. The cost and quality pressures resulted in a major restructuring of operations around small business units (SBUs) in 1980. These SBUs focused on electronics. Each had a free hand to determine how it built systems, whether from in-house supplied or off-the-shelf parts.

In the 1970s, Xerox had a central plant in Pomona, California, which manufactured the bare printed circuit boards for the entire company. The next logical step was to implement a central manufacturing location for the assembled boards consumed by the company. The existing assembly plant in El Segundo was used to manufacture boards for the electronic printing division. Its role was expanded to be the in-house manufacturer of all printed circuit board subassemblies for Xerox electronics products. Forced to compete with outside electronics suppliers, Electronics Division Manufacturing had to develop a strategy to become the low-cost, high-quality supplier of assembled boards to the Xerox SBUs.

2.4.1 The evolution of process automation

The 1980s are a challenging period in the electronics manufacturing industry. Process difficulty is rising exponentially, as is the quality requirement. In Electronics Division Manufacturing in 1981 and 1982, the number of parts per board was between 100 and 200; by 1985 it was approaching 1000. At the same time, the quality target in boards had grown from the industry standard of 1 to 5 bad boards in 100 to 1 in 1000. Because the plant had to serve a large number of internal clients, the number of current designs which could be required to be manufactured at any time exceeded 300. The problems caused by the combination of those trends had only one solution: extensive automation of the assembly process for reliability, repeatability, and flexibility.

In 1982, two major strategic moves were begun. Xerox began to consolidate its buying power by implementing multinational procurement; it negotiated favorable terms with its suppliers for major volume contracts and influenced supplier quality through an aggressive supplier quality control program. At the same time, it undertook the reduction of direct and indirect labor contents of its manufactured products, both of which were higher in proportion to the cost of the product than those of the competition. Xerox decided to address both problems through phased implementation of increased automation and the use of computers in the manufacturing process.

Major automation projects implemented since 1982 in the electronic manufacturing division have included the following:

- Continuous assembly processing (CAP) line 1 installed in December 1982; fully operational in April 1983.
- Equinox project initiated in February, 1983; fully operational in 1984.
- CAP-2 installed in 1984.
- In-line robot test cell integrated with CAP-1 in 1984.
- SMT line installed in 1984; fully operational in 1985.

In 1982, assembly was carried out with a series of independent universal insertion machines. After it was installed, CAP-1 produced 1500 boards per day in a continuous process, including insertion, inspection, wave soldering, and testing. Although some boards still cannot be assembled on CAP-1, the number which can be is increasing. As the board mix moves to all CAP-compatible and the SMT line becomes fully functional, the older individual-insertion devices will be phased out.

At Electronics Division Manufacturing, manufacturing engineering follows boards through the design process to help ensure that they can be built. In 1982, it was decided to make all boards CAP-compatible, and a joint manufacturing and engineering group was created to formulate design rules for board dimensions, components, spacing, and so on, to ensure that the boards were both CAP-compatible and backwards compatible to the universal insertion machines as well.

By 1983, the need became apparent for higher component density per board than present manufacturing techniques could support. With this pressure from engineering, management decided to implement a surface mount device capability in manufacturing to more than double the number of devices that could be placed on a single circuit board. As before, design worked with manufacturing engineering on design rules to ensure that boards would be SMT-compatible.

Information obtained from testing is a key ingredient to overall process control. Past testing was a stand-alone process, not integrated into the manufacturing process itself. With boards growing an order of magnitude in complexity, the strategy of testing became increasingly complex. Acquisition of the appropriate test became a major problem, because test development cost can be several times more than the design cost of the board. The final issue is that boards must work in their system environment, when integrated with other boards built at this and other plants, and still maintain the quality goal of 1 in 1000.

There are four basic types of inspection and tests that can be implemented: *physical*, to ensure all parts are mounted and soldered properly; *component*, to ensure components of the board function as expected; *functional or board level*, to ensure the entire board works; and *system*, to ensure the board works as part of a system. Each assembly inspection and test strategy is determined individually and may include any or all of the four basic types. The goal of the test and inspection process is to push fault identification and correction back up the manufacturing process to where the fault was created.

Working closely together, board designers and test engineers designed a series of tests to be integrated into the manufacturing information system. The increasing amount of joint activity of design and manufacturing engineering on issues of manufacturability and testability is blurring the boundary between engineering and manufacturing and making them parts of the same process.

With the increasingly close working relationships developing among departments brought about by the requirement to cooperate on the integrated manufacturing system, Xerox is implementing Taguchi's most recent theories of manufacturing. It arrived at them independently through first principles of manufacturing in attempting to achieve goals under the shifting constraints of the market.

2.4.2 Information automation

As automation of the manufacturing process began to be implemented, management realized it needed a proactive decision support tool to understand and control quality in its increasingly complex operations. In 1982, a task force studying board quality focused on the failure of shipped product in the field. The results were that although 40 percent of field failures could be attributed to inadequate testing, the majority were associated with environmental or procedural problems in the manufacturing process.

The Equinox project was initiated to address problems in the manufacturing process. The system was designed to address process deficiencies in

- Process control
- Material flow
- Product traceability
- Process performance
- Process error correction
- Factory communication

As Equinox evolved, the information it contained was found equally useful in operations, materials, test engineering, and finance. Thus a system designed for quality is becoming the foundation for CIM. The information system was designed in 1983 and put into initial operation by the end of that year. In 1984 the shop floor database was implemented and verified, to ensure the system would accurately model the factory. The next 6 months was spent installing the system of workstations, bar-code readers, and online testers. By the end of 1984, Equinox was fully functional.

The system breaks the manufacturing process into a series of workstations through which boards are checked in and worked on and important data elements of that work are collected. Another way of looking at the system is as a series of workstation nodes in which each node performs both control and informational functions. Informationally, the nodes may be either sources or sinks. The system requires preestablished routings and criteria for each board. Depending on the results of a test or special problems with it, a particular board may be rerouted by the smart routing features of Equinox.

A workstation transaction consists of two major parts: *validation* and *on-line data logging*. The board is scanned and its position is verified; that is, its history is checked to determine if it has followed the correct sequence of manufacturing operations described in the process plan to get to this workstation. If a board is found to be out of place, the operation will not start. Once permission to work has been generated from the verification, the process is performed and the second part of the transaction, *online data logging*, takes place.

Over 300,000 transactions, with an average of five data elements each, are processed per month. An important feature of the system is that the data collection and informational structure extends to Xerox's (internal) customer sites. That is, remote U.S. sites collect data on boards after the boards have been shipped to machine assembly operations. (This process will eventually extend to the field.) This allows Electronics Division Manufacturing to track board quality to machine assembly and ultimately to the field, thus making those locations an extension of the board-testing system.

Equinox is extensible. As the process and products become more complex, problem detection and removal must be pushed back up the manufacturing process as far as possible through additional tests and inspection. Each year, additional testers are tied to the system. With each new tester, a clearer picture of the process emerges.

To be effective, data collected through the manufacturing process must be reduced to highly usable form. To accomplish this, Xerox has defined the term "informational statements" to identify the highly focused directives which the system provides as action items to management. To convert raw data to these concise action items, Equinox performs three successive levels of analysis, and each level yields a more refined report for management. The first level is to simply sort and summarize the data elements; the results are large, standardized reports of the type most manufacturers use in an attempt to understand their plants. Although these reports are useful for understanding the details of a problem, they are difficult to use to identify and isolate the problem.

At the second level the results of the data reduction are compared with a model of an ideal factory which could achieve the ambitious goals of quality and productivity Electronics Division Manufacturing sets for itself. Performance deviations from the ideal model result in a series of concise statements about the performance of the factory against the model within the sampling period.

Third-level processing extends the single time window comparison with the ideal factory model to multiple time periods, which allows management to view trends in the process and determine if previous actions had favorable or unfavorable results. This level of analysis allows easy implementation of statistical process control techniques. Here are some of the features of the "ideal factory" model:

- The model does not consider nonrandom defects.
- Products undergo only one repair cycle.
- Tester and inspection yield increases the further into the manufacturing cycle the test or inspection is made.
- Everything in the manufacturing process adds value to the product rather than cost.
- Defects are fixed at their source.

By evaluating the data continuously against the model, Equinox can immediately identify problems or opportunities that could not otherwise be perceived. Periodically, management reviews the action item list, usually no more than three pages, determines priorities, and assigns the priorities to individuals. Weakening processes are immediately exposed, and corrective action can be taken.

The Equinox information is active and available worldwide. The history of every board is available, and customers will interrogate and append information through the life of the board. This allows the electronic manufacturing division's customers to avoid surprises—to actually review the manufacturing and test results, as well as field histories of new designs, and anticipate problems early in the process.

No matter how well tested they are—full systems testing of every board in every possible physical configuration is prohibitively expensive—some boards simply don't work in the field. The system monitors the movement of each board from field to repair to field; if any board circulates three times, it is scrapped. In one case, when the system was initially implemented, 100 boards were found in spares inventory which had failed in the field and yet passed every test available to the refurbishing center. The boards were discarded rather than allowing them back into the field. The 100 bad boards would have required 100,000 good boards to be produced to meet quality requirements of the factory; this would have been a terrible burden—particularly in comparison with the simplicity of the remedy.

Some board defects appear only over time. As the boards pass through refurbishment, their repair histories are maintained. Eventually, the sample becomes large enough for meaningful analysis and provides useful information for improving product quality over time.

Designed as a quality information and control system, Equinox does not address factory scheduling, although it does contain much of the required information. Manufacturing scheduling was implemented in 1978 and 1979 by using the Xerox computer services (XCS) form of material requirements planning (MRP-1). In the early 1980s, this system was tailored to the needs of the plant for materials control and centralized purchasing. The modules for master scheduling, capacity planning, and shop floor control tie together; they form portions of a manufacturing resources planning (MRP-2) system tailored to the needs of the facility. Full-scale implementation of an MRP-2 system began in mid-1985.

2.4.3 Computer systems installed

Xerox has IBM machines at XCS which perform the planning and scheduling applications, and it has VAX's in engineering and manufacturing. Sigmas, the legacy of the old Xerox computer mainframe business, are being phased out. A number of 8010s and 820s are used, particularly in design and test, and some Computervision systems are still being used. Networking through portions of the integrated system is via Ethernet.

Future plans are to link the IBM and DEC machines, to upgrade both systems to provide detailed lot tracking, production planning, quality, and a series of other functions. The problem, of course, is the different environments. Where not currently practical, the systems will not be linked, but the architecture is designed to accommodate such linkages in the future.

Electronics Division Manufacturing plans to continue to integrate its information systems through multiple networks. For example, the component layout should come directly from the CADD system to the plant floor, as should the bill of materials for each board. The test procedures and tolerances should come from the computer-aided engineering (CAE) systems where they were developed. Over time, additional integration will take place, and it will result in more effective use of manufacturing resources in the pursuit of quality and productivity.

2.5 Summary

The dream of factories with their operations fully integrated through the use of computers is still alive. Its realization cannot be cataclysmic; rather, it must evolve at the rate that managers and workers can tolerate. The key to acceleration of the process is not technology; that particular cornucopia already offers help at a rate greater than we can cope with. We must focus on adopting useful paradigms, just as Xerox chose its own manufacturing process as its fundamental view of information. These metaphors of information structure give us a framework on which we can build our transition plans. Without them, we can easily head in the wrong direction.

CHAPTER

3

Obstacles to Computer-Integrated Manufacturing

Michael C. Mills

3.1 Introduction

The years since 1960 have seen a long list of manufacturing application products identified as key solutions to the problems existing in the discrete manufacturing industry. The list includes such notables as numerical control (NC), distributed numerical control (DNC), computerized numerical control (CNC), material requirements planning (MRP I), manufacturing resource planning (MRP II), computer-aided design and computer-aided manufacturing (CAD/CAM), robotics, and so on. Computer-integrated manufacturing (CIM) has recently surfaced as yet another key technology to achieve what the above technologies have apparently failed to accomplish. CIM is currently being viewed as everything from a philosophy to a turnkey system which can be purchased off the shelf.

Why have the numerous existing manufacturing application products failed to resolve the problems experienced in the U.S. manufacturing community? Conditions and obstacles that exist in the manufacturing environment must offset much of the benefit derived from those application products. The same conditions and obstacles will, more than likely, have much the same effect on CIM.

Before obstacles to CIM can be adequately identified and addressed, it is important to derive an acceptable definition of CIM. An understanding of factory automation is a necessary prerequisite to defining CIM. For the purpose of identifying and examining CIM obstacles, the following definitions are presented:

- An automated factory or facility is an environment in which a productive balance is achieved through the integration of human resources and automation.
- CIM is defined as the use of database and communication technologies to integrate the design, manufacturing, and business functions that comprise the automated segment of the facility.

These definitions suggest that there is not a generic CIM specification to be used or a standard CIM system to be purchased. Each individual

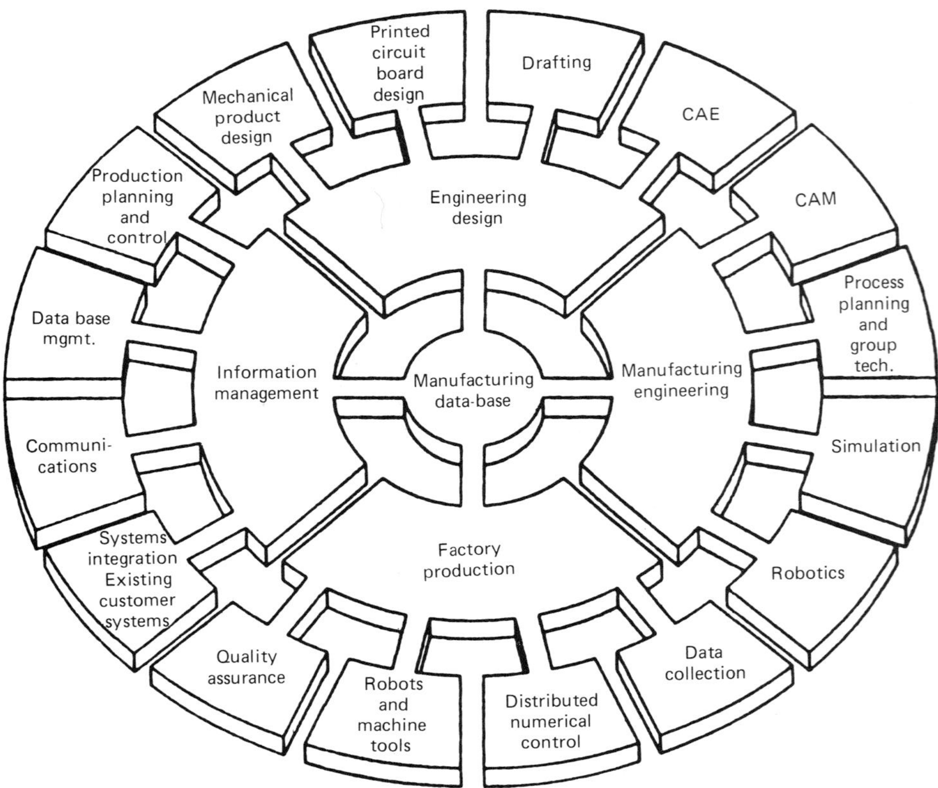

Figure 3.1 Generic CIM architecture.

company will require a different point of balance between the functions which humans perform more efficiently and those which automation performs more efficiently through effective software solutions (Fig. 3.1). The individual balance achieved will reflect the sets of business policies, rules, and methodologies that will maximize productivity and are unique to each company. This viewpoint further suggests that CIM is far more a strategic issue than a technical issue and is specific to each company.

This architecture reflects applications which reside in major systems and are integrated through local area communication networks and a common manufacturing database. At the risk of creating yet another acronym, perhaps computer integrated facility (CIF) would be a more descriptive term to define the automated environment.

An overview of the current situation in both the discrete manufacturing industry and the process manufacturing industry will provide the necessary foundation for the identification and discussion of the conditions and obstacles which will have a dramatic impact on the future implementation of CIM technology. Emphasis is placed on the discrete manufacturing industry, because the process of implementing automation in that environment is far more difficult owing to the inherent flexibility of manufacturing operations. Many of the CIM obstacles apply equally to the process manufacturing industry, however.

3.2 The Current State of the Discrete Manufacturing Environment

The current state of automation in the discrete manufacturing industry spans the range of companies having no implemented automation to companies having a high degree of implemented automation. This discussion of CIM obstacles is targeted at the average automated company profile, but certainly it applies throughout the entire industry spectrum.

3.2.1 Foreign competition

On the whole, U.S. manufacturing is beset with serious problems. It is losing its capability to compete in a worldwide marketplace. Foreign countries can produce higher-quality goods at lower prices. This does not mean foreign competition enjoys a technological edge over the United States. Most of the high technology innovations originated and still reside here. The computer-aided design and computer-aided manufacturing (CAD/CAM) market, for example, grew at an estimated 35 percent average annual rate over 5 years to an estimated 1984 revenue level of $2.4 billion. That figure includes both hardware and software revenues. What has limited the potential productivity gains in the U.S. manufacturing sector is an inability to integrate this high technology.

3.2.2 Partitioned technology

The partitioned state of automation currently existing in industry has been heavily influenced by the following methodology:

> When a problem assumes unmanageable proportions, throw a computer at it and hope it goes away . . . quickly!

This approach has led to many installed application products that do not communicate with each other or, at best, communicate through expensive custom interfaces. Many databases coexist within the same organization, each with a different file structure and containing data with mixed formats. Our free enterprise system does not foster technological standardization which would facilitate the integration of hardware and software products purchased from a wide selection of suppliers.

3.2.3 Condition-driven economy

The discrete manufacturing industry, not unlike most industries today, is undergoing a dramatic transformation. The ambition-driven economy of the 1960s and 1970s is giving way to a condition-driven economy which more than likely will continue well into the twenty-first century. Business conditions are heavily influenced by the state of the economy and by the cost of capital. The recession of the early 1980s saw wholesale personnel layoffs. By increasing the productivity of the remaining workforce and by implementing automation, management observed that it could remain in business, stay competitive, and generate an acceptable level of earnings. As the economy improved, much attention has been directed toward high technology and automated solutions. The U.S. automobile industry is an excellent example. The rapidly emerging robot and robotics industries were born out of the crisis situation in which the auto industry found itself during the late 1970s and early 1980s. The ability to react quickly to a change in the state of the economy is directly dependent on the amount of integrated automation a company has installed.

3.2.4 Current financial emphasis

The initial reaction to changing economics was very survival-oriented, and a major emphasis was directed toward short-term financial performance. That shifted the focus from long-term technical planning and seriously impacted the required long-term investments in high technology required by CIM. Management must realize that, although there is no quick return on investment in integrated automation, the acquisition of CIM technology will not be possible without the requisite long-term financial investments.

3.2.5 Organizational cognizance

Organizations are traditionally structured according to the similarity of functions. Human management theories dictate the relations between the functions and the flow of information through the resulting "systems." Managements view their respective organizations through their formal organization charts. During the past 25 years, these organizational infrastructures have undergone tremendous change. Their form has remained relatively constant, but the way they basically function—the way they handle their data—has changed. Through this period of change, managements have in some cases lost touch with how their respective organizations actually function. They attempt to effect changes in the formal structure while the organization operates according to an informal structure.

3.2.6 Management education

The state of current automation implementation highlights a major problem facing U.S. industry: management education! For the most part, the average company is run by management that has a limited knowledge and understanding of manufacturing automation. In many cases, current management is looking to the supplier community, industry consultants, industry seminars, and the field of journalism to provide the required education and knowledge to select, justify, implement, integrate, and manage automation intelligently. In order to include CIM in a realistic 5-year strategic planning effort, management education in CIM technology is a key requirement.

3.2.7 Manufacturing perception

Computer scientists have spent a great deal of time analyzing and evaluating the classical engineering environment. The result has been the development of many productivity tools such as CAD/CAM and computer-aided engineering (CAE) systems. The success of this effort can be seen in the size of the CAD/CAM market and the number of suppliers to that market. The manufacturing environment, on the other hand, is often considered to be a black hole. Manufacturing has traditionally been empirically driven and at times approached far more as an art than as a science. With few exceptions, computer scientists have stayed away from the area. It has been perceived as a very confusing and uncomfortable environment—one in which the less time spent the better. It is difficult to locate individuals who have a strong data processing background, systems analysis and design experience, and an in-depth working knowledge of the manufacturing process.

3.2.8 CIM information sources

A review of CIM and factory automation literature reveals a wide diversity of fact and opinion. It soon becomes evident that, although at times this source provides valuable information, caution should be exercised in relying too heavily on this medium for input to the management decision process.

3.2.9 Industry consultants

The same situation exists with the use of industry consultants. There are few manufacturing consultants who have the appropriate technical background and are up to date on the latest CIM technologies such as manufacturing-based expert systems, relational database management systems, and communications. There are even fewer who have actually been involved in the specification, design, and implementation of a CIM system. While this new breed of consultant is certainly in the process of evolution, care should be exercised in the screening and selection of "CIM consultants."

Obstacles to CIM will be approached as strategic issues and technical issues as they apply to the following areas:

1. The discrete manufacturing industry level
2. The manufacturing company level
3. The CIM system user level
4. The CIM system vendor level
5. The university and the federal government level

3.3 The Discrete Manufacturing Industry Level

3.3.1 Shared research

Our free enterprise system and its implied competitiveness have negatively impacted technology sharing among most organizations. The majority of the applied research is conducted by large corporations and is considered to be proprietary. Sponsored research, conducted by independent industry organizations, is primarily funded by the same large corporations. The direction and final deliverables of these research projects tend to be self-serving to the sponsors.

3.3.2 Standards

Free enterprise has been responsible for the lack of industry-accepted standards. This can be seen in the lack of intercommunication capability in much of today's installed computer hardware and software. The computer industry on the whole has had a great deal of difficulty in self-regulation.

It has not produced integration standards that are readily accepted by industry suppliers in a timely manner. By the time a standard has been finally agreed upon by industry, the marketplace is saturated with hardware and software products that will not communicate with each other.

Large corporations have a vested interest in influencing proposed standards. If a new standard takes a direction that is favorable to their existing product lines, a great deal of product development expense can be saved. With the rapid advances in technology and the increase in the number of new high technology start-up companies, standards seem to fall further behind. Acceptable standards can be set up and integrated, but to do so will require an industrywide commitment. Without a common communication language, our industry will be about as successful as the Tower of Babel.

3.3.3 Technical relations between large and small corporations

As the large corporation becomes more technically proficient and increases the level of installed automation, a technology gap is created between the corporation and its base of smaller suppliers. It is pointless to install a just-in-time inventory system if the suppliers cannot deliver the inventory at the specified time and in the specified volume. Productivity suffers when the same large corporation, which has designed a product on a CAD/CAM system, is dealing with a supplier that either does not have an installed CAD/CAM capability or has a CAD/CAM system from a different vendor. Part files created on the corporation's system cannot be used on the supplier's system without an expensive interface software package. This interface may be made obsolete by the next system software release from either CAD/CAM system vendor. Data format standards such as the Intermediate Graphic Exchange Standard (IGES) will help alleviate this communication problem, but IGES still cannot pass all the geometric and non-geometric data elements required to transfer a machine-readable part file completely.

3.4 The Manufacturing Company Level

3.4.1 Five-year strategic technical planning

Although a great deal of effort is currently being expended on strategic planning, it does not always include CIM. For strategic planning to be effective, solutions to the many problems existing in the manufacturing environment must be found. For strategic planning results to be realistic and attainable, the planners must have first-hand knowledge of the manufacturing process. Many of the cost reduction goals targeted by strategic planning can be realized through the introduction of CIM technology into the manufacturing environment.

3.4.2 Top management's role in and commitment to CIM

Since automating a facility crosses traditional organizational boundaries, top management must be involved. Organization structures have to be evaluated as to their effectiveness in an automated environment. It might be more effective to structure an organization around a well-designed system than to attempt to integrate a potentially productive system with an unproductive organizational structure. Large amounts of capital have to be budgeted and approved over multiyear time frames. Top management must determine the future policies of the organization, which in turn will drive the CIM specification. Top management must also determine how it wants to monitor the new system and what information it will require from the system. It is imperative that it make known throughout the organization that it wants the new system implemented and will support the effort through to completion. Efforts on the part of the appropriate personnel to facilitate the successful implementation of CIM technology should be noticed and rewarded.

3.4.3 Capital asset justification methodologies

In the past, justifying capital assets has been strictly a formula-driven procedure. Certain tangible attributes were decided upon, an internal rate of return on investment level was established, investment recovery timeframes were set, investment monitoring programs were developed, and accounting procedures were established. The validity of the selected tangible attributes was directly proportional to the understanding of the informal organizational structure and the impact the attributes would have on other organizational functions. To justify CIM technology successfully, additional attributes—both tangible and intangible—will have to be included. A long-term technical implementation plan should not require the investment in CIM technology to be recovered in the traditional 2- to 3-year time frames. Investment monitoring techniques and programs will be a key not only to providing asset recovery assurance but also to constantly monitoring the performance of the asset implementation and management plans.

3.4.4 Existing hardware and software review

A large obstacle to CIM is the current investment in installed hardware and software systems. The monetary value and the amount of online data invested will have to be closely evaluated. Existing systems will have to be analyzed for their consistency with the 5-year CIM plan. A replacement strategy will have to be measured against required interfaces, price/performance ratios, compatibility with new CIM application products, and so on. It might be more cost-effective over the long term to consider replacement rather than attempt to interface a hybrid system or application with a structured CIM architecture.

3.4.5 Manufacturing data

A CIM system will be as productive as the quality and quantity of online data it has to work with. Much of the data required to drive a CIM system is still in the traditional manufacturing database—the file cabinet! Efforts will have to commence in the areas of key data element identification and the loading of this data into a common manufacturing database. This data will become the foundation for the manufacturing expert system. By applying group technology techniques to this data, part and process relations can be established. They will drive the efforts toward maximizing the utilization of manufacturing resources. When one considers the dynamics of a CIM system, one soon realizes that consistency of rules and procedures will be a key ingredient in eventual success. Artificial-intelligence-based manufacturing expert systems will provide this consistency because its output is determined by the rules that are loaded into the system.

3.4.6 Industrial labor unions

A realistic view will have to be taken by labor union executives toward CIM and its relationship to the survival of the U.S. manufacturing industry as we currently know it. Although a CIM system does not take human resources out of the manufacturing equation, it will certainly redefine many job requirements. It is important to realize that CIM systems will not be implemented overnight; they will be installed in stages over a planned implementation schedule. Many people will be required to support and maintain the systems. There will be ample time to develop and execute the necessary employee retraining programs required by CIM technology. A turnaround of the U.S. manufacturing industry will be expedited by enthusiastic planning for change rather than by fighting the technology.

3.5 The CIM System User Level

3.5.1 CIM acceptance

The success of an automated system is very dependent upon its users. CIM systems are going to require interaction with enthusiastic and positive employees if they are to be productive. The introduction of CIM into an organization should be welcomed as an opportunity for personal advancement. Many new jobs will be created through the implementation of the technology, and many of the more mundane and hazardous jobs will be eliminated.

3.5.2 Education and training

Education is not just an obstacle at the user's level. It must be addressed by the university environment, industry service organizations, system suppliers, and the individual companies that are implementing CIM. Although a great deal of effort will have to be put into the development of user-friendly system front ends, it is also mandatory that users do not expect a complex system to run by itself. It is just as important to know why a system is installed to handle specific organizational functions, where the system receives its input data from, and where its output data goes as it is to know which buttons to push in order to turn the system on. Potential CIM users will have to take advantage of all the education and training opportunities they have at their disposal.

3.6 The CIM System Vendor's Level

3.6.1 Experience and education

CIM has found a very congenial home with the CAD/CAM vendors. In the space of 6 months, most of the vendors were advertising their current product lines as CIM systems. There is a big difference between CIM as a marketing and/or an advertising technique and CIM as a deliverable technology. Very few of the announced CIM vendors have enough experience and expertise to be truly classified as bona fide suppliers of CIM technology. It is one thing to develop application software and another thing entirely to assist the user in successfully implementing the software in its intended environment.

System integration and consulting skills are going to be required in assisting the potential CIM customer in planning, selecting, justifying, implementing, and managing a CIM system. For the most part this capability is not to be found in the average CAD/CAM vendor. Supporting CIM technology is a little like having to eat your own cooking when your wife is away—it can be a very humbling experience. There are very few CAD/CAM vendors who have evolved from a manufacturing environment and have a working understanding of that environment and the problems and complexities it represents.

3.6.2 Resource availability

To be a qualified CIM vendor, a tremendous amount of research and product development needs to be accomplished. Manufacturing expert systems must be developed as a forerunner to artificial-intelligence-based CIM technology, and technology must be developed to load these expert systems with the appropriate manufacturing decision rules. Database schemes must be developed to define just what a "common database" is, what it actually contains, and what are its relations with the variety of distributed databases that are certainly not going to go away. Professionals with different expertise are going to have to be added to existing staffs. The commitment to be a qualified CIM supplier is going to take a considerable amount of capital. Realistically, there are few companies that are large enough and that have the resources necessary to take on this commitment.

3.6.3 Existing products

A CIM vendor is going to require product development capacities similar to the following:

1. The ability to develop CIM application software that will run under standard operating systems on standard hardware.
2. The ability to integrate near-term advances in communications and networking technologies into current product lines.
3. The ability to adapt current software applications to a distributed, intelligent stand-alone workstation environment.
4. The ability to implement new structured approaches toward software development in order to assure application modularity.
5. The ability to assure current and future application software compatibility with relational database management systems.
6. The ability to integrate artificial-intelligence-based manufacturing expert system technology with current application software.
7. The ability to make different systems share common data sets. Users must have the capability to build intelligent models under the control of a CAD/CAM system and pass the models to other systems. Applications residing on the other systems must be able to read and write the passed models.
8. The ability to upgrade the current user base to new CIM-level application software without making current applications obsolete.

3.7 The University and the Federal Government Level

The education and training of future users of CIM technology fall to the university environment. Few universities are currently providing manufacturing engineering programs, and few of those that do add the necessary computer science courses to the program to ensure that the graduating student is prepared to deal with CIM-based technology. Evening programs must be developed to provide the working individual with the knowledge necessary to understand CIM technology and vie for the new jobs that are being ushered in with the technology.

A classroom environment for the education of engineering students has worked very well in the past, but the education and training of manufacturing students must have an applied approach. An understanding of the complex manufacturing environment cannot be acquired solely from textbooks. The development of programs that revolve around an actual manufacturing environment will not only better prepare the graduate for successfully functioning in a CIM environment but will also position the universities to compete more effectively for industry-sponsored grants in the field of CIM.

The federal government can be very instrumental in the industry move toward manufacturing automation. The implementation of CIM technology in U.S. industry is vital to the preservation of our economy and is therefore an issue of national importance and is worthy of subsidy. Liberal tax incentives in the areas of investment tax credits, CIM-training credits, revised accelerated depreciation schedules, special CIM technology research tax write-offs, and a complete review of the current OSHA requirements are just a few of the ways the federal government can help facilitate the development and implementation of CIM technology in the United States.

3.8 Conclusions

Although U.S. industry on the whole has developed and implemented many components of a CIM system, it is still lacking the foundational technology and architecture to integrate these islands of automation. Industry is still not focusing on the facility as a whole; it is dealing at the application level in specific areas of the organization. The business functions of an organization cannot be separated from the design and manufacturing functions. Before a company can implement CIM technology successfully, it has to have a very good idea of how it intends to conduct its business in the future. Organizations must view themselves as interconnected networks of functions. All these functions require data to drive them, and the data must be available at specific time intervals in order for the individual functions to be performed in a timely manner. By using structured analytical techniques and application tools, a company must define its structure from the top down, which will allow it to meet the goals and requirements of its 5-year strategic plan. This top-down design will drive the requirements of and specifications for the foundational system architecture that assures the integration of future CIM applications.

CHAPTER

4

Future Trends and Developments

M. Eugene Merchant

4.1 Concept of the Future CIM System

4.1.1 Nature of the CIM system

"Computer-integrated manufacturing" (CIM) is a term coined to represent the full range of the capability potential which the digital computer holds for manufacturing. That potential is threefold. First, the computer has *unique* potential to provide manufacturing with two powerful capabilities, never before available, namely:

1. Online variable program (flexible) automation
2. Online moment-by-moment optimization

Second, and very important, the computer has the capability to do the above not only for the hard components of manufacturing (the manufacturing machinery and equipment) but also for the soft components of manufacturing (the information flow, the databases, and so on). Third, and of utmost importance, the computer has the capability to do the above not only for the various bits and pieces of manufacturing activity but also for the entire system of manufacturing. The computer therefore has tremendous potential to *integrate* that entire system and thereby produce what is called the computer-integrated manufacturing *system*.

This generic concept of the CIM system is shown in Fig. 4.1. It may be defined as a closed-loop feedback system in which the prime inputs are product requirements (needs) and product concepts (creativity) and the prime outputs are finished products (fully assembled, inspected, and ready for use). It is comprised of a combination of software and hardware, the elements of which include product design (for production), production planning (programming), production control (feedback, supervisory, and adaptive optimizing), production equipment (including machine tools), and production processes (removal, forming, and consolidative). It is amenable to being realized by application of systems engineering and has the potential of being fully automated by means of versatile automation and of being made fully self-optimizing (adaptively optimizing). The present major resources for accomplishing this are the computer-related technologies.

It is this concept which provides the background for today's strong technological trend toward development and implementation of computer-automated, computer-optimized, and computer-integrated manufacturing—now being collectively called computer-integrated manufacturing. However, it should be recognized that, as yet, full CIM has not been realized in practice anywhere in the world.

4.1.2 Nature and role of the elements of the CIM system

Figure 4.1 illustrates the five main elements of the future full CIM system: product design, production planning, production control, production equipment, and production processes. There is nothing hard and fast about this particular characterization of the elements of the manufacturing system; others have subdivided it differently or used different names for elements similar to those shown. How the system is composed is relatively

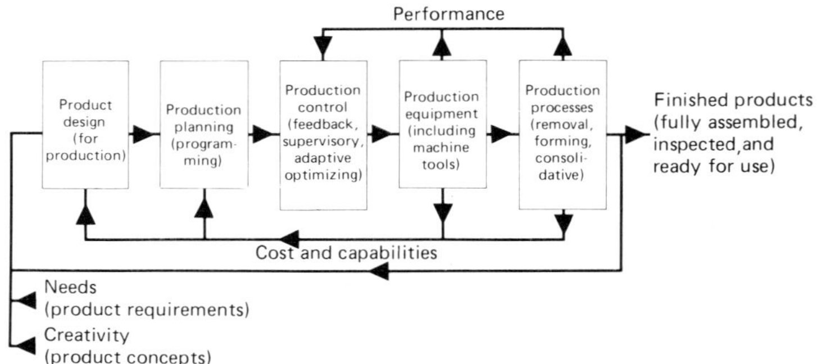

Figure 4.1 Computer-integrated manufacturing system.

immaterial. The important concept to recognize is that all the types of activities, equipment, and processes represented by these terms are, and must be, an integral part of the manufacturing system which is in the process of being automated, optimized, and integrated by applying the computer to these tasks. That is particularly true of the product design element of the system.

The reason for the preceding two statements can be recognized most readily by a simplified description of the roles and interrelations of the five elements shown in Fig. 4.1:

1. *Product design* establishes the initial database for production of a proposed product. In a CIM system, this is accomplished through such activities as geometric modeling and computer-aided design while considering the product requirements and the product concepts generated by the creativity of the design engineer. Ideally, however, in a CIM system the design process should be constrained by the costs that will be incurred in actual production and by the capabilities of existing production equipment and processes.

2. *Production planning* takes the database established by product design and enriches it with production data and information to produce a plan (program) for the production of the product. In a CIM system, this planning process should, ideally, be constrained by the production costs which the plan will engender, and by the capabilities of existing production equipment and processes, in order to generate an optimized plan.

3. *Production control* of the actual activity of producing the product according to plan requires further enrichment of the database with performance data and information about the production equipment and processes. In a CIM system, this requires such activities as modeling, simulation, and computer-aided scheduling of the production activity. This should include, ideally, online dynamic scheduling and control based on the real-time performance of the equipment and processes to assure continuous optimization of the production activity.

4. The *production equipment* further enriches the database with equipment and process data and information, resident either in the operators or the equipment, to carry out production processes. In a CIM system, this equipment consists of computer-controlled process machinery such as computer numerically controlled (CNC) machine tools, flexible manufacturing systems (FMS), computer-controlled robots, flexibly automated materials-handling systems, computer-controlled assembly systems, flexibly automated inspection equipment, and so on.

5. *Production processes* that create the finished product are carried out by the production equipment. This is done under the guidance of the data, information, and knowledge resident in the operator or, for CIM systems, in the equipment and the total integrated database supporting it. These

processes consist of material removal, material forming, or material consolidation (joining, sintering, assembly, and so on), as well as automated quality assurance. From these processes come products which, ideally, are fully assembled, inspected, and ready for use.

The above description of the roles and interrelations of the various elements of a manufacturing system which is in process of being automated, optimized, and integrated by application of the computer serves to illustrate the importance and interdependence of such elements in the overall future CIM system. Also, it serves to clarify the potential technological benefits of such an approach. However, the potential benefits of CIM are not purely technological. CIM promises in fact to bring major long-term socioeconomic benefits to nations which develop and implement it in their manufacturing industries. Thus, it is important to understand the nature of these benefits if the technological trend is to be properly guided and the resulting technology is to be properly used. The nature of these benefits can best be illustrated by examining examples of socioeconomic drivers toward CIM.

4.2 Socioeconomic Drivers toward Future Full CIM

4.2.1 The overall socioeconomic driver

The major, long-term, overall socioeconomic driver toward development and implementation of CIM in the manufacturing industries of industrialized countries is that which stems from the potential of CIM to reduce the cost of creating the primary, real, tangible wealth of such countries. The reason for this is that manufacturing is the principal real-wealth-creating activity of industrialized nations today. Although manufacturing now accounts for about one-third of the gross national product (GNP) of the typical industrialized country, services account for about one-half of that GNP. Services are essential to the support of a high standard of living and quality of life, but they do not create primary, real, tangible wealth. Yet such wealth is, historically, the basis and source of all other wealth in a nation. Therefore, by subtracting the half of the GNP coming from services, it can be seen that manufacturing then accounts for two-thirds of the remainder, or two-thirds of the real-wealth-creating sector of the economy. Thus it follows that manufacturing is responsible for about two-thirds of the primary wealth-creating activity of a typical industrialized country. The remainder of the real wealth creation comes from the extractive (agriculture, fishing, and mining) industries and the construction industry.

To take a specific example, in the case of the United States today, manufacturing accounts for about 21 percent of GNP and the extractive and construction industries account for about 13 percent. On the other hand, the service industries account for about 66 percent of GNP. However, only the 21 percent of GNP represented by manufacturing and the 13 percent represented by the extractive and construction industries result in creation of primary, real, tangible wealth. Thus, in the case of the United States, manufacturing accounts for about 62 percent of the direct real-wealth-creating activity of the country, or about two-thirds.

These facts provide a direct indication of the potential of computer-integrated manufacturing to provide overall benefit to a country and society in general—its effects on employment, quality of life, standard of living, and so on. Since, as indicated above, creation of primary, real, tangible wealth is the basis and source of all other wealth in a nation, increases in the standard of living, quality of life, employment, and the general economic well-being of a country stem directly from decreases in the cost of creating such wealth. With manufacturing accounting for two-thirds of that wealth creation in industrialized countries, it follows that reduction of the cost of manufacturing must be a top priority item in such countries. CIM-based automation, even in its present early stage of development, has already demonstrated far greater potential to reduce manufacturing costs than anything that has appeared on the scene since the onset of the Industrial Revolution. As such, development and implementation of CIM in a nation's manufacturing industries promises to provide a basis for major increases in the standard of living, quality of life, employment, and general economic well-being. This indeed constitutes a major driver toward such development and implementation.

4.2.2 Some specific economic drivers

In addition to the major, overall socioeconomic benefit which CIM promises to bring to nations which vigorously pursue its development and application, a variety of specific economic drivers of the trend toward full CIM have already been discovered and documented, even in these early stages of CIM's realization. Of these, two stand out. They are the potential reduction of the tremendous economic waste resulting from huge quantities of work in process in conventional factories and the potential reduction of the equally great economic waste resulting from the very low percentage utilization of capital equipment in those factories. The significance of these drivers can best be illustrated by a simple example for the production of machined parts. When the time spent by the average workpiece in going through a *conventional* factory is analyzed, it is found that only about 5 percent of the time is actually spent on machine tools and, of that 5 percent, only about 30 percent (or 1.5 percent of the total time) is actually spent as productive time in removing metal. This result is illustrated graphically in

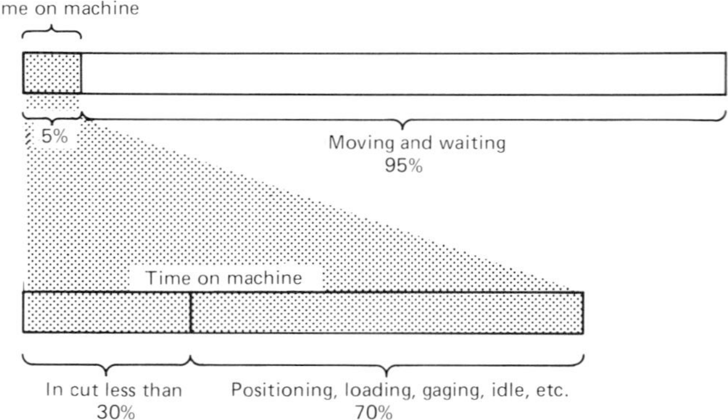

Figure 4.2 Analysis of time spent by average workpiece in moving through a conventional plant.

Fig. 4.2. The situation highlights the two main areas in which by far the greatest improvement in the economy and productivity of manufacturing can be made today. The first of these is a reduction in idle time for parts in process in the shop. Consequently, reductions are achieved both in the extremely high inventory of unfinished parts on the shop floor and in the number of finished parts waiting for others in process so that assembly of the product can proceed. In economic terms, the amount of capital lying idle in those parts in process is often as great as all the capital invested in the plant and its equipment—enough to build and equip *another whole plant*! The application of CIM, even in its present elementary state of development, has already demonstrated the capability to free up most of that idle capital.

The second area of potentially great improvement is in the percentage of machine utilization. The 30 percent machine utilization, when a machine is actually being operated (Fig. 4.2), must be combined with the fact that the average machine is waiting approximately 50 percent of its time for parts on which to work. As a result, the average machine tool in a batch-type shop is being utilized productively (i.e., actually cutting metal) only about 15 percent of the time that it is manned. Its actual utilization may be as low as 5 percent or less because it is seldom manned 24 h/day. This represents only about 5 percent utilization of the capital invested in machine tools. Therefore, the amount of capital lying idle is often greater than the capital required to equip the plant with machinery that more fully utilizes the capabilities of computer-integrated manufacturing! Application of such technology, even in its present elementary state of development, has already demonstrated the capability to equip factories with full systems of shop-floor automation with capital investments no greater, and often less, than that required to equip them with conventional machinery.

4.2.3 Some specific social drivers

The major social driver toward development and implementation of CIM is that already described above, namely, the major increases in the standard of living, quality of life, and level of employment which can come from decreasing the cost of manufacturing by use of CIM technology. However, there are a variety of more specific social benefits which it can impart as well. They derive mainly from the fact that the major long-range social trend affecting the field of manufacturing today is that toward postindustrial society. As described by Bell [1], the trend is from *preindustrial* societies to *industrial* societies to *postindustrial* societies.

Preindustrial societies were (and still are, since they exist in many primitive countries of the world today) structured in traditional ways of routine and authority. The labor force was engaged largely in the extractive industries—mining, fishing, forestry, and agriculture. The unit of social life was the extended household, and productivity was low because existence was a contest against nature.

Industrial societies were, and are, goods-producing societies, the environment of which has become technical and rationalized. As Bell points out, the machine predominates and energy has replaced raw muscle to provide the power that is the basis of productivity. The watchwords are maximization and optimization. Life is a contest against fabricated nature. Skills are broken down into simpler components, and the artisan of the past is replaced by two new figures: the engineer and the semiskilled worker.

A *postindustrial society,* as described by Bell, is based on services and hence is a contest between persons. What count are not raw muscle power and energy; what count are information and knowledge. The key person is the professional, who is provided by education and training with the kinds of skills most needed in such a society. The key consideration in this society is not so much whether productivity is high or low as whether the attainment of improved productivity is properly balanced against the quality of life which will result.

To help clarify the meaning of the concept of postindustrial society, Bell specifies five dimensions, or components:

1. *Economic sector.* The change from a goods-producing to a service economy
2. *Occupational distribution.* The preeminence of the professional and technical class
3. *Axial principle.* The centrality of theoretical knowledge as the source of innovation and of policy formulation for the society
4. *Future orientation.* The control of technology and technological assessment

5. *Decision making.* The creation of a new "intellectual technology" (of which the computer is the prime tool)

This continuing, long-term trend toward becoming primarily a service-oriented economy—a postindustrial society—which is particularly marked in the United States, has generated a variety of associated long-term social drivers toward development and implementation of full CIM. Three such social drivers stand out today—three sets of trends in attitudes toward manufacturing:

1. On the part of workers, an increasing reluctance to work in manufacturing
2. On the part of employers, an increasing effort to improve worker satisfaction and environment
3. On the part of governments, a change from a somewhat passive to a more active role in requiring improved working conditions

Concerning the first of these trends, there is a steadily increasing preference of workers in industrialized countries for employment in the service industries instead of in manufacturing. This trend is dealt with directly by Bell. He points out that this same type of trend in work preferences occurred earlier in the transition from the preindustrial to the industrial society in industrialized countries. For example, in the United States, as the manufacturing industry developed, the percentage of the work force employed in agriculture declined from 90 percent in 1790 to 4 percent in 1976 and is down to less than 3.5 percent today. Meanwhile, the percentage of the work force employed in manufacturing rose correspondingly during the nineteenth century. In recent years, however, it has begun to decline: from 30 percent of the work force in 1947, to 24.6 percent in 1970, and to 21.7 percent in 1978. A Rand Corporation forecast projects that, by the year 2000, only 2 percent of the labor force in the United States will be employed in manufacturing. It is probably unrealistic, however, to expect the figure to be that low at such an early date. Bell's own estimate of this figure is a more conservative 10 percent. Nevertheless, in all the industrialized countries of the world there already is, on average, an increasing shortage of willing, capable manufacturing workers as a result of this trend.

The migration of the labor force into services puts a direct social pressure on the manufacturing industry to increase its productivity at a rate sufficient to more than compensate for the shrinkage in its labor force. Fortunately, computer-integrated manufacturing offers considerable benefit here and has already demonstrated major potential to reduce the pressure. It has done so by demonstrating tremendous potential for automating nearly all aspects of manufacturing so that highly productive manufacturing can continue with a smaller and smaller percentage of the labor force in more and more highly computer-integrated factories.

Concerning the second of these social trends, that of the changing employer attitudes, the employers are now clearly recognizing the human need for the nature of work to be such as to assure the worker of deep satisfaction from performing it (as well as freedom from potentially dangerous or unhealthful conditions). Thus, much attention is being directed to methods of accomplishing that end. Here the pioneering work on job enrichment by such investigators as Herzburg [2] is proving most useful. Herzburg's significant finding, illustrated in Fig. 4.3, is that, although the so-called hygiene factors of a job (i.e., company policy and administration, supervision, work conditions, salary, and so on) can cause *dis*satisfaction if they are not satisfactory, they can do little to provide ongoing job satisfaction. Instead, job satisfaction derives from the adequacy of the so-called motivator factors of a job: opportunity for achievement, recognition, responsibility, advancement, growth, and so on. The major feature of jobs which provide such opportunities is participation in decision making, or participative management. Here again, computer-integrated manufacturing offers considerable benefit. It can substitute challenging, satisfying, participative-

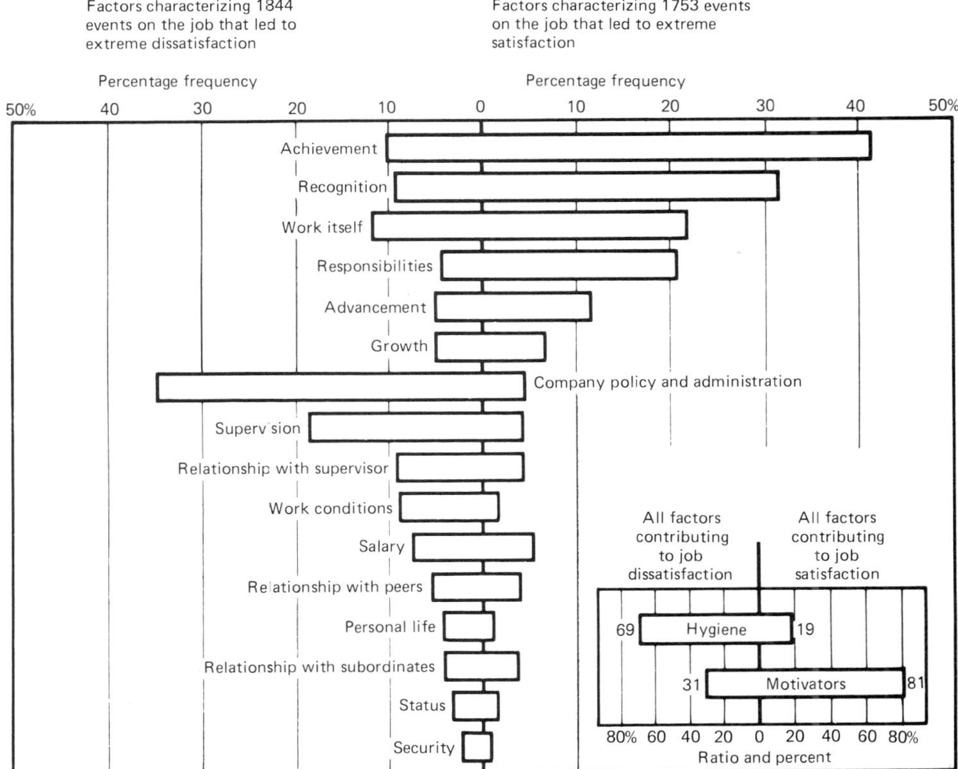

Figure 4.3 Satisfaction and dissatisfaction in jobs.

management-type jobs (which interactive computer systems offer) for the unpleasant, physically taxing, or even unsafe and unhealthful jobs found on the conventional factory floor. Thus, CIM is already proving to be a powerful tool for management to respond positively to (1) the trend toward increasing efforts to improve worker satisfaction and (2) the need to provide jobs in manufacturing better suited to the educational level of today's labor force, which would help in attracting willing, capable workers.

Concerning the third of the foregoing social trends, that of government attitudes throughout the world, governments are changing their approach to freeing workers from potentially dangerous or unhealthful conditions. In most of the industrialized countries of the world, government is no longer playing an essentially passive role by requiring that, as technology to accomplish improved working conditions is developed, it be put to use. Instead, governments are now playing very active roles by requiring that such technology be developed. For example, in the United States, the Occupational Safety and Health Act is, in effect, requiring such action on the part of industry as:

1. Development of technology to eliminate the necessity for a worker ever to insert a hand, arm, or any part of the body into a potentially dangerous area of a machine (such as a press)
2. Development of technology to keep the average noise level to which a factory worker is exposed over an 8-hr period below 90 dBa, or even below 85 dBa

Here again, CIM-based flexible automation offers considerable benefit. It can provide management with a prime tool for responding positively to such requirements; whereas without this technology, such requirements can be met only with the greatest of difficulty, if at all. In fact, CIM-based flexible automation will allow the worker not only to be released from operation of dangerous machines but also, eventually, to spend very little time each day on the noisy factory floor.

4.3 The Long-Term Goal: Realization of Full CIM Strategies for Accomplishment of the Goal

4.3.1 Strategies for accomplishment of the goal

As a result of today's powerful technological trend toward computer-integrated manufacturing and its perceived large economic and social benefits, realization of full CIM has now become a major long-range goal of many industrialized nations and of much of the manufacturing industry throughout the world. As a result, many industrialized nations and many manufacturing companies are today pursuing large programs of research, development, and implementation of CIM to hasten the realization of fully computer-automated, computer-optimized, and computer-integrated factories. Both national governments and manufacturing companies have realized, however, that advancing from today's industrial methods, know-how, and installed equipment to the CIM goal requires an *evolutionary* process—rather than a revolutionary one.

The strategy being followed, therefore, is to develop and to implement a series of viable, economic steps in the form of shorter-range programs of research, development, and implementation. Each of these programs should have two main characteristics:

1. Potential for sufficient economic return to justify the program by itself and to generate or free up capital to support development and implementation of the next step
2. Compatibility with the eventual attainment of the goal of full computer automation, optimization, and integration of manufacturing

As a corollary, the strategy being followed for the technological aspects of the evolutionary development is as follows:

1. Plan from the top down (to assure compatibility of each step with all others)
2. Implement from the bottom up (to permit nondisruptive stepwise building of the total system)

Such a strategy clearly requires that a company develop a *master plan* for research, development, and implementation of CIM if it is to be successful in attaining its goal. This master plan should be a dynamic plan that is updated at regular intervals to take into account changes in technology, business conditions, and so on. It is interesting to note that those companies throughout the world which are achieving the most success from their CIM programs are following the above strategy and, in most cases, are using comprehensive, dynamic master plans.

4.3.2 Nature of R&D efforts worldwide

In pursuit of the strategies described above, many industrialized nations and much of the manufacturing industry throughout the world are today pursuing large programs of research, development, and implementation of CIM. The nature and geographical distribution of these programs and activities vary greatly. However, the following items seem to be receiving a major amount of attention worldwide:

1. Development of *integrated manufacturing software systems* through development of individual software modules which can, in the long run, be readily interfaced with each other to assemble full software systems

that are appropriate for various types of manufacturing applications. This work is proceeding rapidly in the COMECON (Council for Mutual Economic Assistance, a trade organization for East European countries) countries, as well as in West Germany and, to a lesser extent, in Japan. The U.S. Air Force has carried out a major program of this type known as ICAM (Integrated Computer-Aided Manufacturing). In addition, CAM-I (Computer-Aided Manufacturing—International) has several programs in progress for developing integrated manufacturing software.

2. Development and application of *group technology and cellular manufacturing* as a required base for application of hierarchical factory computer systems and the evolution of flexible manufacturing systems. Such work is proceeding most rapidly in the Netherlands, Japan, West Germany, Norway, Britain, and the COMECON countries. The U.S. Air Force ICAM program has had this goal as one of its features, as has the CAM-I program.

3. Development and application of broad *computer control of machine tools and manufacturing processes* through use of numerical control (NC), computer numerical control (CNC), direct numerical control (DNC), and hierarchical factory computer systems. Such work is proceeding most rapidly in Japan, the United States, and West Germany.

4. Development and application of *computer-controlled robots* for automating workpiece and tool handling, machine operation, and various manufacturing processes. Such work is proceeding most rapidly in Japan, the United States, West Germany, Norway, Sweden, the USSR, and Bulgaria.

5. Development and application of *computer-controlled flexible manufacturing systems* which are, in effect, rather fully automated group technology cells. Such work has proceeded most rapidly in Japan, the United States, and East Germany, and more recently in West Germany, the USSR, Bulgaria, Czechoslovakia, Hungary, Norway, and Britain.

6. Planning and development of actual *prototype computer-integrated factories.* Such work is proceeding most rapidly in Japan, which undertook a major national program to develop a prototype minifactory. In addition, several Western European and COMECON nations, including the USSR, have general plans and supporting programs aimed at this objective.

4.3.3 Insight provided by FMS as a microcosm of future shop floor integration

As yet, as stated in the preceding section, development and implementation of full CIM has not been realized anywhere in the world. Thus, there is as yet no significant performance data which can document the benefits of full integration. However, on the factory floor, integration has proceeded further than in any other sector of the manufacturing system. It has taken the form of flexible manufacturing systems (FMS). Their importance stems from the fact that such systems are not stand-alone pieces of equipment; they are fairly sophisticated *integrated* systems of a number of software and hardware modules. They include integrated automation of not only such "hard" functions as workpiece and tool transport, loading, unloading, and changing but also of such "soft" functions as planning, scheduling, and control of workpiece and tool flow and workstation load. Thus, in a sense, they are microcosms of a portion, at least, of the future computer-automated, -optimized, and *-integrated* factory. Therefore, a look at the performance benefits already being demonstrated by such systems can provide some insight into the nature and magnitude of the still greater potential benefits to be reaped eventually from integration of the total system of manufacturing. In addition, such a look reveals how amazingly large some of the benefits already being obtained with these microcosms of a portion of the future total CIM system really are.

A sampling of some of the benefits obtained from experience with actual operation of FMS systems in various countries will serve to illustrate their nature and magnitude. The sampling set forth in Table 4.1 represents a composite of performance results, reported in the literature, of three modern

TABLE 4.1 Typical Performance Benefits Experienced with Modern Flexible Manufacturing Systems*

	Quantified benefits†	Percent
Reduction in:	Lead time for product	40
	Lead time for parts	53–75
	Required number of machine tools	53–81
	Required personnel	53–92
	Labor costs per part	90
	Required machining hours	65
	Required floor space	42
	Tooling costs	30
	Total annual costs	24
	Capital investment cost	10
	Inventory of work in progress	90

Nonquantified benefits

- Improved quality:
  - Higher accuracy and reproducibility
  - Lower rework costs, scrap rates, and quality-assurance costs
- Closer adherence to production schedules:
  - No order chasing
- Improved working conditions:
  - Decreased accident risk and physical labor
  - Increased challenge
- Increased flexibility:
  - Increased independence of batch size, types of parts, and production quantities

*Composite of results from one German, one Japanese, and one American system.
†Comparison between stand-alone machines and integrated systems.

FMSs—one West German, one Japanese, and one American. The West German system, reported on by Dronsek [3], is operating at Messerschmitt-Bölkow-Blohm (MBB) in Augsburg, machining a large variety of aircraft parts. The Japanese system [4] is an unattended (unmanned) FMS operating at the Internal Combustion Plant of Niigata Engineering Company in Niigata, machining a variety of cylinder heads for diesel engines. The American system [5] is operating at the Vought Corporation in Dallas, Texas, machining some 600 different aircraft parts.

The nature and magnitude of the typical performance benefits compiled in Table 4.1 speak volumes concerning the tremendous power of CIM, even in its present elementary and incomplete form, to improve the capability and cost-effectiveness of manufacturing both today and in the future. However, the last two entries in Table 4.1 are of particular and very great significance. It has been found that they are quite representative of the experience being obtained with the great majority of installations of modern FMSs. What they indicate is that the implementation of computer-based integrated factory automation can be essentially *free of capital cost* to the user *if properly utilized*! The FMS, as a microcosm of the future computer-integrated and automated factory, demonstrates this.

First, the reduced capital investment in FMS workstations (due to the much smaller number required because of *greatly* increased utilization when compared to stand-alone, unintegrated workstations) more than pays for the required additional supporting, integrating facilities. These facilities include not only such entities as the control computer and the work-and-tool-transport equipment but also the software system.

Second, the drastic reduction of work-in-process inventory and stock waiting to be assembled (virtually to zero because of the capability of these flexible systems to produce just whatever mix of parts is required for immediate assembly) frees up sufficient capital (previously lying idle in inventory) to more than pay for the total FMS. Another way of looking at this latter fact is that the just-in-time production, which is made possible by truly flexible automation, allows a plant to turn over its total inventory *many* more times per year than is possible in a conventional plant.

4.4 Technological Forecasts

As stated earlier, the full development and realization of CIM is still in its relative infancy. The question therefore arises as to the nature and timing of its future technological development and implementation. Although no one has hard-and-fast answers to this question, the art of technological forecasting has been able to shed some light on this matter.

A number of technological forecasts of the future of manufacturing have been carried out in recent years by using the Delphi technique. (The Delphi technique of technological forecasting makes use of the anonymous intuitive consensus opinion of a panel of experts as to the nature and timing of future technological events.) These forecasts have given a progressively stronger indication of the future advent of computer-integrated factories. The first of these forecasts, conducted by the author [6] was carried out by the International Institution for Production Engineering Research (CIRP) in 1971 by using an international panel of manufacturing research experts.

Out of 94 forecast events on which there was good consensus, 24 (or over one-fourth) of the experts strongly forecast the coming of computer-integrated factories. The early perceptions about the nature and timing of this major development can be summarized by 3 of the 24 events forecast at that time. These are:

By 1980 (*median date*)

A computer software system for full automation and optimization of all the steps for manufacturing a part will be developed and in wide use.

By 1985 (*median date*)

Full online automation and optimization of complete manufacturing plants, controlled by a central computer, will be a reality.

By 1990 (*median date*)

More than 50 percent of the machine tools produced will not have a "stand-alone" use but will be part of a versatile manufacturing system featuring automatic part handling between stations and control by a central process computer.

In other Delphi forecasts that followed a still higher percentage of the forecast events strongly foreshadowed the future advent of computer-integrated factories. Then, in 1977 and 1978, the Society of Manufacturing Engineers produced three forecasts [7,8] on the future of manufacturing that covered material removal, manufacturing systems, and assembly. Of 133 events forecast, 99 (or 74 percent) events strongly forecast the coming of computer-integrated factories. The selected events that are cited below give an excellent feel for the expected timing and direction of this highly important technological development.

By 1985 (*median*)

Assembly operations will be integrated with other manufacturing operations so that a CIM system might result. Computer software systems for automation and optimization of all steps in manufacturing planning (selection of machining sequence, machine tools, clamping methods, sequence of operations, tools, and optimum cutting conditions) will be used by at least 25 percent of the firms represented by the panelists (a cross section of U.S. industry).

By 1987 (*median*)

Approximately 15 percent of all machine tool production will not have a "stand-alone" use but will be a part of a versatile manufacturing system featuring automatic parts handling between stations and control by a central process computer.

By 1990 (*median*)

The development of sensory techniques will enable robots to approximate human capability in assembly. Computer-aided design (CAD) techniques will be used for 50 percent of the new assemblies designed.

By 1995 (*median*)

Of the direct labor in final assembly of automobiles, 50 percent will be replaced by programmable automation.

It can be seen that these forecasts portend the rather full realization of the computer-integrated factory before the end of this century.

The 1977 SME forecast and another produced in 1980 [9] also provide some interesting insights into the expected trends in the development of the technology associated with flexible manufacturing systems as microcosms of a portion of the future complete CIM system. A selection of these forecasts reflects the nature of these trends, as follows:

By 1987 (*median date*)

1. In-process adaptive control of surface roughness in machining and grinding will be developed and will be in wide use.
2. Noncontact, high-speed, high-accuracy inspection systems for 100 percent inspection (online) with closed-loop feedback to the machine control will be in wide use.
3. Approximately 24 percent of the new machines will employ online gaging of machined parts. (Such machines should be capable of feedback control compensation of the process to maintain a prescribed accuracy.)
4. Approximately 38 percent of the total manufacturing systems will be equipped with diagnostic sensors and associated software for implementing the indicated changes or action.
5. Approximately 30 percent of current part scrappage will be reduced because of improved accuracy control and online sensing.
6. Automated fixturing and holding devices will be applied on 25 percent of the machine tools.

By 1990 (*median date*)

1. Of all manufacturing industries, 25 percent will be using online inspection systems that sense and correct deviations from the part definition through a feedback subsystem.
2. Of the industrial machine tools sold, 20 percent will be designed to eliminate the need for an operator at each workstation through automatic

part loading and unloading, automatic sensing of tool wear and changing of cutting tools, and completely computerized process monitoring.

4.5 A Technological Vignette of the Factory of the Future

The foregoing consideration of the trends, benefits, and forecasts relating to the advanced technology of the developing field of CIM naturally raises the question as to what the factory of the future ultimately may be like, when the full potential of CIM is realized. Even though it is surely not possible to give a definitive answer to this question, the potential power of the technology is already sufficiently clear to permit creation of a glimpse-like technological vignette of the future factory. It might be somewhat as follows.

The process of designing a product will be carried out by iterative communication between humans and computers, the human supplying the design concepts and requirements and doing the creative work and the computer supplying stored and standardized information and carrying out the design calculations. During this design process, the computer can, through the internal loop labeled "cost and capabilities" shown in Fig. 4.1, be constantly calling for and taking into account information on the manufacturing costs and capabilities (equipment and process costs and capabilities) required to produce, on the true production elements of the available system, each of the alternative features conceived by the designer. The computer then uses that information to find a design which not only satisfies the product requirements but also can be manufactured at minimum cost and with maximum producibility.

Almost simultaneously, the production planning part of the system is using this information to set up an optimized production plan for the production of the product and choosing proper equipment and processes, sequence of operations, operating conditions, and so on. This numerical information is in turn used to control the array of automatic machines and equipment which do the actual production and assembly of the product. These are machines and equipment capable of setting themselves up, automatically handling parts, selecting their own tooling, and carrying out automatically a variety of fabrication processes of the removal, plastic flow, and consolidation types, including assembly of the product. Self-optimizing, they feed information back to the control system through the other internal loop labeled "performance" in Fig. 4.1. This system, as it constantly receives information as to the actual performance of the equipment and processes, compares the information with the "ideal" performance planned in the earlier phase. Then, as it finds performance beginning to depart from the planned optimum, it overrides the original plan and performs dynamic scheduling, adjusts operating conditions of the machines and processes, and so on, as necessary to maintain optimum, minimum-cost performance.

Meanwhile, the machines and equipment carry on self-diagnosis of their condition and, when an impending malfunction is detected, carry out appropriate corrective action including automatic replacement of defective modules in the system. Further, the machines also carry out automatic in-process inspection of the product at each stage of manufacture so that any impending deviations from the original specifications are automatically corrected and held within prescribed tolerances. Thus, the final assembled product is turned out fully inspected and in full conformity with the original design concept and requirements.

Such is the prospect for future manufacturing technology as the full power and potential of CIM approaches realization!

REFERENCES

1. D. Bell, *The Coming of Post-Industrial Society: A Venture in Social Forecasting*, New York, Basic Books, 1973.
2. F. Herzburg, "One More Time: How Do You Motivate Employees?" *Harvard Business Review*, vol. 46, no. 1, 1968, pp. 53–62.
3. M. Dronsek, "Technische und wirtschaftliche probleme der fertigung im flugzeugbau," *Proceedings, Produktionstechnisches Kolloquium Berlin 1979*, Munich, Carl Hanser Verlag, 1979, pp. 107–115.
4. Editorial Report, "Paying a Visit to FMS Plant," *Metalworking Engineering and Marketing*, vol. 4, no. 1, 1982, pp. 72–76.
5. Editorial Report, "$10 Million FMS at Vought to Build Fuselage Components," *The Production Engineer*, vol. 63, no. 1, January 1984, pp. 39–40.
6. M. E. Merchant, "Delphi-Type Forecast of the Future of Production Engineering," *Annals of the CIRP*, vol. 20, no. 3, 1971, pp. 213–225.
7. B. Colding, L. V. Colwell, and D. N. Smith, "Delphi Forecast of Manufacturing Technology: Manufacturing Systems, Material Removal," Society of Manufacturing Engineers, Dearborn, Michigan, 1977.
8. B. Colding, L. V. Colwell, and D. N. Smith, "Delphi Forecast of Manufacturing Technology: Assembly," Society of Manufacturing Engineers, Dearborn, Michigan, 1978.
9. R. Wilson, L. V. Colwell, and D. N. Smith, "Delphi Forecast of Manufacturing Technology: Computer-aided Manufacturing," Society of Manufacturing Engineers, Dearborn, Michigan, 1980.

PART 2

Components

CHAPTER

1

Product Definition: The Role of CAD/CAM in CIM*

Peter A. Marks

1.1 Why Is a Strategic Direction for CAD/CAM Important?

Is computer-aided design/computer-aided manufacturing (CAD/CAM) a strategic issue in CIM? Should it command a significant share of corporate resources? There is, of course, no single answer that applies to all companies; these are, however, important questions for nearly all companies to be asking. Although CAD/CAM technologies are not a strategy in themselves, they are among the most powerful tools available for implementing various competitive strategies. In the hands of a competitor, CAD/CAM tools become a threat; managed competently within an enterprise, they represent a competitive opportunity.

There are at least four reasons why company management should approach CAD/CAM from a strategic perspective. The first is that CAD/CAM

*This article is an edited version of two articles in *CAD/CAM: Management Strategies*, Auerbach Publishing, Pennsauken, N.J., July 1984.

is likely to represent a significant investment whether or not it is treated as a significant strategic issue.

The second reason is that many CAD/CAM investments will fail to provide any real strategic advantages. They may in fact be nothing more than the industrial equivalent of keeping up with the Joneses. For example, efforts to reduce drafting costs (which are usually less than 1 percent of total costs in a typical design and manufacturing company) may be good tactics, but they do not result in a strategic edge for most companies.

Third, several companies have been able to turn CAD/CAM capabilities into an important competitive advantage. These enterprises, including relatively small manufacturers as well as Fortune 100 firms, are outperforming their competitors largely because of their effective management of this technology. For these companies, there is little question: CAD/CAM is a strategic issue.

Fourth, proper CAD/CAM implementation is a prerequisite for effective CIM. Although CAD/CAM might be considered as the product definition component of a total CIM plan, an integrated strategy is essential to maximize the benefits of both CAD/CAM and CIM.

This chapter describes CAD/CAM's major role in designing and implementing a company's business strategy, and it offers guidelines for determining the most effective use of CAD/CAM capabilities in a specific business environment.

1.1.1 The status of CAD/CAM technology

In some respects, the implementation of CAD/CAM systems is much like the implementation of management information systems (MISs) in the 1970s. Many companies entered the promised land of MIS with great expectations and a long wish list of capabilities. Ten years later, many were still being buried under an avalanche of MIS information. Others, however, concentrated on critical needs and can now demonstrate that better control of key information provided a strategic edge.

In the same way, too many companies are approaching an investment in CAD/CAM with great expectations and a wish list of capabilities in hand. Pursuit of capabilities by these companies is often driven more by the desires of functional managers than by the strategic needs of the businesses. It is natural and desirable for functional managers to want the best possible tools; but given the costs and importance of the CAD/CAM investment, more strategic direction is needed.

As a basis for understanding the strategic importance of CAD/CAM planning and management, it is helpful to view CAD/CAM as any useful combination of the following four technologies:

- Database management
- Computer graphics

- Mathematical modeling (analysis)
- Data acquisition and control (physical prototypes, production processes)

A typical turnkey CAD/CAM system, for example, includes a database system to store and retrieve drawings and part attributes. It includes computer graphics for drawing creation and display, and it often contains companion programs for electrical or mechanical analysis (e.g., circuit simulation or finite-element analysis). Finally, a typical turnkey CAD/CAM system may be provided with programs to control production processes (e.g., numerical control, offline robotic programming). Thus, it could include all four of the basic technologies. A material requirements planning program, as another example, is based primarily on only one of the basic technologies: database management.

The important notion behind this four-part definition is that each of the four computer-aided technologies often represents an alternative approach for modeling and predicting the outcomes of various product decisions. Selecting the best overall combinations of problem-solving tools, computer-aided or otherwise, is an important tactical decision.

Furthermore, the slash (/) in CAD/CAM implies an integration of capabilities. The alphanumeric database created in design might become a bill of material in manufacturing. Part geometries might become the basis of tooling design, but without recreating the entire geometry. Analysis could include a look at questions about producibility and manufacturing quality. This interplay and integration of various CAD/CAM technologies is an important source of potential benefits. However, such integration of capabilities cuts across functional disciplines. It requires better coordination of the entire development process; the tools alone are not sufficient to achieve the coordination. The slash in CAD/CAM is a reminder that a commonly understood strategic direction is needed for integration to occur.

1.2 CAD/CAM's Impact on Resource Allocation

Strategy answers the question, "How should we allocate our resources?" In companies without a clear strategy, many people allocate corporate resources, but not in a coordinated manner. In such companies, one might find quality control managers identifying scrap and production managers shipping it, value engineers taking out product costs and design engineers putting costs back.

There is a direct relation between strategy and the design function. Design activities have a significant effect on the allocation of a company's resources. Design affects decisions about the products a company chooses to make, how those products will respond to trade-offs between cost and performance, the costs of materials and manufacturing processes, ultimate product reliability, and product life cycle costs. Therefore, although the typical manufacturing company allocates only about 5 percent of its resources to design and engineer its products, those activities have a tremendous impact on product price and performance. Once the product has been designed, about 80 percent of its ultimate costs and performance have already been determined. The graph in Fig. 1.1 contrasts these typical patterns of expenditure vs. commitment of resources over a product's life cycle. Clearly, the 5 percent or so of corporate resources spent on design and engineering can have tremendous leverage on competitive success.

CAD/CAM implementation, in turn, can have tremendous leverage on the 5 percent or so of resources allocated to design and manufacturing engineering. Whether by freeing up additional time for engineering tasks, by helping examine more options, or by conducting sophisticated simulations and evaluations, CAD/CAM investments can make engineers more effective. Effectiveness is the key concept; the goal is not so much to reduce the 5 percent of costs spent in design and manufacturing engineering as to make better decisions about the 80 percent of product costs and performance which are determined by design and engineering.

This relation among corporate strategy, product design and manufacturing engineering, and CAD/CAM may seem fairly obvious. However, most companies and vendors are still approaching the technology as if the primary goal were to reduce engineering costs rather than to improve product value and competitiveness. To return to the earlier example about drafting, the majority of drafting systems are justified on the basis of drafting productivity ratios. Even if drafting costs were eliminated, the product would not necessarily perform any better, and it would cost only 1 percent or so less. More important, this emphasis on producing more drawings in less

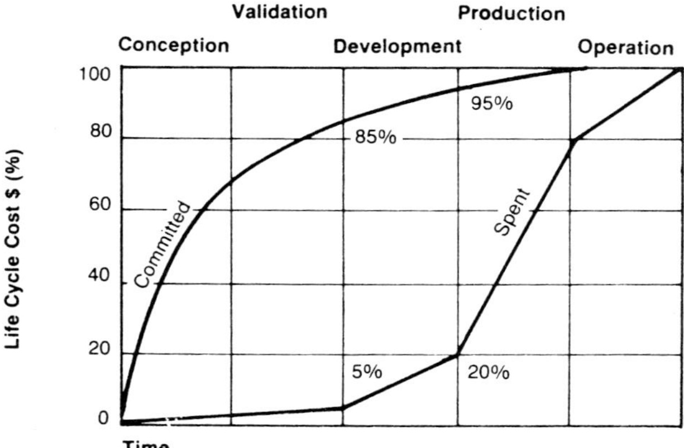

Figure 1.1 Product life cycle costs (*G. Jacobs, "Designing for Improved Value,"* Engineering, *London, February 1980.*)

time rewards the wrong objective. From a competitive standpoint, drawings are a somewhat necessary evil, not a measure of design quality. Companies generally do not need more drawings any more than they needed telephone book–size MIS reports. CAD/CAM systems can have an impact on design quality, but only if they are implemented with a clear set of strategic design objectives.

1.2.1 Strategies for improving product value

All things considered, the customer always tries to buy the "best" product. What keeps more than one competitor in business is that one customer's definition of "best" differs from that of the next. One customer needs immediate delivery; the next buys for price. Price is no object for yet another customer. Therefore, each company must find its own competitive edge and market niche. Not everyone can have the lowest-cost design, the highest-quality design, or even the design best suited for a narrowly segmented group of customers. Other strategies, such as concentration of sales or marketing resources, should also be examined.

The key issue is that a company that does not incorporate into its product design an appropriate set of values is not likely to win many customers. (Included in product design are the decisions about materials and manufacturing processes that set the agenda for production.) Strategy determines an appropriate set of product values for each competitor. CAD/CAM, in turn, can help shape and improve product values in ways such as the following.

Faster response Companies that are currently losing orders because of extended delivery times are often able to win a larger market share by compressing the design and manufacturing cycle. For example, CAD/CAM has provided a competitive edge to companies involved in the manufacture of special machines and the design of high-technology products.

The use of CAD/CAM will not necessarily result in shorter development times unless the automated functions are on the critical path. If faster response is a strategic priority, a first step should be a detailed audit of the current development process. Depending on the organization, drafting may be a significant source of delays in one case, prototype testing in another, and the product change notification process in yet another. Once the critical functions have been identified, a CAD/CAM tool can probably be found to speed the process.

Automation of critical steps in the existing development process is a reasonably good beginning. Even better, however, is a fresh look at the entire process. For example, if product testing is on the critical path, computer-aided testing can help speed it up. However, it could prove even more useful to significantly reduce the amount of testing needed by early use of computer simulation. Examples of this include circuit simulation as a partial substitute for test and debug cycles in electronics or structural simulation (i.e., finite-element analysis) before proof and fatigue testing of mechanical products.

Making such a fundamental change in the existing product development organization can shorten the development process by months or even years. It is, however, a significant cultural transition for most companies. The organizational part of this transition may itself require several years. For that reason, it is often practical to automate parts of the current process in the short term, within the context of a longer-term plan to fundamentally reorganize development functions to make better use of CAD/CAM.

Greater design and manufacturing flexibility Production environments have traditionally been based on either batch production using general-purpose machines or mass production using dedicated machines. Batch production allows considerable flexibility, but at the penalty of high per-unit costs for such operations as setup, materials handling, and shop floor management. Mass production lowers per-unit costs, but only if production volumes warrant the initial investment and loss of flexibility.

With CAD/CAM and flexible manufacturing, companies can often provide a much better balance of custom-design performance benefits and mass-production cost benefits. In a traditional design and manufacturing environment, parts and model complexity tends to directly increase costs. When computers are used to manage this complexity, through the grouping of similar parts in databases and the production of families of parts in flexible manufacturing systems, costs can be partially decoupled from complexity. A significant initial investment is still required, but the resulting design and manufacturing process can accommodate a wider range of similar parts. The benefits can include better satisfaction of customer performance needs, timely production and delivery, fewer design and production errors, and lower inventory costs.

Those planning the use of these technologies, however, should keep in mind that advanced design and manufacturing methods will not revive obsolete product lines. Thus, a prerequisite for any investment in CAD/CAM for flexible manufacturing should be a serious attempt to anticipate customer requirements, applicable new technologies, and competitive thrusts three or more years in the future.

Improved product cost or performance Product victories are often measured in fractions. A slight edge can mean major differences in market share and profitability. Companies that manufacture a narrow range of products in mass quantities can often afford significant investments in design to gain a slight competitive edge because the cost will be amortized over a large production volume. In some cases (e.g., the design of medical or military systems), costs may actually be unimportant relative to the desire for a slight performance advantage.

CAD/CAM technologies can play roles in better understanding and optimizing product performance to gain that slight competitive edge. For example, solid-modeling programs are becoming highly effective tools for achieving improved fit, finish, and appearance. Other related CAD/CAM applications can aid in the achievement of such objectives as reduced cost, lighter weight, simpler mechanisms, increased structural reliability, and better user ergonomics.

It should be emphasized that the pursuit of a competitive edge, while potentially rewarding, can be difficult and costly even with computer aids. Furthermore, CAD/CAM systems and sophisticated analyses are not substitutes for good designers and engineers.

This means that companies must carefully choose the areas in which a competitive edge will make a difference to the customer and then provide workers with the best tools for the job. For an example of the difference a sharply focused strategy can make, compare General Motors' X car and Chrysler's K car. The cars were relatively comparable and the companies had access to similar CAD/CAM technologies. GM had access to vastly superior resources: more money, more technology, more people. Chrysler, however, concentrated most of its resources on a few key product values: good appearance, good economy, and the promise of reliability. Even with fewer available resources, Chrysler made the K car a relative success by seeking a competitive advantage where it mattered most.

Reduced need for physical prototypes Most companies design and build a product several times in order to gain the experience necessary to produce satisfactory results. Often, several to several dozen physical prototypes are involved in this process. Many companies even let their customers do some product testing.

Products like buildings, bridges, satellites, and heart pacemakers, however, need to be built correctly the first time. Other products, like automobiles and airplanes, are very expensive to prototype. The four basic CAD/CAM technologies can often be used to eliminate or reduce the need for traditional prototype testing. Databases of past performance and best practices are one approach. Graphics simulation is another. More insightful testing of fewer prototypes, through computer-aided testing, is a third approach. Perhaps the most powerful approach, though, is mathematical simulation—building and analyzing a mathematical prototype in the computer.

Computer simulation can be much faster and cheaper and nearly as accurate as testing the actual product, and sometimes simulation is the only workable approach. For example, test conditions may be impractical to replicate in the lab (e.g., a hurricane or a nuclear power plant failure), or the product may be so costly to produce that it must be done correctly the first time. A related benefit is that simulation often forces engineers to understand the physics behind product performance. This design insight is much more valuable than trial-and-error prototyping.

Naturally, simulation is limited to problems that are understood well enough to be modeled mathematically. These problems include geometric interferences, mass properties, static and dynamic structural performance, kinematics, heat transfer, many fluids problems, most digital and analog circuit evaluation, and many problems in process and inventory optimization. Because of the trends toward lower computing costs and the attention currently being given to simulation, new applications are continually being added.

When an effective simulation technique is to be applied to a key product performance concern, it is often beneficial to use both physical testing and mathematical analysis capabilities in tandem. The role of testing, however, will increasingly be to check assumptions about simulation inputs and modeling techniques. Such a partnership can be extremely effective in producing the product correctly on the first attempt or at least in reducing the time and cost of prototype testing.

Better performance at the leading edge Companies at the leading edge of a technology are often faced with no strategic alternative but to employ whatever tools are needed to design and build the next generation of products. The possibility is not so much for a slight competitive edge as for order-of-magnitude improvements. When such improvements are technically imaginable, the use of CAD/CAM technologies may be a prerequisite for success. The use of CAD/CAM systems in the design and manufacture of integrated circuits is an often-cited example. Other modeling technologies will probably prove essential in such diverse areas as controlled-fusion energy research, the design of ultraprecise machines, advanced structural applications for composite materials, or even genetic engineering.

Performing at the leading edge is a high-risk, high-payback strategy that depends on the insight of key research and development people. One CAD/CAM implementation strategy is to first conduct studies to determine probable sources of breakthroughs and then provide computer aids that are likely to be essential for support.

More efficient use of scarce expert knowledge The availability of special skills and knowledge can be a limiting factor on growth or profitability. Every profession seems to be in short supply at one time or another: teachers, toolmakers and diemakers, airport controllers, and engineers. It is often feasible to leverage scarce knowledge resources either by improving experts' productivity or by incorporating experts' knowledge about best practices into computer systems.

The potential payback of eliminating or automating routine tasks can be high. The typical engineer spends only about two hours per day doing actual engineering. The rest of the time is spent on activities such as attending meetings, writing reports, gathering information, traveling, answering the phone, and trying to stay current on technology developments. While these functions are certainly a necessary part of the job, intelligent automation of the six hours of office routine each day can often yield an extra hour or two to devote to engineering or project management.

Productivity tools applied on an individual basis can typically provide 2- or 3-to-1 efficiency gains for each automatic task. The overall gain, since not all tasks will or can be automated, might be as much as 50 percent for a typical professional. While this is a notable improvement, sometimes much greater improvements are needed. What if all 20 of a company's die designers are about to retire or the head widget engineer is suddenly faced with 100 widgets to design?

In such cases, the strategy may be to incorporate the expert's knowledge about best practices in a system others can use. For example, standard design elements are fairly easy to build into a CAD database (assuming that appropriate CAD software is selected). If the company's most experienced designer builds the CAD database, other less experienced designers can use it with a reasonable expectation of success. Although the use of expert systems and artificial intelligence also comes to mind, it seems reasonable to anticipate that people capable of building expert systems will be in just as short supply as the experts whose knowledge is to be captured. The practical approach today is to employ existing CAD/CAM technologies, especially database programs, to provide flexible access to information about best practices. Even microcomputer spreadsheet programs can prove quite useful if several technical information users must use information provided by one expert. This might, for example, be a good way of communicating information about allowable configurations or probable costs and performance to distributors or sales engineers. A considerably more powerful approach is to link geometric information from a CAD or solid-modeling system to a relational database of design, manufacturing, and service attributes.

Regarding implementation, the effort required to create even a relatively simple database of best practices is easy to underestimate. The problem is not so much picking appropriate database software or entering the data as making sure that the contents do, in fact, reflect best practices and that they are organized for easy retrieval. In general, however, the cost of establishing such databases can easily be justified for key functions, as identified in the strategic business plan.

Additional strategic benefits There are many other potential CAD/CAM strategies and related benefits, all variations on the often-heard promises of better, faster, and cheaper. CAD/CAM tools can support product objectives that focus on reliability, energy efficiency, speed, initial cost, appearance, or delivery time. These potential CAD/CAM benefits are real, but their achievement depends on linking a competitive product strategy with the best available people, processes, and tools to get the job done.

1.3 Directing the Planning Process

Given the many opportunities that CAD/CAM presents, where should a company start in planning CAD/CAM investment? Like most questions about strategic planning, the answer involves both top-down and bottom-up components. In some enterprises, the top and bottom have yet to meet. For example, some top-management mandates to modernize and computerize have been barely able to budge the organizational inertia called resistance to change. Conversely, many R&D staffs and engineering departments have energetically pursued the state of the art, often without a good answer to the question *why.*

Even in companies in which business and technical managers are not frustrated in their attempts to communicate, there is often a missing element—the customer. Any investment in CAD/CAM must ultimately be measured by the customer's willingness to buy one company's product rather than another's. Both general and technical managers can and should find common ground through a customer-oriented view of what each investment in CAD/CAM might provide.

Thus, the process for evaluating and planning for CAD/CAM from a strategic perspective should ideally start with a top-management mandate to develop a strategic profile of the business—to view its strengths, weaknesses, and opportunities through the eyes of its customers. Such a profile is most useful when it is the joint product of the enterprise's key disciplines, including research, design, production, marketing, sales, and finance. It then provides a common ground for communication.

Even if top management has abdicated its responsibility for technology management (as often happens given the rapid and confusing pace of change), product engineering and manufacturing managers should not take upper management's inaction as a license to avoid the issue. Many jobs can be performed better or faster with CAD/CAM tools; the key is to view the priorities through the customer's eyes.

1.4 Implementing the Strategic Planning Process

The current implementation of CAD/CAM is quite often backwards. Reports of exceptional paybacks and an interest in the technology lead many companies into an immediate search for the "best" CAD/CAM system. This is premature if they have not done their strategic homework. They must first set appropriate product objectives, prepare their people to accept and manage the new technology, and examine whether their current forms of organizations include barriers to integration and overall product and process improvements. If these steps are not taken before CAD/CAM is introduced, the results will be long delays before the targeted productivity improvements are finally achieved. Such delays are costly. Worse yet, when the goals are finally achieved, it often turns out that the companies have achieved Pyrrhic victories over trivial (or at least nonstrategic) problems.

This portion of the chapter examines the strategic planning process and identifies four basic elements of a strategic CAD/CAM plan. A proven method for strategic CAD/CAM planning is presented, and its use in preparing all four elements of the CAD/CAM plan is described.

1.4.1 Alternative implementation approaches

User experience to date has not yielded a universal pattern for the level at which CAD/CAM technologies are investigated and their implementation is planned. Some companies have proceeded on a functional basis, with leadership from within the functional department (e.g., drafting, analysis, testing, assembly, quality control, and machining). Others, particularly autonomous divisions, have proceeded on a product-by-product basis. Still others have a corporate mandate that may span several functions, divisions, and product lines.

Implementation has tended to occur at the lowest organizational level able to justify a system. For example, when 16-bit turnkey computer-aided drafting systems cost $250,000 to $500,000, investigation was almost always at a corporate level in small organizations and at a functional level in large organizations. Until the latest generation of engineering workstations, a company needed to have about ten engineers, analysts, or drafters in one place to justify a turnkey minicomputer-based system. The critical mass of people in a function would naturally be larger for a dedicated mainframe installation.

Perhaps the best approach for most companies in linking CAD/CAM with corporate strategies is to proceed on a product-by-product or family-of-products basis. Function-by-function implementation tends to be too parochial in large companies and below critical mass in small companies. Corporate implementation dominated by central staff, particularly in large multiproduct organizations, tends to be too far removed from both customer needs and operational considerations.

An obstacle to product-oriented implementation is that the organization of many companies is so specialized that no one except the top managers really has overall product responsibility. Nevertheless, it is necessary to conduct CAD/CAM planning from a customer viewpoint to avoid many implementation traps. Many companies find themselves with a hodgepodge of incompatible systems because of function-by-function and department-by-department implementation. Establishing a corporate staff function to manage CAD/CAM implementation has not been successful either; the staff function tends to be too far removed from the customer or too intent on reinventing technology already in the commercial domain.

Some examples may help make this recommendation for product-oriented CAD/CAM planning more concrete. A company like 3M Corporation might begin by looking at the role of CAD/CAM in coatings technology: for such a company, a family-of-products approach would probably best support corporate strategy. A company like General Motors Corporation, in contrast, needs to approach the technology on a market-by-market basis. Even though all cars have engines, drivetrains, bodies, and suspensions, the customer who buys a Corvette has different performance and cost expectations than the buyer of either a Pontiac or a Cadillac. These differences will influence the CAD/CAM strategy.

1.5 A Suggested Planning Methodology

CAD/CAM technologies can serve many objectives, but no company can afford to pursue all objectives with equal vigor. Concentration is thus the key to success. Four basic elements are vital to a planning process through which strategic priorities are established:

- A focus on selected strategic product objectives
- A plan for how the process of design and engineering should evolve to achieve product objectives
- A high priority for the preparation and training of people for a CAD/CAM environment
- The selection of software and hardware tools

1.5.1 Defining product objectives

The starting point for the strategic planning process is to establish a targeted set of product values, including costs, that must be achieved for the enterprise to win a profitable share of business in the years ahead. Product performance goals should be based on a thorough evaluation of the competition; customer needs; and corporate strengths, weaknesses, opportunities, and threats. Neither too vague nor too detailed, the ideal is to set no more than about eight customer-oriented goals that are quantifiable and that can be addressed by using the best available technology.

Setting a good strategic direction requires a blend of creativity and analytical thinking. Above all, it requires a comprehensive understanding of the business and its competitive environment. Yet, it is difficult for any one individual to know the breadth of a business in sufficient detail to set strategy, and it is even more difficult to communicate that strategy clearly enough that the entire enterprise can act on it. When the same problem is given to various executives with backgrounds in marketing, sales, finance, engineering, and production, each tends to see it as a marketing, sales, finance, engineering, or production problem. The resulting "strategic" allocation of resources depends almost entirely on who is doing the allocating. The depth of knowledge required to make strategic use of cutting-edge CAD/CAM technology serves only to complicate the situation.

A strategic planning tool The Product Performance Profile^{SM} is a management tool used to gather an organization's best thinking about strategic options and communicate the results in a powerful, graphic manner. The profile method provides a framework particularly suited to examining the strategic impact of CAD/CAM technologies on a product-by-product or family-of-products basis.

The framework, as shown in Fig. 1.2, is basically a set of eight customer-oriented product attributes designated on an eight-axis graph. The initialism *$APPEALS* formed by the eight attributes ($ cost, availability, packaging, performance, ease of use, assurances, life cycle, and social acceptance) is intended to serve as a reminder that each product attribute should be examined from the standpoint of its customer appeal.

Each of these generic product appeals must be carefully tailored to the product or service at hand—to see the product as the customer sees it. An electronics manufacturer's management, for example, may see itself as being in the cathode-ray-tube display business. However, customers may

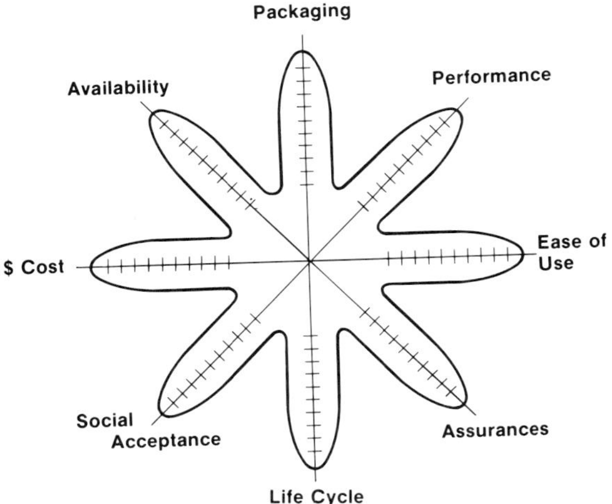

Figure 1.2 Generic product performance profile.

TABLE 1.1 Product Performance Parameters (SAPPEALS)

$ cost	Packaging	Ease of use	Life cycle
As influenced by: Design Producibility Technology Materials Suppliers Production Fabrication Labor cost Facilities Quantities Experience Scrap Inventory Growth trends Installation Financing Selling cost Overheads Automation Simplicity Shipping Distribution	Form both implies and permits function. Mechanisms Layout Surfaces Size, quantity Modularity Styling Geometry Interferences Fit and finish Color, texture Framework Identity Inside, outside	Users include buyers, managers, operators. User friendliness Controls, displays Graphics Ergonomics Training Manuals and documentation User-machine interface Operating conditions Noise and vibration Handling Feedback Simplicity Sensory input and output Anthropometry Illumination Effort Environment Timing Human factors Cognition	Foresight and quality often mean lower life-cycle costs. Lifetime Uptime, downtime Fatigue, wear, and corrosion Deterioration Maintainability Parts and supplies availability Service Component interchangeability Backups Monitoring Upward compatibility Futures Obsolescence Standardization Depreciation and appreciation Energy costs Operating costs Installation costs Disposal costs Life-cycle costs

Availability	Performance	Assurances	Social acceptance
Providing what customers want and why, when, where, and how they want it. Marketing Sales Distribution method Stock Lead times Advertising Ordering Locations Configuration Sizes Quantities Options Pricing Demonstrability Awareness Positioning	The primary functions may be tangible or intangible. Function Appeal Specifications Speed Capacity Power Payload Size Accuracy Repeatability Productivity Versatility Performance vs. constraint (power vs. weight, capacity vs. cost, miles vs. gallons)	Assured performance. Reliability Quality Safety Margin for error Structural integrity Safety factor Stability Redundancy Error detection and correction Failproofing Foolproofing Diagnostics Certification Testing Guarantees	Often, there are hidden costs for builders, buyers, operators, and bystanders. Indirect influencers Status Effect on bystanders Effect on environment Social approval Legal concerns Government regulation Deregulation Tax incentives Restrictions Safety Product liability Pollutants Health effects Insurance Resource consumption Security Side effects Toxicity Customs and taboos Special interests Workers and workplace Political impact

NOTE: Computer-aided systems can help optimize most of these parameters.

see the product as any display technology meeting their needs (e.g., for cost, resolution, and color).

The customer must be carefully defined as well. The customer may not be the end user, but a purchasing agent or a distributor. Individual product attributes ($APPEALS) must then be defined as the customer views them. Assurances, for example, could mean the safety factor to one customer and mean time between failure to another. A list of alternative product parameters for each generic category, as shown in Table 1.1, is helpful in this stage.

The completed profile is a valuable strategic planning tool. Some of its most useful features are that:

- It represents the buyer's view and not that of engineering, manufacturing, sales, or any other isolated function. It overcomes the product, engineering, and marketing myopia common in most companies.
- Each parameter is quantifiable; a 10-point scale is used along each axis, with zero at the center of the graph representing the worst condition and a perfect (and generally unattainable) 10 at the periphery. Although the parameters may start out as qualitative measures, the profile is not complete until there is a quantitative measure for competitive success.
- The competitive positions and appeals of various competitors can easily be overlaid for comparison.
- The market share leader is readily identifiable by a cluster of high scores in the upper left quadrant, as shown in Fig. 1.3.

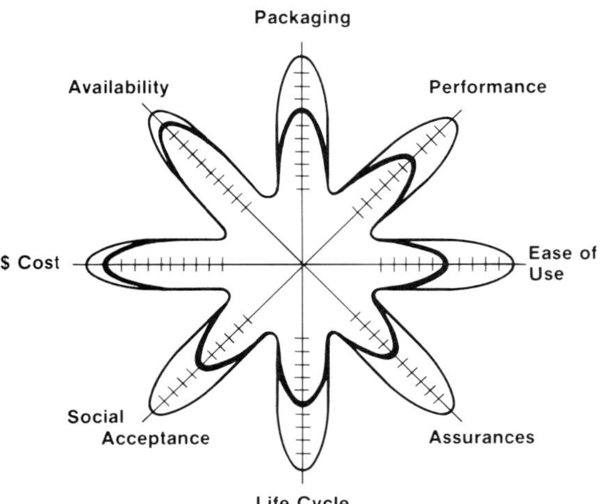

Figure 1.3 Market share–oriented product performance profile.

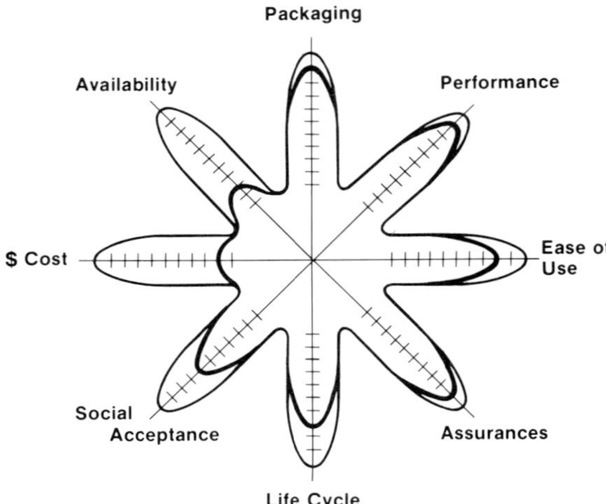

Figure 1.4 Quality-oriented product performance profile.

- The quality leader is identifiable by a cluster of high scores in the right quadrants as shown in Fig. 1.4. [The ability to identify market leaders and quality leaders from product attributes is particularly useful to companies that follow strategies, such as the profit impact of market strategy (PIMS),* that emphasize trying to maximize profits through either market share or quality leadership.
- Day-of-sale attributes are clustered in the upper quadrants, and future attributes are clustered in the lower quadrants.
- The graphs and competitive overlays become quickly recognizable as patterns of competitive position. Strengths, weaknesses, problems, and opportunities are highlighted for strategic response. The amoebalike appearance is intended to emphasize that products must struggle for the marketplace equivalent of an ecological niche.
- The profile forces users to focus and allocate resources in a disciplined manner. As a brainstorming aid for new-product strategies, it focuses the effort without dampening the output of useful ideas. It also helps set performance objectives for departments and individuals, even in areas like research and development.

While other features of the profile are helpful, listing any more would begin to violate one of the principles on which the profile is founded.

---

*A database of business experience collected by the Strategic Planning Institute, Cambridge, Mass. It attempts to correlate various business strategies and conditions with their past successes.

Human factors research tells us that people—and this certainly includes both customers and strategic planners—can deal with only about seven, plus or minus two, variables. Beyond that, they discard whatever is perceived as lesser information. The initialism *$APPEALS,* and hence the concept behind it, is much easier to remember and work with than most planning methods. The power of the Product Performance Profile is that it focuses the strategic planning effort on product attributes that customers keep in mind.

Using the Product Performance Profile to define product objectives The profile should be used as a framework and bookkeeping device in information-gathering and planning sessions. It is important to involve the appropriate people in a series of profile sessions, at times including a cross section of internal viewpoints as well as external consultants, distributors, and customers.

Using the Product Performance Profile to link a CAD/CAM strategy with a corporate strategy generally requires a three-step process. First, the profile must be customized to reflect the business and strategic environment accurately. Each of the eight axes must be defined for the product under consideration, ideally by a group including typical customers. Competitors' relative strengths and weaknesses are compared on the profile axes. If the profile does not accurately correlate with market realities, the initial profile group reviews its assumptions about the customer appeal of the selected product attributes.

Next, an explicit strategy for competitive success must be determined. This involves a decision that the product or family of products will be better than its competitors' with respect to one or more profile parameters and equal or nearly equal in certain other parameters, with minimal resources devoted to those that remain. The company must be superior on at least one parameter. Usually, high performance on two or three related parameters is the basis of a competitive strategy.

For a reliable reckoning, each parameter must be at least approximately quantifiable to distinguish between competitors. Parameters like cost and performance are easy to quantify and rank on a 10-point scale. However, quantifying parameters like appearance—one of the packaging variables—may involve focus panels or other forms of market research. In the early stages, relative rankings are usually satisfactory to understand competitive strategy. Later on, more precise quantification of competitive performance is desirable to make the profile into an operational set of objectives for engineering and manufacturing.

The third step requires a new profile team staffed with experts in CAD/CAM technologies, particularly the technologies with a high probability of profitable application to the profile attributes in which the company intends to excel. For example, if step 1 identified packaging and appearance as a key attribute, then an expert on solid modeling could probably make a helpful contribution. On the other hand, if structural integrity was highlighted as a customer concern, then expertise in finite-element modeling and analysis might be more appropriate. The objective is to consider the best possible technologies to achieve the desired profile and to feed that information back to the first (strategic) profile team. Some members typically participate in both profile teams; the use of a facilitator knowledgeable in both the profile technique and the relevant technologies can significantly reduce the overall effort. In any event, one advantage of the profile is that it immediately focuses the experts on the relevant kinds of problems and opportunities. Even when all three steps are combined in a single session, new insights typically come to light.

The Product Performance Profile is not a cure-all. Applying the profile as a complete management tool requires hard work and intelligence, and special considerations often arise. A large company with multiple products must decide what to profile, whereas the choice for a small, single-product company is obvious. The corporate environment and internal politics can make the profile method impractical, because it works best when there is a spirit of open cooperation throughout an organization. International and multinational companies may approach different customers with the same product. As a result, efforts to achieve worldwide standardization on CAD/CAM networks may require careful thought regarding cultural and regulatory differences.

Companies might view their product profiles as snapshots of the products they must design and build to remain competitive in the near future. With the snapshots of tomorrow's products in hand, design, manufacturing, and marketing can begin to plan for and implement the processes, people, and tools needed to turn those snapshots into profitable realities.

1.5.2 Defining process objectives

Once a workable product strategy is in place, the next step is to translate product objectives into process objectives. For CAD/CAM in particular, a plan for the organization of the product and manufacturing engineering functions is needed. A key advantage of computer-aided methods—allowing design and manufacturing engineers to consider more options and manage more information—is lost if the new methods are unduly constrained by the bounds of the old organization.

Today's manufacturing company is as much involved in handling information as it is in handling materials. Despite this, the design and manufacturing engineering process is often much the same as it was decades ago. One example of changes that might be needed involves bringing design and manufacturing closer together so that information about producibility and manufacturing costs can be considered at the design stage. Such a process objective might be an appropriate way to achieve product goals for cost reduction. However, those goals will not be achieved simply by automating the transfer of electronic blueprints from design engineers to manufacturing engineers. Furthermore, establishing a local area network between the two groups will not ensure communication if it hasn't existed in the past. Rather, design and manufacturing engineers may very well need to report directly to the same manager and meet face to face at frequent intervals. Such organizational changes often take time, yet they are essential to the best uses of the new engineering technologies.

Another possible change in methods might involve replacing physical prototypes with mathematical prototypes. Such a plan, however, raises the organizational issue of what happens to the old testing department. Closer collaboration of marketing, engineering, and customers might be another process objective, one that could be met through the use of computer-generated models early in design. First, though, it will be necessary to decide who in the organization will take the lead in such a joint effort.

As these examples are intended to illustrate, any company that has made a major investment in CAD/CAM and left its organizational chart unchanged is either phenomenally foresighted or is failing to take advantage of the opportunity. Any strategic plan for implementing CAD/CAM technology must also plan for the evolution of the design and manufacturing engineering organizations. The key is to establish priorities for information needs and then design an effective organization for the new process of design and manufacturing engineering. Only then will a company be able to take full advantage of the increased capability for information control offered by computer methods.

1.5.3 Planning people-related strategies

Most companies say something to the effect that people are their most important asset. Fewer plan and budget for personnel transitions with the same intensity that they give to selecting, justifying, installing, and maintaining capital assets like CAD/CAM systems. This often means that the people responsible for managing and for using costly tools are less than fully enthusiastic, trained, and competent. Yet, skilled people multiply the return on any investment in CAD/CAM, whereas unskilled people or an unfavorable organization climate can undermine anticipated benefits. It doesn't make sense to tie up thousands or millions of dollars in a CAD/CAM investment and then wait months or years for paybacks because personnel were not adequately prepared.

The product strategy is supported by process strategies, and these in turn must be supported by people-related strategies. People-related issues that must be addressed include hiring, training, overcoming resistance to change, creating new and relevant standards of performance, participating in cooperative programs with universities, and determining the role of internal and external consultants. In particular, the company must realize that although it may be able to secure delivery on an item of CAD/CAM hardware within weeks or months, acquiring or developing skilled and enthusiastic people will take longer. Ideally, the company plans to "take delivery" of both skilled personnel and hardware and software tools in the same time frame. In practice, this means that work on the personnel part of implementation must be at least parallel to and probably in advance of the usual selection and justification of tools. For example, it can take 6 to 9 months to build a rudimentary database and train a fully productive user on a typical turnkey CAD system. Much of this learning can be done through such agencies as service bureaus, vendor training programs, and independent training organizations, before the system arrives.

Employees should be involved in selecting their own tools, whether on the CAD or CAM end of the business, with full appreciation of the company's product, process, and personnel strategies. Workers should also be involved in establishing standards of performance. Performance measures should reflect the organization's progress toward meeting customer needs and its position relative to its competitors. For example, one appropriate measure of a CAD operator's productivity might be the reduction in defects that result from misinterpretation of drawings in production rather than the number of drawings produced. The customer does not care whether it takes one drawing or a hundred drawings to define a product for production. However, the CAD manager who feels responsible for minimizing errors is likely to support the linking of CAD to CAM to further reduce the possibility of misinterpretation.

Implementation should be planned to minimize unnecessary disruptions. If people are accustomed to filling out forms manually, for example, the new database screen display can be configured to look like the existing forms. The individual's perception of his or her role in the company should also be addressed. For example, drafters who take pride in craftsmanship and attention to detail may not feel kindly disposed to drafting automation unless they have an opportunity for similar recognition and satisfaction with the new system.

It is helpful to create ways to try out the new technology that will minimize the perceived risks of failure. This is especially important in convincing an older generation of knowledgeable engineering personnel to make the transition. These employees and their product knowledge are essential; they must not be left out of the CAD/CAM implementation process. Creating opportunities to try the technology may involve advance training at a service bureau, becoming acquainted with the user groups before buying the software, working on the first job hand in hand with an experienced engineering consulting company, or setting up an internal CAD/CAM facility on a no-cost or low-cost basis to start.

Workers also need to see how the technology benefits them. Fast terminal response, top-management support, and pay raises or bonuses to match productivity and skills increases are some examples of benefits workers will respond to favorably.

1.5.4 Selecting tools

Once product goals have been translated to organizational objectives and workers are prepared to manage the technology, the next step is to select the most effective software and hardware tools to support product, process, and people-related needs.

Naturally, there will be a set of questions about integrating systems, verifying software and hardware vendor claims, ensuring backup, and related concerns. However, the most troublesome questions, from a strategic standpoint, involve decisions regarding an uncertain future.

Computing capability advances that can quickly render today's hardware and software obsolete are occurring within 2- or 3-year periods. For example, in the mid-1970s, few managers predicted that superminicomputers would displace many minicomputer and mainframe applications. More recently, networks of full 32-bit virtual memory microcomputers have been replacing superminicomputers. Similarly, solid modeling is displacing or augmenting many three-dimensional wireframe graphics applications, and relational database structures are replacing hierarchical structures. The cost of main and peripheral memory continues to drop, and central processing units combined with specialized math, graphics display, and database coprocessors are continuing to offer more capability at lower prices.

Once again, the best strategy should be to aim ahead of the moving target. Those who are familiar with the CAD/CAM and computer industry can generally outline the probable positions of companies, capabilities, and standards 2 or 3 years in advance. The key consideration is to allow for upward compatibility—a growth path—in the years ahead.

The cost of conversion from one set of systems to another can be particularly high if past data cannot be transferred or if massive reeducation is needed for employees to learn a different set of software and procedures. For this reason, it is particularly important to look at the inherent flexibility within the database and graphics architecture of any capability contemplated for immediate purchase. For instance, it would be a mistake for most integration-minded companies to commit a vast database of engineering drawings to a CAD system that managed the attribute data only in sequential files (rather than hierarchical, network, relational, or fully associative forms). Both the vendor and users will require a more flexible CIM database architecture to add new applications in the future. Similarly, one should be aware of emerging standards, such as more complete Product Definition Data (PDD) standards as the probable successor to the Initial Graphics Exchange Specification standard for transfer of design information.

The specific examples are meant only to illustrate the need for predicting likely software, hardware, and data communications trends as a part of strategic CAD/CAM implementation. One way to do that, without creating an in-house bureaucracy, is to convene a focus group of outside technology experts and inside engineering managers once a year or so. A manager with broad technical interests, knowledge of the product, and respect within the organization is needed to manage this process. The group should work from a strategic planning tool, such as the Product Performance Profile, in order to determine how new technology trends might affect company capabilities in strategically important areas. It is an educational experience for the internal participants and a low-cost way to survey new technologies. A related benefit is that it is also a good way to screen potentially useful outside experts and familiarize them with the business's operations and objectives. If a new and potentially important technology emerges, a further internal investigation might then be planned.

From a strategic standpoint, there is little to add about tool selection—only the caution that, despite vendor claims and user hopes for easy solutions, the belief that it is possible to buy a "strategic edge" right off the shelf from the local CAD/CAM vendor simply is not true.

1.6 Management Responsibilities

CAD/CAM technologies are having a significant impact on corporate strategy; they often help to make possible that which was previously impossible. Despite this proven potential, managers often abdicate the responsibility for CAD/CAM planning and implementation to more technical subordinates. As a result, the majority of current CAD/CAM success stories are successes in a limited tactical and functional sense rather than a strategic sense. Many companies are not even achieving the 3:1 productivity improvements they originally anticipated, and, worse yet, the technology is being applied to functions for which even a 3:1 improvement will barely affect the product's appeal to the customer.

Top-management commitment and guidance are essential to the best use of CAD/CAM technologies. Without such support, many opportunities will elude the enterprise. There must be a commitment not only to invest in the tools but also to make a comparable investment in product understanding, people, and the changing process of design and manufacturing.

Perhaps the key commitment is to structure an environment in which the strategic issues can be examined at appropriate levels throughout the organization. In such an endeavor, the Product Performance Profile or its equivalent can provide a framework for communication and objective setting. If companies simply implemented the best available technology one year earlier, there would be no productivity problem. Management can use an organized approach to strategic planning, such as the one outlined in this chapter, to implement available technology within a company more rapidly and effectively.

So, is CAD/CAM a strategic issue in CIM? It should be clear by now that the answer for each company will be determined—whether by action or inaction—by management.

CHAPTER

2

Group Technology

Alex Houtzeel

2.1 Background

Beginning with the Industrial Revolution, a great deal of research and development has been devoted to the optimization of design and manufacturing in mass production. When production rates of hundreds of thousands or millions of units are involved, very small gains in efficiency can lead to significant production or cost savings. There have been many improvements in mass production technology over the years, and the twentieth century would have been a much different time without them.

Most manufacturing is not done with endless assembly lines or other mass technologies, however. Even in the United States, approximately 75 percent of all manufacturing is carried out in smaller batches—in lots ranging in size from one or two to a few thousand. Until recently, very little has been done to improve batch manufacturing, despite its importance. The seemingly random nature of both design and manufacturing activities in the batch environment have made it difficult to even define approaches to optimization. There were approaches to queuing and sequencing programs in batch manufacturing and to standardization of certain design and manufacturing activities, but those efforts were somewhat limited and never had a great impact. A basic problem is that it is difficult to invest large sums for research and development into operations which may produce a few washers here and a few shafts there or a half-dozen gears. It is one thing to shave a fraction of a penny from an item which is to be produced in the tens of millions; it is another to justify efforts to save pennies on the cost of producing an item made in a lot size of two or three.

The problem, of course, is that improvements made to batch manufacturing must be generic if they are to be cost-effective. They cannot focus on the production of specific products; instead, they must deal with principles which can be applied to all, or at least almost all, the items produced in a batch manufacturing environment. Considering the random nature of batch manufacturing, solving this problem has not been easy.

There have been scheduling programs, sequencing programs, and, more recently, material requirements planning (MRP) and just-in-time systems which have been developed to help improve manufacturing productivity. All of these types of systems are ultimately geared to the production of parts to be finished at the right time for final assembly. Although this is a worthwhile objective, it is often accomplished at the expense of other things.

Constantly meeting deadlines does not mean that assets such as people and machine tools are effectively utilized. In fact, such systems can increase production costs, because additional people and machines are required to fulfill their objectives. It is not difficult to do more with more, but it is not easy to do more with less, especially in batch manufacturing environments.

Since World War II, there has been an increased awareness of the problems of batch manufacturing. The general feeling, however, has been that more and more items would be made in mass production environments and that the importance of batch manufacturing would decrease. That has not been the case. We are now beginning to see that the trend is to smaller and smaller lot sizes, in mass production as well as in batch manufacturing. Government regulations, energy costs, the demand for more personalized products, marketing considerations, and other factors are resulting in smaller lot sizes, even of basically standard items. An automobile may be a mass-produced standard item, but when the producer specifies a certain type of stereo system, air conditioning, a special steering wheel, and other factory-installed options, the car is ultimately produced in a very small lot size. As specialized features become more integral to basic products, the trend will continue.

All of this could have a tremendous impact on future costs. An automobile manufacturer can produce a car at a cost of $3000 to $8000. The same car would cost a job shop $500,000 or $1,000,000 to manufacture. If more and more things move from mass production to small series production, the impact on the costs of consumer goods and inflation rates could be dramatic, not to mention the need for more production personnel.

Skill levels are another factor. In mass production operations, the skill levels required for standardized jobs are relatively low. In batch manufacturing, however, much higher skills are required. With an increasing shortage of skilled machinists and others, this could be an important problem. Whatever the reasons, recent years have seen sharply increased recognition of the problems of batch manufacturing and greater efforts to help overcome them. Advances in computer technology have been an impetus.

2.1.1 The search for similarities

It is apparent that when a company makes a particular series of a type of product, there inevitably are underlying similarities among at least some of the parts in production. For example, a company producing centrifugal pumps may manufacture many different types of such pumps. In doing so, however, it will make many similar impellers, housings, packings, and so on. The same is true in the manufacture of airplanes, machine tools, and all other mechanical, electrical, or electronic products.

The fact that there are such similarities is obvious and has long been recognized. Until recently, however, it was difficult to take advantage of the similarities. A supervisor on the shop floor might organize the sequence of parts awaiting production on a given machine tool to help minimize the setup times required to produce the parts, and there have been other types of informal attempts to organize parts by similarities. Very recently, flexible manufacturing systems based on part similarities have been developed. These systems tend to deal with larger lot sizes—less than mass production, but more than small batch quantities.

There have been other approaches to the utilization of underlying similarities in the batch manufacturing environment. In the manufacture of machine tools, for example, certain types of similarities are obvious. Lathes may be manufactured with different horsepower ratings, feeds and speeds, and sizes, but many spindles, gears, and shafts are really very similar, even though the main frame may be different. These types of perceptions were the basis for work carried out in the 1940s and 1950s. The basic principle was that if numerical values which reflected similarities could be assigned to parts, then it might be possible to take some advantage of the similarities. This was done through classification and coding systems, which were means of classifying parts according to similarities and assigning code numbers which reflected those similarities. The code number could then be used as a tool to find classes of similar parts.

The initial applications for these concepts were in design retrieval. If a designer had a means of knowing whether the same or a similar part had been designed in the past, unnecessary designs and duplication could be avoided. In the 1950s, a number of systems were developed to facilitate design retrieval. They became popular in Europe, especially in the United Kingdom and in the eastern bloc countries, and in the United States. Essentially, they were systems which permitted the designer to describe a part in numeric or alphanumeric values. When a new part entered into the

Figure 2.1 Classification and coding and design retrieval.

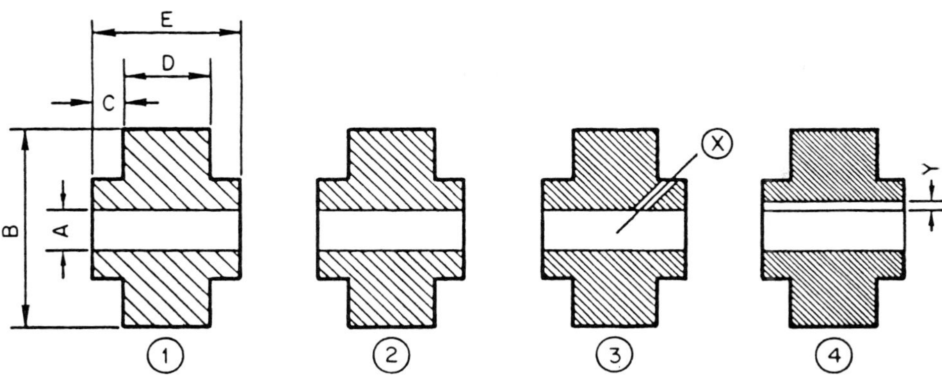

Figure 2.2 Similar parts based on shape.

design process, the designer could code it and, with the code designation, retrieve the same or similar drawings from a file cabinet.

The early systems were manual. The designer had a handbook illustrating various design features and their associated code numbers. He or she would manually code the part by using the code handbook as a reference. The drawings themselves were stored in manila envelopes or folders in file cabinets (Fig. 2.1). These systems were quite useful, but they had their limitations. For the most part, they were tied to the particular products of specific users. They were highly customized to meet individual needs.

A second problem was size. Retrieval with a code number is relatively easy when the file cabinets contain a few thousand drawings. When the number of drawings gets into the hundreds of thousands, however, complications set in, both in the coding process and in the retrieval process.

Another significant problem grew out of the design application itself. Design features are not the only criteria for similarity. A group of parts may look somewhat the same (Fig. 2.2), but if part 2 is made of tungsten carbide, part 3 of stainless steel, and part 4 of cast iron or if lot sizes or other factors differ significantly, different manufacturing techniques will be required to produce each part. Therefore, although they may look the same, they are not the same from a manufacturing point of view.

At the same time, parts that look different may in fact be the same from a manufacturing point of view. Figure 2.3 illustrates a group of apparently dissimilar parts. All of them are castings, however, and the only operation to be performed on each of them is a holing operation. From the manufacturing point of view, they are indeed similar.

2.1.2 Meeting design and manufacturing needs

In the Federal Republic of Germany, at the University of Aachen and elsewhere, the emphasis shifted from classification and coding for design retrieval purposes to the development of classification and coding systems which addressed similarities in manufacturing processes. This was a very important step, because it led to a realization that the grouping of manufacturing elements could lead to an inexpensive method of mass production, with groups of machine tools especially set up to manufacture families of similar parts. In other words, work cells of dedicated machine tools could be formed to produce families of parts with similar manufacturing requirements. This was the beginning of group technology for manufacturing (Fig. 2.4).

In addition to work in the Federal Republic of Germany, efforts were made in the United Kingdom, France, and the eastern European countries to optimize batch manufacturing by using group technology principles. The question was not only how to find parts with similar manufacturing requirements but also how to balance the manufacturing load requirements with available machine tools to make the parts. The work-cell concept could not be effective unless loads were balanced so that existing resources—both personnel and machine tools—were effectively utilized.

Experiments in work-cell implementation were successful primarily in

Figure 2.3 Similar parts based on manufacturing process.

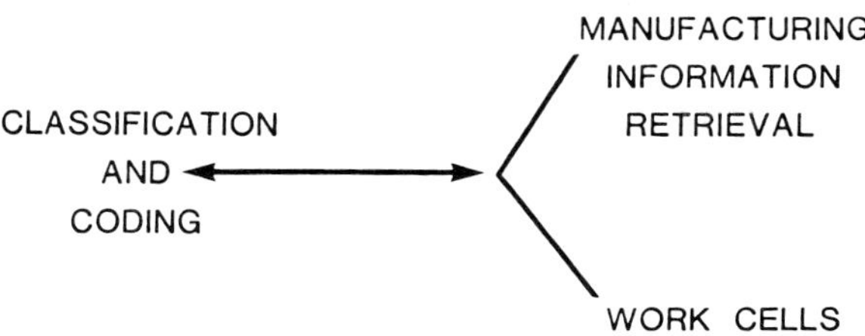

Figure 2.4 Classification and coding and manufacturing information retrieval and work cells.

small European companies, where load balancing and grouping could be done manually. In the United States, where batch manufacturing was often on a much larger scale, the approach was not very successful.

A major breakthrough occurred in the late 1960s in the The Netherlands. The first classification and coding system which served both design and manufacturing needs was developed, and the system was implemented on a computer rather than operated manually. New computers which could interface with users interactively rather than work in batches helped to make this possible. It marked the beginning of the relationship of the computer to classification and coding, and it was a major step forward in the development and use of computer-aided manufacturing (CAM) software. This was also the first use of classification and coding as a tool to interface with a computer design and manufacturing information database (Fig. 2.5).

Figure 2.5 Classification and coding and interface to parts database.

2.2 Classification and Coding

With its application to both design and manufacturing, classification and coding came of age. It had evolved from a means of manually indexing drawings for retrieval purposes to a compact vocabulary for the storage, recall, and use of large amounts of design and manufacturing information. With the developments in The Netherlands, classification and coding embraced designs, manufacturing information, cost data, and much more.

In classification and coding, code numbers are used to identify parts by specific characteristics. By doing so, they make it possible to classify parts—group them according to their characteristics. In a sense, this is a sifting of all of a company's products and detailed parts by similarities. It is a major step in the translation of the seemingly random nature of batch manufacturing operations into a rational order.

When handled manually, the length of a code number was significant. The longer the number, the more tedious the coding process and the larger the chance for error in the transference of a code number from one piece of paper to another. With the computerization of classification and coding, code length was no longer as critical. Thus the earliest truly comprehensive classification and coding system for design and manufacturing application was a matrix of 300 positions—a "sieve" with 300 holes (Fig. 2.6). Within this system, a code number could be as long as 30 digits, and each digit had a possible value ranging from 0 to 9. This meant that 30 different parameters could be coded, each with as many as 10 different attributes.

The growth in code length somewhat paralleled the evolution of computer technology. The computer not only was able to handle the lengthier codes with ease but also made it much easier to assign the codes to parts. In the interactive mode, the computer would ask a series of questions requiring yes, no, or numerical answers and would then generate the code number. As the process became easier, different departments became interested in using classification and coding. What began with design and then spread to manufacturing now began to find applications in quality control, testing, purchasing, and other areas. The code length grew from the original 7 or

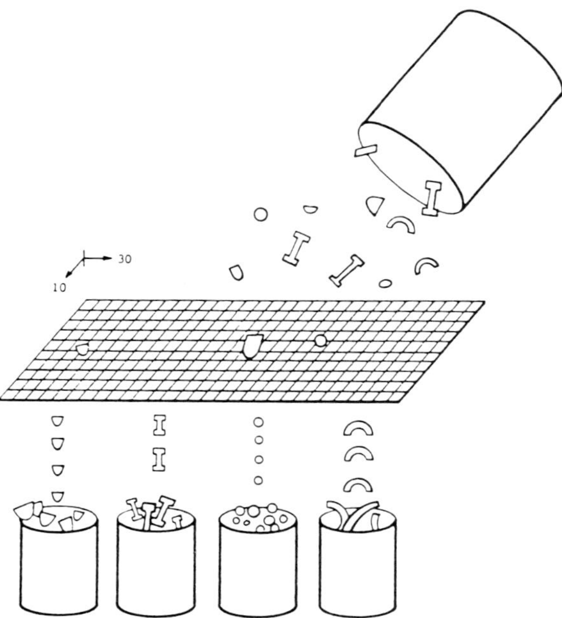

Figure 2.6 The sieve.

TABLE 2.1 Part Attribute Information Required by Various Departments in a Company

Data	Department			
	Purchasing department	Design office	Work preparation	Manufacturing planning and scheduling
Main shape	—	×	×	×
Shape elements	—	×	×	×
Type of material	×	×	×	×
Rough shape	×	—	×	×
Dimensions	×	×	×	×
Dimensional accuracy	—	—	×	×
Batch size	×	—	×	×
Production time	—	—	—	×
Processing sequence	—	—	×	×

8 digits used for drawing retrieval in customized systems to more generalized codes of 18 to 25 or more digits.

2.2.1 What do you code?

In the development of a classification and coding system, the most basic question is "What should I code?" Different departments require different types of information (see Table 2.1). For example, the purchasing department is interested in information about type of material, rough shape, dimensions, and, of course, batch size. The design office, on the other hand, deals with such things as main shape, shape elements, and dimensions. Batch size is much less important. Process planners need still more information. It is very easy when developing a multiple-purpose system to attempt to capture so much information that the entire system becomes cumbersome and unworkable. The key is to be able to define the information that is really needed.

An important issue in the development of this definition is the number of different shapes and manufacturing methods which the system will have to embrace. There are some basic principles involved, which most people have difficulty in accepting at first. The experience of a centrifugal pump manufacturer is illustrated in Figs. 2.7 and 2.8. The company had between 65,000 and 75,000 part numbers and produced parts in batches ranging in size from 1 to 50. Management's assumption was that everything was different and that all parts would have to be included in the classification and coding database in order for the system to be truly effective. A random sample of parts going into production was coded—roughly 200 parts a week for a period of 10 weeks. At the end of the 10-week period, 2100 parts had been coded. As Fig. 2.7 indicates, the number of different shapes leveled off at 900. All others entering into production were essentially of the same shape. In other words, instead of 65,000 to 75,000 different shapes, the company actually had 900.

As illustrated in Fig. 2.8, the same type of phenomenon occurred with respect to new process plans. The number of distinctly different process plans leveled off at about 1200. All of the other process plans created by the company were essentially reinventions of the wheel.

One may argue that all centrifugal pumps are essentially the same and that the results cited above are not surprising. Since then, however, the author has been witness to the same type of experience at nearly 150 other installations. Within 10 to 30 weeks of the beginning of the coding of representative samples of parts going into production, the curves for both the number of distinctly different shapes and the number of distinctly different process plans will flatten out. Even in the very largest of organizations, the total number rarely, if ever, exceeds 4000 to 8000 different shapes or process plans. Thus, a company may have 500,000 part numbers or process plans but no more than approximately 8000 of them will be different. Invariably, company personnel will argue that these principles do not apply in their organization. An actual test, as indicated above, will always demonstrate otherwise.

There is still further argument, however. "What happens next year, when we change models?" Unless a company drastically changes its business, from machine tools to airplanes, for example, the mix of parts will stay about the same. Washers, gears, spindles, and other components that are basic to large classes of products do not change easily.

Thus the size of the classified universe for many companies can be defined by looking at all the produced parts over a period of about 10 to 30 weeks. In that time, it is possible to identify all of the design and manu-

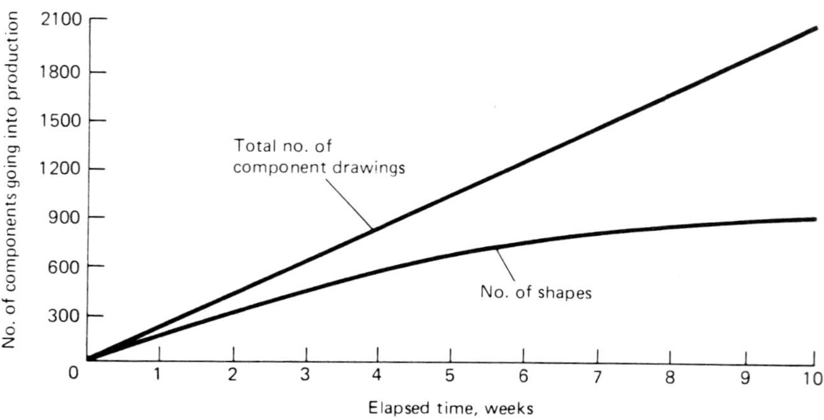

Figure 2.7 New component shapes for manufacture.

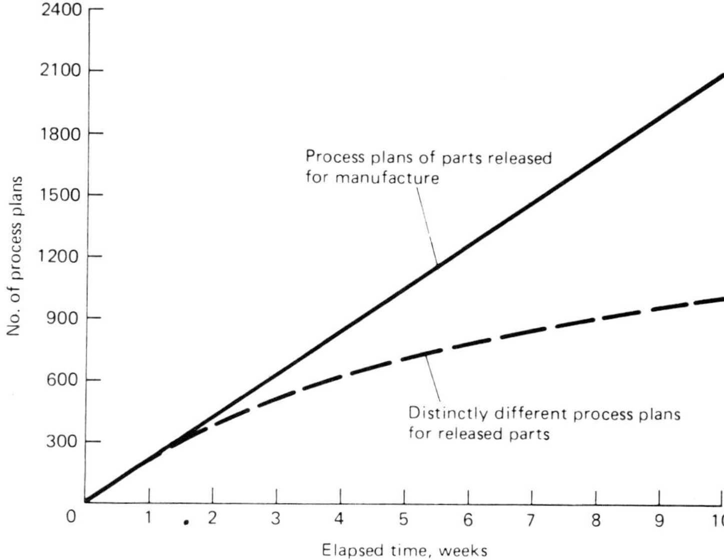

Figure 2.8 New process plans for manufacture.

facturing attributes of importance. This assumes, of course, that one has started with a coding system which is capable of capturing the different attributes. Assuming that this is the case and that roughly 10 to 30 weeks of production has been classified and coded, it is possible to analyze the coded data to begin the development of relevant design and manufacturing families of parts for standardization and retrieval purposes. The results can be impressive, even in the early stages of implementation.

2.2.2 System criteria

A classification and coding system is a tool to capture relevant design, manufacturing, and other features of parts. It is also a tool to analyze and retrieve parts by design and manufacturing features, and it is a communication tool with an information database which can ultimately be used to link different computer-aided design and computer-aided manufacturing (CAD/CAM) systems.

A basic issue in the creation of a coding system is its form—should it be numeric, alphabetic, or alphanumeric? In the development of postal zip codes, telephone numbers, and other coding systems, a great deal of research has been conducted on the effectiveness of various ways of creating a code number. It has been found, for example, that blocks of no more than four digits are most effective for lengthier codes. It has also been found that numeric rather than alphanumeric systems are least prone to errors. With each digit being decimal-based (10 possible values), the numbers are easy to understand and work with. There appears to be no doubt that the best type of code designator is a series of decimally based numeric values expressed in blocks of four digits.

What should one look for in a classification and coding system? What characteristics correlate with ultimate effectiveness? A very basic requirement is that the system be product-independent. If it is customized to specific products, it will not last. The day a somewhat different product is developed, the system will fall apart if it is too product-dependent. A product-independent system will meet all needs for now *and* the future. It is also essential that the system be capable of classifying all parts. While its application may begin with machined parts or some other specific type of product (such as purchased parts or assemblies), it should ultimately be capable of handling assemblies and subassemblies, tools, purchased parts—any and all components which may at some time enter into the manufacturing stream.

Reliability is yet another essential component. There should be checks within the system to help assure accuracy and minimize coding errors. One of the functions a classification and coding system serves is to capture experience so that it is not lost when personnel leave or retire. Reliability is essential to the retention of this experience.

The system should also be extendable. As indicated above, its first applications may account for only a small portion of its ultimate uses. It should be capable of ultimately capturing quality control information, for example, or testing data. In line with this, it should be conceived with the current and future needs of all departments in mind. In order for different departments of a company to get the information they require, the code should capture information about manufacturing operations as well as about shape and other part attributes.

Potential interfaces with other types of systems must also be considered. For example, when part families are established, it should be possible to create a master numerical control (NC) tape for a family. Such a tape would require only changes in dimensions when used to manufacture parts within the family and would consequently save considerable time in NC tape preparation. The capability of interfacing with computer graphics is another important feature. The system should allow graphic systems to access information about design, NC requirements, manufacturing requirements, and so on.

In terms of these applications and others, the classification and coding system should not create a duplicate database. It should interface with existing parts information databases or provide the framework for a new comprehensive database, i.e., relational database if none exists.

2.2.3 Computerization

As indicated previously, classification and coding systems were initially manual. The design engineer compared the sketch of a new part to drawings of part features contained in a handbook, derived a code number, and then searched for a similar drawing in a file cabinet.

Although classification and coding were first linked to a computer in The Netherlands, with the advent of convenient interactive modes of computer accessibility, its real growth and the evolution of computerized classification and coding took place in the United States as the computer found its way into the manufacturing environment.

Manufacturing companies first began to use computers as accounting tools. In the early 1970s, when interactive capabilities were still relatively primitive, design and manufacturing engineers had low priorities in the allocation of computer time. The introduction of the minicomputer was an important turning point. The minicomputer freed design and manufacturing engineers from the need to access corporate mainframe computers. Its evolution coincided with the development of increasingly sophisticated interactive capabilities.

The introduction of computer graphics systems in the 1970s also helped to simulate the use of computers in the design environment. By the end of the 1970s, minicomputers were no longer unusual in design or manufacturing offices. Computerization greatly enhanced the effectiveness and efficiency of classification and coding systems. It not only significantly reduced error rates, particularly when coding was done in different locations, but also greatly speeded the retrieval of design and manufacturing information.

A modern minicomputer takes the user through a decision-tree process by asking questions appropriate to the identification of relevant part features and manufacturing requirements. The computer can do this very well and then can integrate the results with applications such as computer graphics, NC tape generation, and computer-assisted process planning.

2.2.4 Code structures

There are two basic types of code structures: hierarchical, or monocode, and attribute, or polycode. In a hierarchical code, each position relates to the preceding position. Thus if the first digit defines a main shape, such as rotational, the second digit will define a feature related to a rotational part, and the next digit will define a feature related to the feature defined in the second digit. This is the basic philosophy of a hierarchical code such as the Dewey decimal system, which is used in libraries.

Figure 2.9 illustrates a typical hierarchical code structure. As the diagram illustrates, each digit is directly related to the preceding digit. Thus the first digit, 3, may define a rotational part. The digit in the second position may then define the size of the outer diameter of that rotational part. Thus a 1 in the second position may refer to a size of $\frac{1}{2}$ in or less, a 2 may refer to a size of $\frac{1}{2}$ to 1 in, and so on. In the third position, if a 1 is preceded by 2, then a 3 may define an inside diameter of a particular dimension; or if a 1 is preceded by 1, then a 3 might have a totally different meaning.

As the process continues, parts may be grouped into families with increasingly specific feature definitions. Each digit fine-tunes the information contained in the preceding digits. The meaning of any number in any position (a 2 in position 3, for example) will rarely, if ever, be the same.

Hierarchical codes are compact—it is possible to include a great deal of information in a few digits because each digit is an exponent of the preceding digit. The hierarchical approach is very useful in highly customized applications for retrieval only. It also works well in a manual system.

While a hierarchical code is good for design retrieval, it is not useful for much else. Because the same digit in the same position may have different meanings at different times depending on the preceding digit, hierarchical codes do not lend themselves to computerization or to easy analysis. Their usefulness is thus limited when analysis and comparisons of data are desired.

Attribute or polycodes are based on the total population of different

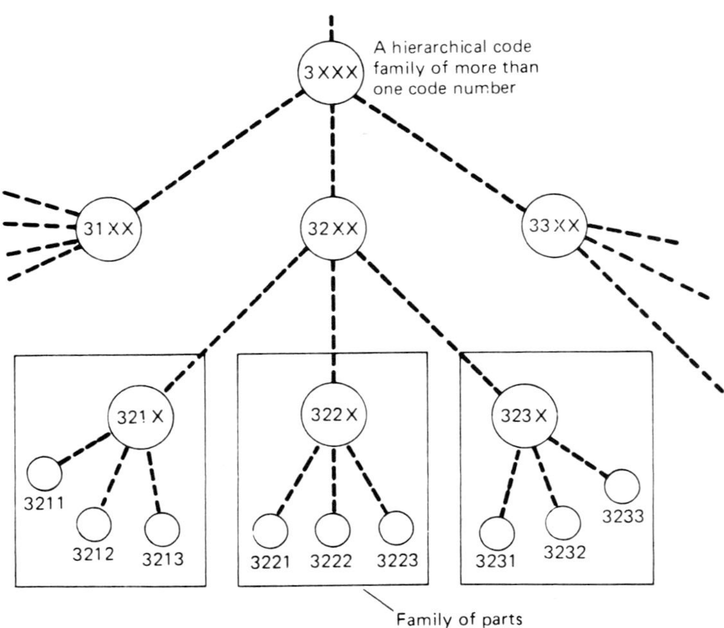

Figure 2.9 The structure of a classification and coding system based on hierarchical principles.

Digit	Class of feature	Possible values of digits							
		1	2	3	4	5	6	7	8
1	External shape	$Shape_1$	$Shape_2$	$Shape_3$	—	—	—	—	—
2	Internal shape	None	$Shape_1$	—	—	—	—	—	
3	≠ holes	0	1-2	3-5	5-8				
4	Type holes	Axial	Cross	Axial and cross					
5	Flats	None	Internal	External	Both				
6	Gear teeth	Spur	Helical	None					
7	Splines								

Figure 2.10 Attribute code structure.

attributes. A typical attribute code structure is illustrated in Fig. 2.10. As the figure illustrates, each digit and each position (value) of each digit has a specific meaning. Thus, for example, the first digit might always describe external shape; the second, internal shape; the third, the number of holes; the fourth, the type of holes; and so on. A complete code number will identify all of the part's major attributes. As a result, the meanings of digit values for any given part will be consistent in the system.

This opens the door to detailed computer analysis and the definition of the specific operations required to make all the parts in a specific group. If, based on the Fig. 2.10, the user needs to know how many parts require cross holing, for example, all the computer needs to do is to check on the number of times the value 2 appears in the fourth position. The same is true of much more complex comparisons.

The problem, of course, is that there are a great many possible attributes in the total universe of parts, and so a pure attribute code number would have to be very long to accommodate all of the possibilities. The solution to this problem is a hybrid structure—a combination of a hierarchical code and an attribute code structure. The first digit in such a structure could refer to the main shape configuration (Fig. 2.11). An attribute matrix is then tied to each main shape. Since attributes are different for different types of parts (e.g., sheet metal or turned parts), computer analysis is still possible. The hybrid code offers the benefits of attribute coding but in a much more compact form. Most contemporary classification systems are based on the hybrid approach.

2.2.5 Attributes

Experience has demonstrated that there are certain attributes which must be coded in order to accomplish the design and manufacturing objectives of a classification and coding system. Some of them are universal—they seem to apply to every company in every industry; others are more industry-

or company-dependent. The universal elements include:

- *Main shape.* Rotational, box form, sheet
- *Shape elements on the main shape.* Cones, holes, slots
- *Element position.* Where are these shape elements?
- *Dimensions.* A main dimension, an auxiliary dimension, and the ratio of dimensions
- *Accuracy.* How critical are the tolerances?
- *Material*

Beyond that, there are a number of pieces of information which are more industry-, factory-, or operation-dependent. They include such things as lot size, piece time, setup time, the relations among necessary operations, and other specific factory information (relating to such things as testing, assembly, and quality control).

2.2.6 Developing your own system

In the preceding paragraphs, we have discussed the types of attributes to be captured in a classification and coding system. Many companies have chosen to develop their own classification and coding systems rather than purchase a system from a vendor. The results have ranged from moderate success to absolute disaster. It is a process which can be described in relatively simple terms but which is much more complex in actual operation. To begin, one must consider that there are parts which are designed and manufactured in-house, others which are designed in-house and man-

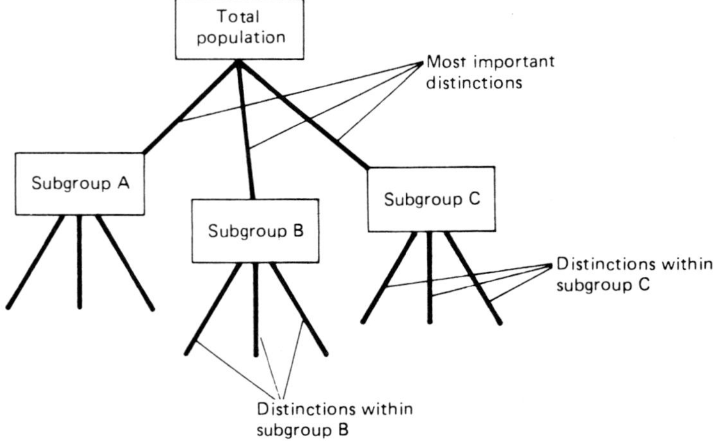

Figure 2.11 Hybrid code.

ufactured elsewhere, and still others which are purchased. Ultimately, the system should accommodate all of these parts.

The first step in the development of the classification and coding system in-house is to sample the existing parts database. The size of the sample will vary with the size of the total base, of course. It may be 20 to 25 percent of a small database of perhaps 2000 active parts and a considerably smaller percentage for much larger active part files. Generally speaking, the sample might range from 1000 to 8000 parts.

The next step is to assemble the sample drawings into families. This, to say the least, is not easy. It requires a good-size room, tables on which to stack drawings, sharp eyes, and a great deal of design and manufacturing experience.

Basically, one should look for classification "buckets" defined by different features. Industry standards may be a useful guide for the definition of these features. The characteristics which are cited in industry standards or specifications are worth considering. One must identify the frequency of occurrence of the chosen features in the drawings contained in the selected sample, work out a rational hierarchy of those features based on the sample, link them to the machine tool capabilities of the company in question, and evaluate the results. It is a continuing process of examination, analysis, review, adjustment, reassembly, and reanalysis. It is a very elaborate activity which requires many labor-years of hard effort. In some companies, the work started with perfectly cylindrical parts and initial results came quickly. Unfortunately, as these companies discovered, perfectly cylindrical parts do not reflect the complexity of the total task. Thus, projects which started off with high optimism ended in disappointment.

It is possible to develop successful classification and coding systems in-house, and that has been done. For most companies, however, doing so is not really cost-effective, and the usefulness of such systems has been limited.

2.3 Manufacturing Applications

Up to now, we have looked at the need to find similarities and how to classify them and at the nature of code designations. We have also briefly mentioned design retrieval as the basic application of coding and classification systems. There are other, more important applications of classification and coding and group technology which emerged in the 1970s and are now the focus of these technologies. The use of classification and coding in group technology applications in manufacturing was the biggest step forward in the evolution of classification and coding systems.

It was one thing to discover ways of codifying similarities in part features and using those features for design retrieval. As was pointed out earlier, this type of approach had limitations, particularly with respect to computerization. With the introduction of the attribute type of structure, it became possible to begin to look at the total design and manufacturing picture in different ways.

Group technology is a means of increasing productivity in design and manufacturing by taking advantage of underlying part similarities. Obviously, classification and coding could relate synergistically to it. Classification and coding, as part of a group technology system, could ultimately impact the total design and manufacturing process in a number of significant ways. It could greatly improve batch manufacturing efficiencies and lower batch manufacturing costs by standardizing design and manufacturing processes and by defining the most practical cost-effective methods of producing each and every part.

Standardization refers not only to the elimination of unnecessary duplication but also to the selection of the "best" way of doing things. The best way involves the optimal utilization of personnel, time, materials, and equipment. This in turn requires a thorough understanding of the flow of parts through the manufacturing facility, the load on each of the machine tools within that facility, and the relative costs and cost trade-offs of production on those machines.

The development of highly efficient and relatively inexpensive computers and of sophisticated software programs helped to make these objectives attainable. At the same time, however, the "ultimate" ingredient remained the knowledge of experienced design and manufacturing personnel.

2.3.1 Analysis for standardization

Perhaps the most important use of modern classification and coding systems is in the analysis of data based on similarities of parts and the use of that information for design and manufacturing standardization. This process can be illustrated with an example.

The management of a large machine tool company felt that the introduction of classification and coding would provide a good vehicle for the retrieval of existing designs and process plans. The parts were coded, but it soon became apparent that there were significant duplication problems. For example, it was found that there were 521 gears with the same or a similar code number; the differences among them were minimal. If a design engineer were to use the classification and coding system for retrieval, he or she would have found 521 gears. Instead of checking each one of the gears, it would have been easier to design gear 522. In fact, that is what happened.

It was also found that the 521 gears had 477 different manufacturing process plans associated with them. If each process plan were four to five pages long, a process planner using the system for retrieval would have to deal with 2000 pages of manufacturing information. Again, obviously, it would be easier to develop a new process plan. When these and other problems became known to top management, a vigorous standardization program for both design and manufacturing was undertaken. After a good deal of work, the 521 gears were reduced to 30 standard gears with 71 manufacturing process plans (taking into account different manufacturing methods for different lot sizes). All of the other designs and process plans were eliminated.

2.3.2 The standardization process

Design and manufacturing standardization requires a considerable amount of effort. The first problem is how to deal with a database of 50,000 to 100,000 parts or more. It is virtually impossible to look at all the designs and all the manufacturing process plans. It is here that the principle of similarity of designs and manufacturing methods becomes valuable.

As pointed out earlier, the number of truly different part manufacturing process plans in any manufacturing environment is limited. All of a company's distinctively different parts attributes will be captured by coding only a few thousand parts, usually somewhere between 2000 and 8000, depending on the company and its products (Fig. 2.12). Once the curve of different shapes and manufacturing process plans (Figs. 2.7 and 2.8) flattens out, there is no need to code any more for analysis purposes.

The next step is to analyze the code numbers to define families of parts. These families may be design- or manufacturing-oriented. By looking at individual part attributes and the frequency with which they occur, it is possible to begin to determine the major families. For example, families may be defined by certain rotational parts within certain diameter ranges and within certain tolerance ranges.

The product mix analysis is a major tool in the differentiation of parts and their division into families. Once the major families are defined, the

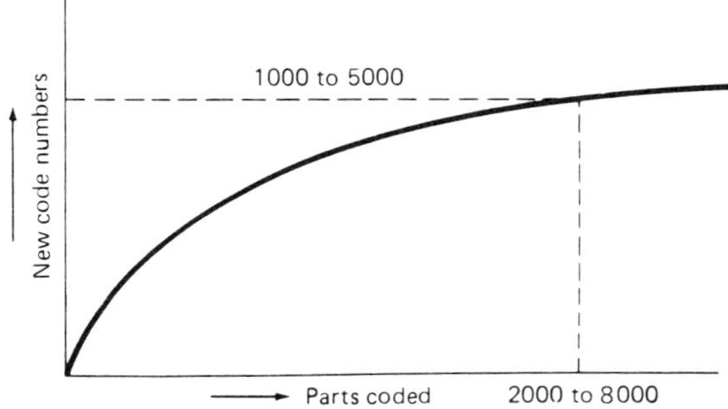

Figure 2.12 Coding of parts to capture distinctly different attributes.

process continues with the breaking down of the major families into smaller and smaller families—each with code numbers describing more detailed aspects of each part and its attributes. For example, from a manufacturing point of view there might be a major separation into rotational and non-rotational parts. The rotational parts might be then divided into those which are turned only and those which are turned and milled. Further separations might depend on such things as grinding and special features like hardening, surface hardening, and surface treatment. Thus by separating a representative sample of the parts in production, manufacturing families of parts can be established.

After the establishment of families, experienced engineers and designers are required to consider each particular family and decide on the standards required for optimal design and manufacturing of that family. Following the optimization by family, the impact of standardization from a manufacturing point of view must be considered, particularly with respect to the availability of machine tools. The optimum manufacturing methods may not be appropriate to the available machine tool capacities. It is obviously essential that loads be balanced so that capital equipment—machine tools—is effectively utilized.

2.3.3 Load balancing

For proper load balancing, the code numbers should include such factors as lot size and number of releases per year. With the inclusion of such information, it is possible to relate each family and each part within each family to load demand over a period of time. One must next look at the individual machine tools available for the production of families of parts. Information about the parts and information about the machine tools available to manufacture them can be combined in matrices.

The principle of a code number matrix is that as one can identify the features of a part by code number, one can also use the same code number to identify the attributes which can be manufactured on a particular machine tool. For example, two different parts are identified in Fig. 2.13. One is a shaft type, with a collet on it, and the other is a shorter flange type. The first part has code number 1220 3251 1141; the second is 1210 3911 1132. Could both these parts be manufactured on a specific lathe?

The part attributes which can be manufactured on that lathe are expressed in the code number matrix. A part with a code value of 1 in the first position could be manufactured on that lathe because in the matrix there is a meeting at that position. However, a part with the code value of 2 or 4 in the first position could not be. Figure 2.13 thus is a "nameplate of capabilities" of the lathe expressed in the matrix.

Looking at the first part, one can see that the code number matches the machine tool matrix, and the part could thus be manufactured on that lathe. Looking at the second part, one can see that the sixth position, with a value of 9, does not match the machine tool matrix. Since this particular position defines dimensions, it is apparent that the part in question is too large to be manufactured on the particular lathe. This process opens the road to matching families of parts with code numbers to dedicated machine tools or groups of machine tools, as illustrated in Figs. 2.14 and 2.15.

2.3.4 Practical application

In day-to-day operations, this is how the results of the standardization process are applied (Figs. 2.14 and 2.15): The design engineer, working with a sketch of a flange, codes it interactively and gives it the number 1772 3231 3144. Going into the parts database, the engineer finds an existing design, part number 2576707. This is the design-retrieval process of the 1950s and 1960s.

The second step is the most important one. The computer system automatically checks the code number to determine if it fits into a matrix represented by the capabilities of machine tools as identified in Fig. 2.15. In this particular case, the code number fits the matrix, as identified in the

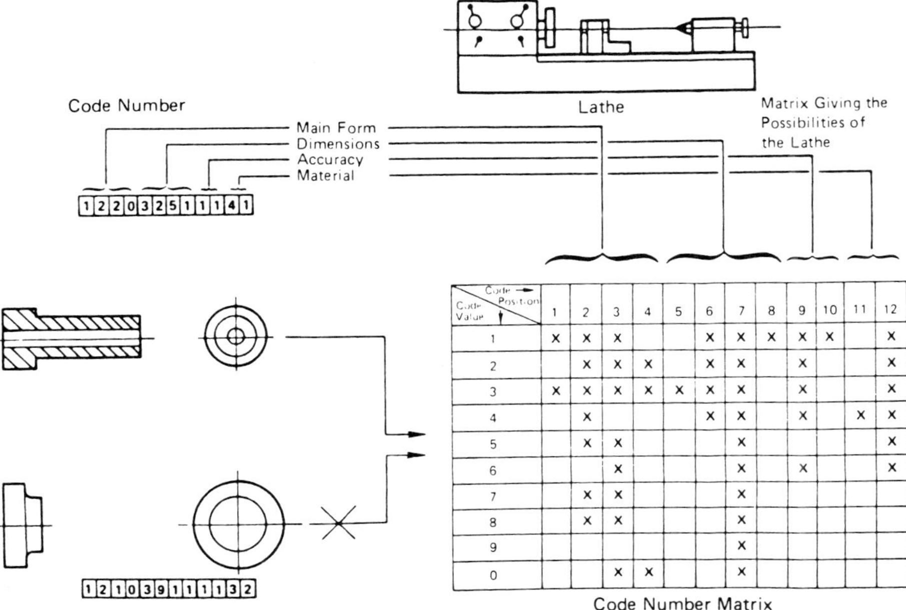

Code Value ↓ / Code Position →	1	2	3	4	5	6	7	8	9	10	11	12
1	X	X	X			X	X	X	X	X		X
2		X	X	X		X	X		X			X
3	X	X	X	X	X	X	X		X			X
4		X				X	X		X		X	X
5		X	X				X					X
6			X				X		X			X
7		X	X				X					
8		X	X				X					
9							X					
0			X	X			X					

Figure 2.13 Matrix of capability for one machine tool expressed in code numbers.

same figure, and the designer is given a set of preferred standards for the part which appears on the screen (Fig. 2.14) in order to fit on the machine tools as identified in Fig. 2.15.

If the design engineer accepts the material, tolerance, surface finish, and other standards, then the code number will lead him or her through the same matrix to a set of machine tools which has been specifically set up to make that part. In other words, the design and manufacturing standards are coupled together so that the classification and coding system is used as the key to the implementation of optimal designs and manufacturing methods.

Once this is done, it is possible to go further and link the manufacturing process plan with automated time standards, cost calculator modules, and shop floor control programs. Even in its earliest stages, standardization can lead to significant results.

2.3.5 An example

The way the process works can be illustrated with the following example. In a manufacturing facility, 150 very similar parts were routed over 51 tools. The parts were so similar that differences were hard to find. There were 87 different process plans for these very similar parts, however (Fig. 2.16).

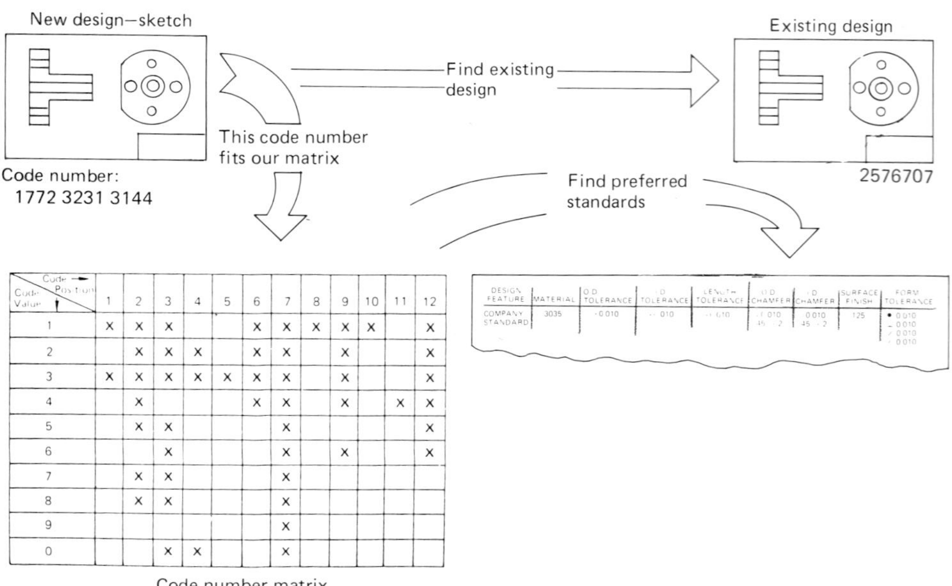

Code Value ↓ / Code Position →	1	2	3	4	5	6	7	8	9	10	11	12
1	X	X	X			X	X	X	X	X		X
2		X	X	X		X	X		X			X
3	X	X	X	X	X	X	X		X			X
4		X				X	X		X		X	X
5		X	X				X					X
6			X				X		X			X
7		X	X				X					
8		X	X				X					
9							X					
0			X	X			X					

Figure 2.14 Design standardization and retrieval.

There were so many different process plans because planners have a tendency to route parts over machines with which they are most familiar. It is not at all uncommon, of course, for two or even more machines to be capable of performing the same operation on the same part. Through group technology analysis and standardization, it was possible to look at all of these parts and the manufacturing requirements for each, including such factors as releases per year and lot size. The total load requirement was determined and analyzed.

Simulation and analysis programs were used to create a group technology work center which could produce the parts in question. It was found that the 150 parts could be manufactured on only 8 tools by using 31 standard routings. (See Fig. 2.17 and Table 2.2.) This type of analysis and cell formation was made possible with new computers and new group technology software.

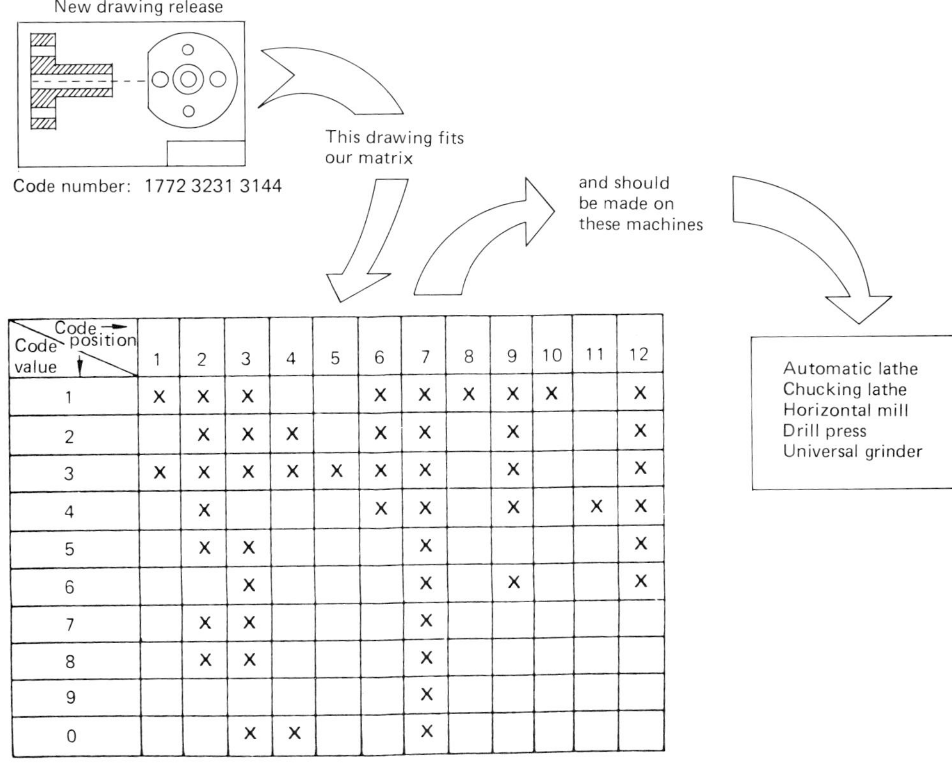

Code value ↓ \ Code position →	1	2	3	4	5	6	7	8	9	10	11	12
1	X	X	X			X	X	X	X	X		X
2		X	X	X		X	X		X			X
3	X	X	X	X	X	X	X		X			X
4		X				X	X		X		X	X
5		X	X				X					X
6			X				X		X			X
7		X	X				X					
8		X	X				X					
9							X					
0			X	X			X					

Figure 2.15 Manufacturing process planning.

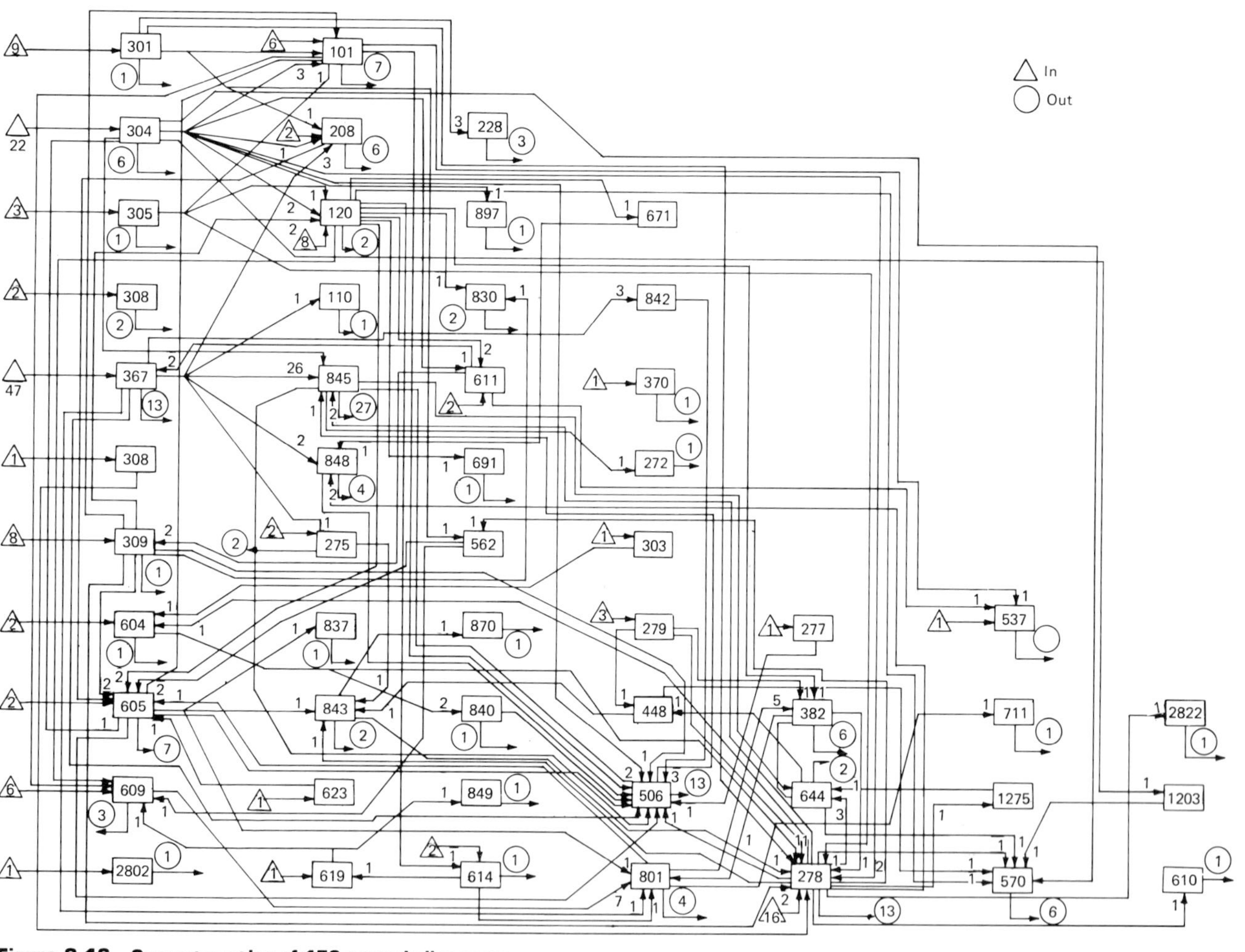

Figure 2.16 Current routing of 150 very similar parts.

TABLE 2.2 An Example of Current and Standardized Methods of Load Balancing

	Current method	Standardized method
Machines	51	8
Flows	87	31
Parts	150	150

2.3.6 Computer-integrated manufacturing

CAD/CAM became a fashionable term in the 1970s. The introduction of minicomputers with their superior interactive capability and ease of use made it possible to dedicate computers to new specific tasks. At the same time, the introduction marked the beginning of the physical spread of computers from the central data processing department to various operations scattered throughout a company.

Recognizing the problems of batch manufacturing inefficiencies, manufacturing management looked to these new computers and the software systems which accompanied them as the potential solutions to its problems. In fact, specific problems were solved, but overall manufacturing efficiencies did not increase very appreciably. Throughput time did not diminish significantly.

The phenomenon was somewhat analogous to expanding the diameter of a pipe in a few places (Fig. 2.18). The creation of 4-in-diameter sections

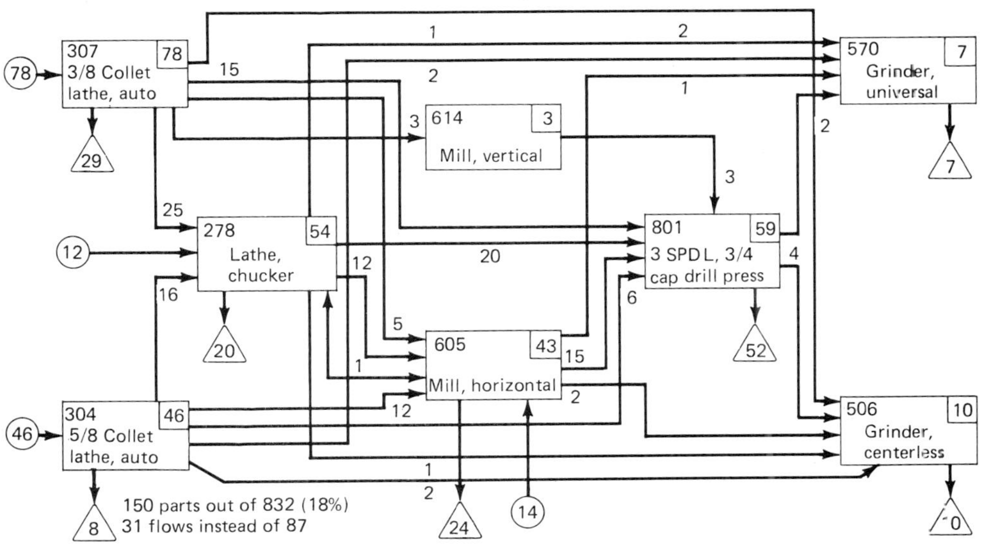

Figure 2.17 Routing of 150 very similar parts after group technology analysis.

in a pipeline of 2 in does not increase overall flow efficiencies. To the contrary, it only creates more turbulence, particularly where the pipeline returns to its normal diameter. In other words, if you want to increase the flow, you have to expand the diameter of the entire pipe. In terms of CAD/CAM, this means that various computerized modules must be connected. As long as they operate in isolation, they will not increase total efficiency very much.

Computer-integrated manufacturing is the term used to describe the interrelations of all of a company's computerized design and manufacturing activities. Group technology is the glue which can bring about the integration of CAD/CAM into CIM (Fig. 2.19) because it provides a common vocabulary for users, and because it is the vehicle for the creation of a database which includes design and manufacturing attributes, processes, and machine tool capabilities. Perhaps even most important, a group technology–based classification and coding system also provides the means for extracting data in usable formats for a variety of different users.

The ability to input data to a database which serves all departments within a manufacturing organization, and to extract that data is a key to CIM.

Companies are now taking advantage of this capability. In one large manufacturing facility, for example, the material requirements planning (MRP) system feeds orders for new parts directly into the computer-assisted process-planning system. This system then produces manufacturing instructions and tooling specifications. Tool requirements and NC programming needs are automatically forwarded to tool designers and NC programmers. The system is also used for the automatic printing of books of job tickets which accompany all parts through the manufacturing cycle.

In another company, parametric design descriptions, defined by a group

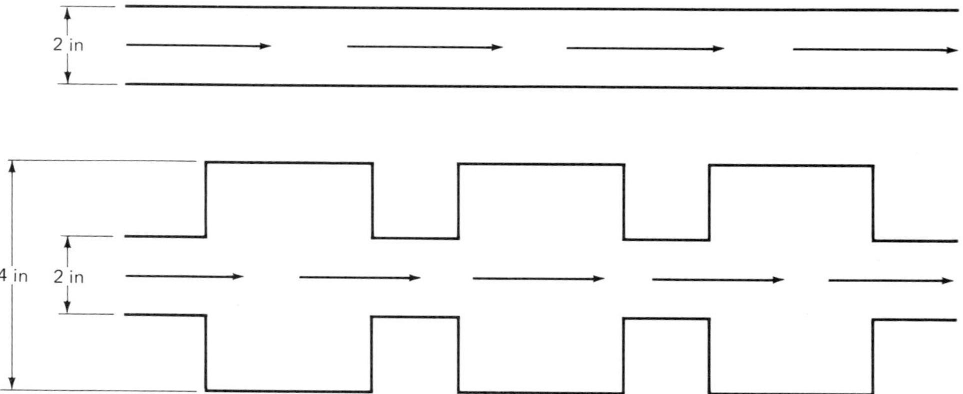

Figure 2.18 Independent computer modules do not necessarily improve productivity.

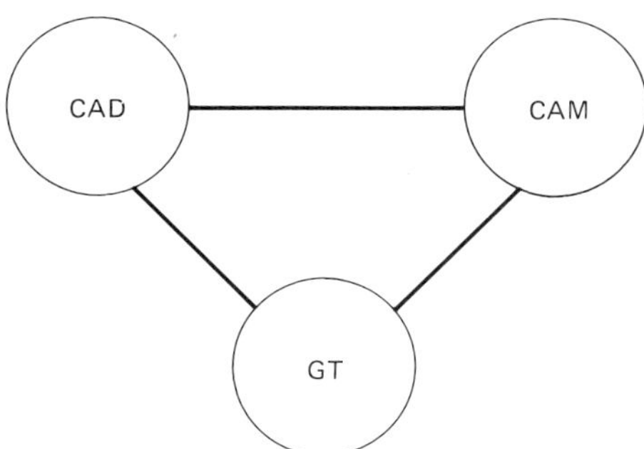

Figure 2.19 Group technology links CAD and CAM.

technology classification and coding system, are used to retrieve same or similar designs and to generate new drawings. As each new part enters the manufacturing cycle, additional data is input to the system. This data, added to the design data, makes it possible to create process planning, NC and robotics programming, and other documentation required for manufacturing, assembly, and quality control.

The group technology classification and coding system provides the common vocabulary for all these activities because it can hold design, manufacturing, and functional and other descriptive data in a consistent format. It should be pointed out that the code number is often transparent to the user. In a design retrieval system, for example, the user can search files by using keyword descriptors. These descriptors are translated into code numbers within the system, and the search process is carried out on the basis of those numbers. The system then provides the user with drawing or part numbers or with actual drawings (when used with graphic capabilities). The code number is never apparent to the user.

As CIM becomes more of a reality, methods of addressing computerized systems will continue to become less complex, so that code numbers are likely to disappear almost entirely from the user's perception. Within these integrated systems, however, the communication links will be the numeric or alphanumeric codes that capture information about the attributes necessary to describe every part or other component from design and manufacturing points of view. To meet the needs of CIM systems of the future, group technology classification and coding systems have become much more flexible in recent years.

2.3.7 Decision-tree-handling systems

Until recently, group technology–based classification and coding systems were geared to detailed parts, particularly mechanical parts. When one considers a user-oriented interface to a computerized database of information, however, it becomes clear that the system should include other elements, such as electronics, assemblies, subassemblies, and purchased parts.

Older computerized coding systems were "hard-coded." Such systems were relatively inflexible and could not be extended to encompass different components without major software changes. This led to a new approach: the design of entirely new decision-tree-handling systems which could handle different coding structures depending on user needs.

Ultimately, a coding system is an interactive means of leading the user through a decision tree to arrive at a string of digits which can be used to retrieve information and for other purposes. With a generalized decision-tree-handling system, it is possible to use one basic structure and software system instead of a number of hard-coded systems with their usual software problems. This type of approach also gives the user the opportunity to short-circuit decision trees after major standardization has been accomplished. Figure 2.20 illustrates a decision-tree-handling system with multiple applications.

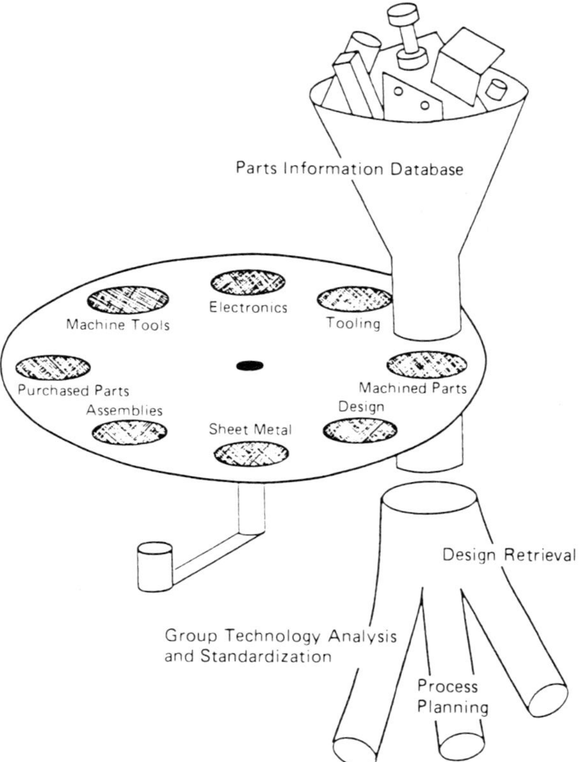

Figure 2.20 Multiple classification and coding system and applications.

Such systems take full advantage of state-of-the-art computer hardware and software technology and also recognize the different needs of different elements of any manufacturing company. The hardware makes it possible to work interactively, easily, and quickly, and the software makes it possible to create coherent, usable code numbers and classifications regardless of the specific attributes under consideration. In looking at a piece of equipment, a manufacturing engineer may be interested in the surface hardness of the main shaft and bearing tolerances; the design engineer may be more concerned with horsepower, rpm's, and mounting devices; and the purchasing agent may look at horsepower and shaft diameter. The new systems meet all of these needs in a single decision-tree structure. They can alsc be tailored to meet increasingly specific needs to the point of providing the retrieval vehicles for generative process planning.

2.4 Applications of Group Technology

2.4.1 Design retrieval

Preparation for manufacturing a typical part costs between $2000 and $12,000 for tool design, jigs and fixtures, process planning, NC tape preparation, and other elements of the preproduction process. Important savings can thus be derived by avoiding the unnecessary design of new parts. A "new" part is often created when the designer varies a tolerance or creates some other design variance which is not absolutely necessary. Also, if the designer does not know that exactly the same part was previously designed, an unnecessary new part may be created.

Design retrieval and the standardization of the designs are very important applications of group technology and classification and coding. Even in a batch production facility, the time spent on such standardization and classification and coding can easily be amortized within a 1- to 2-year period.

Within the design activity itself, 2 to 15 percent of design costs can be saved with an effective design retrieval system. The real payoff, however, as the above demonstrates, is in the manufacturing area. It is therefore important that design and manufacturing understand each other's needs. The system must be implemented in the design area, where the savings are less significant, in order to effect savings in manufacturing.

In CAD applications, an effective design retrieval system can be even more useful. Because designing a new part on a computer graphics terminal is relatively easy, there is less motivation to spend time attempting to retrieve and review existing designs. Unnecessary part design can therefore proliferate rapidly.

Group technology classification and coding and design retrieval have been linked to CAD systems as well as applied to conventional design operations. The use of these technologies with computer graphics will grow and will dramatically increase the usefulness of computer graphics systems.

2.4.2 Process planning

Manufacturing costs are not only determined by the design of a part; the process planner plays a major role in the cost of production. It is the process planner who determines the tools and manufacturing methods. A 2-in shaft can be made on a small lathe or on a five-axis milling machine—the choice is the planner's. The manufacturing cost varies greatly. Manufacturing costs for the same part can vary by factors of 2 to 5.

There is a shortage of experienced process planners which is likely to become more severe in the coming years. As planners retire, their experience could be lost, unless it is captured in some way. The importance of the process planning function, the growing shortage of process planners, and the need to retain the experience of retiring process planners are all arguments for computer-assisted process planning.

Computer-assisted process planning systems are becoming increasingly common in the manufacturing industry. With a proper system, payback is almost immediate. Even in its simplest form—as an "electronic pencil," or word processor—such a system can increase process planning efficiencies by 20 to 30 percent. This results from the elimination of the manual labor associated with writing or typing process plans.

A computer-assisted process planning system should be much more than a word processor, however. With a group technology–based system utilizing classification and coding, code numbers can be used for retrieval of existing and "preferred" manufacturing information. The storage, retrieval, and editing of process plans from a computer-assisted process planning database greatly increase process planning efficiencies. Going an important step forward, however, it is also possible to standardize process plans and to develop preferred routings for part families. The retrieval is done by code number or code number matrix, as explained previously. This preferred routing is, in fact, the optimal routing for the part; it is based on the experience of process planners and the tools available to do the job.

Cost information can be included in a computer-assisted process planning system. Cost calculation modules help the process planner and others to realize the cost implications of a decision and also make it possible to predict costs. This can be done most effectively when an automated time standards (ATS) system is linked to the process planning system. Companies using such systems have reported that it is possible to predict the manufacturing cost of a finished part within 5 to 10 percent, based on a code number. In other words, code numbers can be used for cost estimating.

A group technology–based computer-assisted process planning system linked with ATS and MRP can be very effective in improving shop floor productivity.

2.4.3 Pictorial process planning

Although a process plan can be a very simple document, it is often lengthy, with many steps and many details accompanying each step. Interpreting lengthy instructions is not always easy, especially in this age of visual communications. People are reading less and looking at pictures more. It is now possible to link computer-assisted process planning and computer graphics to create pictorial process plans which contain not only verbal instruction but illustrations as well.

The text of the process plan itself is developed on an alphanumeric or graphics terminal. It is then formatted on the graphics terminal. For example, all text may be on the left side or right side, and the graphic information, such as setup, machining details, jigs, and fixtures, is created on the remaining part of the graphics screen. (See Fig. 2.21.) Furthermore, the tool path can be visualized to prove out the operation before going to the shop floor.

The finished process plan, with verbal and visual instructions, is then sent to the shop floor for machine operator use. As common databases are used for both graphics and process planning, the use of this type of approach will undoubtedly increase.

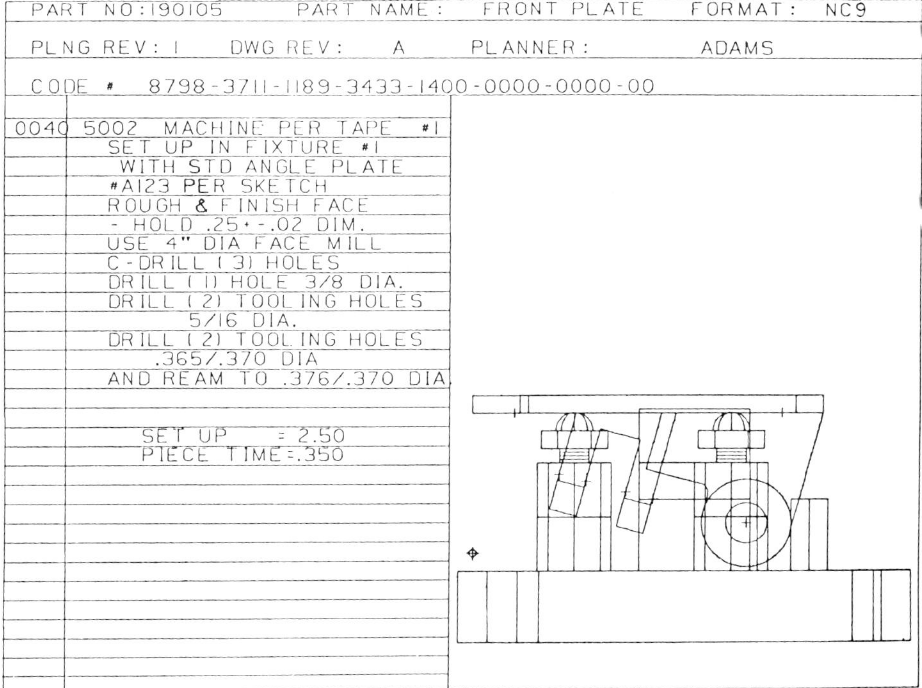

PART NO:190105 PART NAME: FRONT PLATE FORMAT: NC9

PLNG REV: I DWG REV: A PLANNER: ADAMS

CODE # 8798-3711-1189-3433-1400-0000-0000-00

0040	5002 MACHINE PER TAPE #1
	SET UP IN FIXTURE #1
	WITH STD ANGLE PLATE
	#A123 PER SKETCH
	ROUGH & FINISH FACE
	- HOLD .25+-.02 DIM.
	USE 4" DIA FACE MILL
	C-DRILL (3) HOLES
	DRILL (1) HOLE 3/8 DIA.
	DRILL (2) TOOLING HOLES
	5/16 DIA.
	DRILL (2) TOOLING HOLES
	.365/.370 DIA
	AND REAM TO .376/.370 DIA
	SET UP = 2.50
	PIECE TIME=.350

Figure 2.21 Pictorial process planning.

2.4.4 Generative process planning

There are two types of computer-assisted process planning. In a variant system, a process plan is retrieved by part number, code number, or some other key. If the retrieved plan is for exactly the same part entering into production, it can be used without change. If the part is in any way different, the retrieved plan can be varied to meet its specific requirements. In a generative system, the part under consideration is described to the computer, and the computer immediately produces the plan to manufacture the part exactly based on manufacturing logic available.

In the variant mode, standard or prepared process plans are stored for future use (Fig. 2.22). In a generative mode, manufacturing logic is stored to create process plans on "the fly." In general, all systems used today are variant. Although the terms are often used loosely, there are very few, if any, real generative process planning systems. Work is being carried out on the development of such systems, however.

One approach to generative process planning is artificial intelligence. By using artificial intelligence, the computer would look at the attributes of the part in question and, by using decision rules programmed into it, would decide exactly how to manufacture the part. Artificial intelligence is a very desirable means of generating a process plan, but current technology falls far short of its realization because there are very few established and accepted manufacturing rules.

Another more practical approach is through the use of multiple classification and coding systems and group technology. This approach begins with the development of a manufacturing database which can be accessed by part number, nomenclature, or code number. By using group technology,

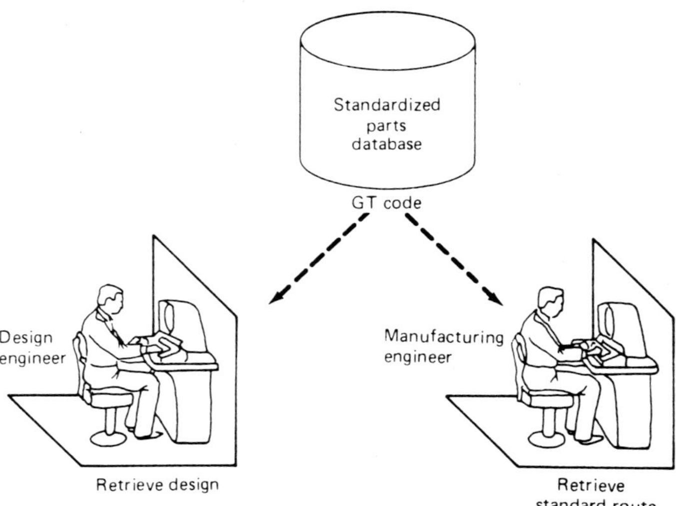

Figure 2.22 Storage and use of process plans.

optimized processes can be developed for each part attribute and part family. Design and manufacturing considerations can be optimized to the point where all factors are considered so that the plan provides the best design and manufacturing technology for the specific facility in which the part is to be produced.

By using a multiple classification and coding system, these optimized process plans can be fine-tuned to a high degree. Matrices, such as those discussed earlier, can be developed for both parts and the tools available to manufacture them. Tight parameters must be devised for the specific facility to meet exacting needs. In other words, in going from variant process planning to design and manufacturing optimization by using classification and coding and group technology, one arrives gradually at a manufacturing rationale (logic) for individual part families. The rationale is specifically geared to an individual production facility.

The system can be continually refined to the point at which it would be possible to ask the computer to generate a plan for such a thing as a gear or a shaft. The system would ask the user several questions to further define the part and then search through its matrices and generate a process plan which would describe the optimal method to manufacture the part. The group technology multiple classification and coding approach to generative process planning is possible with today's technology, and it will lead to the rationale to generate process plans (and designs) automatically. (See Fig. 2.23.)

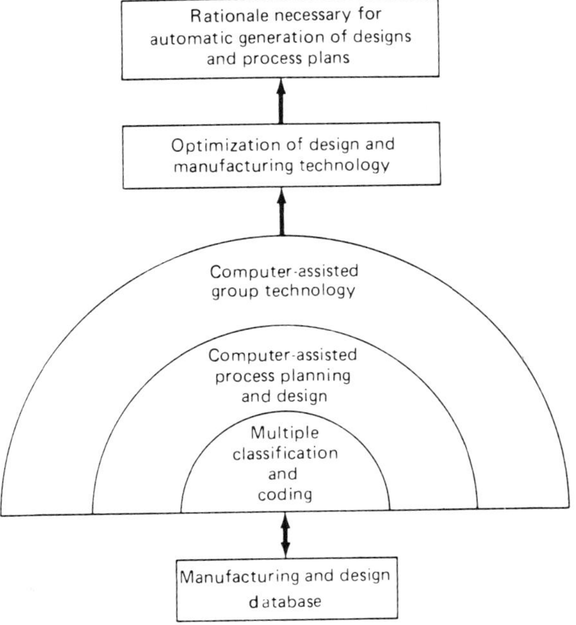

Figure 2.23 Evolution of generative process planning.

2.4.5 Numerical control

Group technology can be used to effect significant cuts in costs and time for the preparation of NC tapes. The principle is simple. In a properly implemented system, the code number matrix for each family of parts contains all pertinent design and manufacturing information. Each member of a family falls into a particular design envelope. Standardized NC programs (tapes) can be developed to produce parts within the family. These macro tapes require only the addition of small amounts of specific information (shaft diameter, for example) to be put to use.

This application of group technology broadens the potential uses of NC equipment. Lot sizes of 15 to 20 are necessary to amortize the cost of preparing an NC tape in the conventional way, but by using prepared macro tapes, the preparation costs can be amortized with 1 or 2 parts. In other words, parts which would otherwise be made manually can now be made on NC tools.

2.4.6 Manufacturing requirements planning

The purpose of an MRP system is to have the right part in the right place at the right time. Parts are scheduled through the manufacturing process, either on an order-launching basis or through all operations. This can be an elaborate system involving the scheduling of all manufacturing operations of all parts. Consequently, proper process plans are required—plans which will assure that parts are manufactured on schedule. In order to meet delivery schedules, MRP schedules which parts should be manufactured on what machine tools and in what sequence. The problem is that the batch manufacturing environment is nonstandard with many different production methods for the same part.

In conventional operations, the supervisor picks similar parts from a queue to reduce setup times. In an MRP system, however, parts are scheduled in terms of delivery and the supervisor does not have that option. Since the supervisor has no choice, he or she cannot optimize setup times. As a result, setup times increase and the time available for metal removal per machine tool decreases. There is thus a loss in production capacity and a decrease in asset utilization.

There is ample evidence that group technology and MRP can be implemented together and complement each other's success rate. Through group technology analyses, a part family can be defined and optimal routings developed. The group technology system tells the planner what parts to send out on what machines. The key is that families of parts are routed consistently to dedicated groups of machine tools. The MRP system recognizes this group of machine tools as one machining center (a black box) where parts enter one side of the center and leave in finished form at the other side. The scheduled flow of materials inside the center is governed by group technology considerations. MRP is related only to the schedule of material and parts to and from the center. Some companies call these centers GT1, GT2, etc.

It is essential to implement group technology first. With group technology analysis, it is possible to deal with issues such as optimized family production, production flow analysis, and load balance by families. Parts can then be produced in families for maximum efficiencies without interfering with MRP operations in any way. This has been done in several companies with high rates of success.

2.4.7 Machine tool selection and shop layout

Since the code number reflects a preferred design or process plan, frequencies of occurrence of code numbers in certain time periods can be related to manufacturing requirements and thus to the machines needed to manufacture the parts over the course of time. In other words, based on the occurrence of code numbers, it is possible to predict with a high degree of accuracy exactly how many hours of time will be required from each machine.

Since group technology analysis can also be used as a simulation tool, it is possible to examine variables in future manufacturing operations and look at future machine tool use with those variables. In short, it is possible to predict machine tool use, and thereby make purchasing decisions, on the basis of current manufacturing patterns or expectations of future changes.

With group technology it is also possible to form work cells, of course. With the definition of work cells and production flow analyses, it is possible to determine specifications for an entire manufacturing facility and to base the layout of the facility on the specifications. This has been done with great success. Asset utilization has improved dramatically, and the investment budget for new machine tools has decreased.

The same principles can be applied to the purchase of robots. In many cases, robots as such are not tools, but they are materials handlers. Based on production flow analyses and the positioning and production expectations of work cells, it is possible to determine where and how robots can best be used.

What has become an increasingly attractive method of selecting machine tools and laying out machine shops is also applicable to the selection of flexible manufacturing systems (FMS). The manufacturing database has all the information available to retrieve the parts (by code number) to be run on an FMS. It can thus be utilized to lay out the specifications of an FMS to be used (or not to be used) most precisely.

2.5 The Future

There is no doubt that computers will be increasingly important in manufacturing in the future. The parameters for future use of computers in manufacturing environments have been established by experience to date. Certain trends appear to be firmly established, and technology is evolving consistently with those trends. For example, computer-integrated manufacturing will require a single integrated database of design and manufacturing information. This database will contain not only conventional design and manufacturing data but also information on cost, testing, quality control, and other elements of the process from initial design through assembly, testing, and shipment. The key to most effective retrieval of this information is to use sophisticated multiple classification and coding structures that interface the user directly with relevant information for the computer database.

Since the activity of classification and coding remains an uninteresting task to perform, voice recognition systems should be considered as a means of imposing efficiency, even extending it into process planning and development.

As mentioned earlier, classification and coding should encompass all parts of the design and manufacturing environment. This is required to perform the necessary group technology analysis for optimization. Once the analysis is done, keyword coding should be introduced to make the classification and coding process more digestible for the daily user. The keywords are based on groups of features in families of parts.

Concurrent with this, microcomputers will play an increasingly important role in batch manufacturing. Microcomputers on the shop floor and in the design and manufacturing engineering offices can handle local processing needs at the workstations and interface with larger computers which will be used to process and store data. Linked systems will accommodate both alphanumeric and graphics hardware and software.

A major breakthrough will be the implementation of solid modeling in computer graphics. Once a computer can actually recognize a part, not as a wire diagram but rather as a physical entity, it will be possible to generate code numbers automatically. These code numbers, invisible to the user, will become the means of tying all database elements together automatically. Interactive communication will become even more efficient.

With a group technology classification and coding system as the key, it will be possible to create designs automatically, link them to the process plans required for manufacturing, produce NC tapes as needed, program robots, and oversee all other aspects of the total integrated manufacturing process. When this is done—and it will not be done quickly—the concept of the automated factory for batch manufacturing will be well on its way to realization. It should be realized that these features are not long-term expectations but rather realities to be expected within the 1980s.

CHAPTER

3

Process Planning

Joel Orr

Saving money and time in the manufacturing process is a universal goal; process planning is an area of manufacturing that can yield rich fruits in pursuit of that goal. By wide observation, parts in process in discrete manufacturing spend 95 percent of the time required to make them in waiting. There is certainly something to be gained by considering ways in which that time might be reduced.

Process planning deals with the manufacturing process. This chapter examines the nature of process planning: its history, the role of the computer in its improvement, its current state, and its future.

3.1 What Is Process Planning?

To understand process planning, we must first define *manufacturing.* Thomas Faulhaber gives it three dimensions:

- The physical function of generating products
- The economic function of generating profits
- The social function of generating change

We trace our current forms of manufacturing most directly to the Industrial Revolution, a tidy name for a very untidy group of events in the 1700s that resulted in the replacement of sweat by steam and streams—the exchange of animal for nonanimal sources of power. Other important changes generally attributed to the Industrial Revolution include the shift of large segments of the population to towns and cities, the widespread introduction of interchangeable parts, the development and refinement of assembly lines and mass production, and an attack of unprecedented violence on the family as the fundamental social institution. The impact of all those issues on the factory, as well as on its place in society, is evident today.

Several notables of that time, including Arkwright, inventor of the spinning jenny, and James Watt, known best for steam engines, are recorded as having had some awareness of the importance of planning and record-keeping in the development and implementation of manufacturing processes. Perhaps they had a glimpse of the complexities that arise from the peculiar mix of standard and custom products and processes that are the lot of the discrete manufacturer of today.

Process planning is the developing of a logical sequence of manufacturing operations for a given part. The physical aspect of those operations is the most obvious, but the economic and social aspects are the fields in which the determination of their "logicalness" is made.

Consider a process for making gears. The gear blank is sawed from round stock; its teeth are cut; and its inside and outside diameters are machined. It is then heat-treated, after which all its surfaces are ground (Fig. 3.1). The job of the process planner is to determine the sequence of operations and interpose the appropriate inspection and testing processes while taking into account such factors as availability of machine tools, deadlines, and present mix of jobs in the shop.

The plan is usually expressed in the form of a *process sheet* which lists the sequence of operations, the machines on which the operations are to be performed, the estimated time needed for each operation, and the required tooling. The process sheet is often illustrated, and it may be accompanied by setup and work instructions and routings.

A variety of considerations, including delivery schedules, machine utilization, and economic processing quantities, as well as material availability and work-in-process (WIP) inventory, must be balanced in the production of a process plan—in addition to the simpler physical issues of determining the appropriate machine and tooling for achieving a particular effect. But this means, of course, that process plans for a particular part may be entirely different on different dates because of variations in the job mix, in the availability of machines, and even in the planner.

3.2 The Process Planner

Experiments have shown that four process planners, under identical circumstances, will usually produce four different process plans for the same part. In fact, there is strong evidence that the *same* process planner will produce different plans for the same part and conditions at different times. There appears to be no clear way to identify an optimum plan.

At the same time, process planners are highly respected and valued by their employers. The difficulty in analyzing and quantifying their craft adds to their value. They share with bus schedulers and chess grand masters the uniquely human quality of being able to tame a complex problem and talk about the process but not be able to state which parts of the activity are essential.

Process planners become expert within particular industries and in certain types of parts. They approach each new job by analyzing it for similarity to previous jobs; often, a planner uses generic process plan forms and customizes their contents for each new job. But like many other professions in which advancement is based on long experience, process planning is finding few new recruits. As a result, salaries in this field are rapidly increasing, much to the concern of management, but higher pay is not attracting young people in sufficient numbers to keep pace with the needs of industry.

3.3 Methods

Process planning belongs to the category of problems mathematicians call *scheduling problems.* As a class, they have long eluded general solution. One side of the difficulty is the quantitative issue: For n jobs and m machines, there are $n!^m$ schedules. Thus for 5 jobs and 6 machines, there are $5!^6$, or 2,985,984,000,000, schedules—a number that defies exhaustive examination. A close look at traditional ways of dealing with process planning manually has yielded a useful approach to simplification of this matter.

Basic to process planning systematization attempts is the recognition of the need for part classification. Relatively few completely new parts are produced in any industry, and process planners rely heavily on their ability to identify the similarity of a new part to a part for which a plan already exists. The new plan can then be created by copying and modifying the old one. A good classification scheme can facilitate the entire process to a surprising extent.

The combinatorics of processes and machine tools, as noted above, are such that relatively small numbers of processes and machines lead quickly to large numbers of possible schedules. For that reason, the decomposition of the problem into a number of smaller problems has generally been fruitful in reducing the total number of schedules to be considered. In the example given above, 5 jobs and 6 machines yield 2,985,984,000,000 sched-

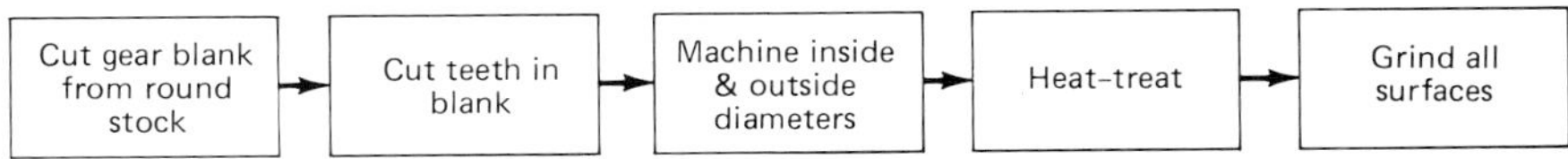

Figure 3.1 Making a gear.

ules. But if the jobs could be grouped as one group of 2 and a second group of 3 jobs, the same formula yields a total of 64 + 46,656 = 46,720 schedules—still a large number, but astoundingly smaller than the first. Clearly, if any of the machines could be eliminated from consideration for any of the jobs, that would further reduce the total number of schedules to be considered. Decomposition strategies have therefore been extensively explored by analysts.

3.4 Group Technology

The awareness that parts with similar geometry and processing requirements are more efficiently produced in groups than separately has led to the birth of a manufacturing philosophy called group technology (GT). GT offers benefits to the designer as well as to the manufacturing engineer; by making it easy to identify designs with particular characteristics, GT saves the designer from creating many parts that are functionally identical. Design time, cost, and documentation can thus be reduced.

The driving principle of GT is *part classification and coding.* Parts are grouped by similarities of various sorts—e.g., size, machining processes, materials—and an alphanumeric code is used to describe the part in terms meaningful to the designer and the process planner. The code is typically quite long—30 or 40 digits— and the digits are grouped in sets to represent the characteristics of the part (Fig. 3.2).

The benefits of GT are many and varied, and they are adequately discussed elsewhere in this volume. With respect to process planning, the power of GT is that it implicitly performs the first step in process planning: classification. Even without computers, GT is thus of inestimable value to the process planner. It is possible, and even desirable, to code machine tools, operations, tooling, and customers, as well as parts; this coding is a prerequisite to the introduction of computers into the process planning activity.

3.5 Computer-Aided Process Planning

There is much that can be done with the help of computers to simplify the process planning activity and render it more efficient—and more effective. The earliest attempts at the introduction of computers to process planning developed into a genre now called *variant process planning.* This use entails the storage of process plans in digital form and allows the planner to select and modify appropriate plans as needed. The computer serves only word processing and simple file management functions in this application; nevertheless, it can render the clerical aspects of process planning more efficient.

In *generative process planning,* descriptions of the parts, the manufacturing processes, the machine tools, and the tooling (in the form of codes, as mentioned above) are entered into the computing system, and the system develops a new process plan. It makes no reference to prior plans, and it is not—theoretically—limited to that which has gone before.

Among the purveyors of computer-aided process planning (CAPP) systems there are some who argue that a generative system throws the users too much on their own resources by requiring them to develop the complex decision sequences required to make planning determinations. They liken today's generative systems to empty shelves whose contents must be determined and acquired by the user. There is some justice to the accusation. CAPP vendors have yet to discover an underlying principle of sufficient generality to produce turnkey generative systems. It is often easier, in the short run, to demonstrate productivity increases with sophisticated variant systems.

A system called DCLASS, created and marketed by Brigham Young University (Provo, Utah), gives the user a methodology for capturing decision-making logic in hierarchical decision trees. Along with software supplied for managing and traversing the trees, DCLASS can be used to create a generative CAPP system. Its application, however, relies strongly upon the understanding of the user, who must become expert in the system as well as being expert in process planning.

Generative systems are, of course, orders of magnitude more complex than variant systems. There are not many generative systems in operation; one of the most notable is the Genplan system, developed and implemented at Lockheed under the direction of Joe Tulkoff. In an interview in *American Machinist* magazine, Tulkoff reported that Genplan automatically determines the sequence of operations, selects proper machine tools, and calculates machining times on the basis of manufacturing logic included in its database. The process planner assigns a code based on part description; Genplan then quickly analyzes the data, evaluates alternatives, and makes planning decisions. Only minor editing of the plan is then done by a human planner. This results in process plans that are consistent in methodology, sequence, format, and terminology and that incorporate the latest technology. Tulkoff says, "The system captures both the art and science of

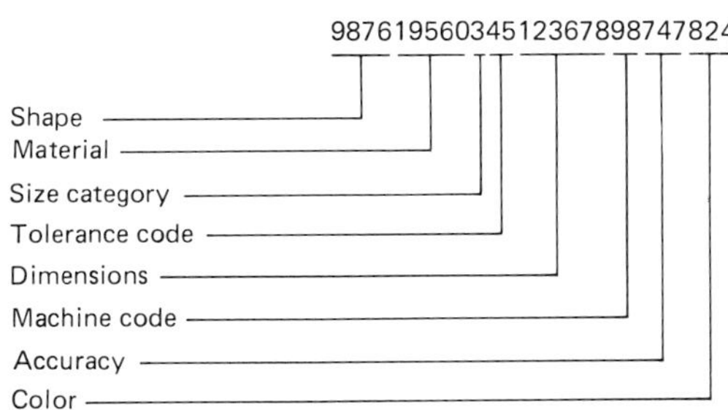

Figure 3.2 A classification code.

manufacturing." However, even Genplan requires a trained process planner, and thus it can legitimately be labeled semigenerative.

But even variant systems, in their relative simplicity, have productivity gains to offer. Over 30 percent of a process planner's time is spent actually preparing process plans; a good variant system can reduce that time by 20 percent or more. The ability to review, compare, and edit process plans quickly and produce them with a minimum of new data entry and in a consistent format is implicit in the variant CAPP system.

Both variant and generative CAPP depend on GT, to a great extent, for high productivity gains. Even in a CADD environment, in which geometry is created and stored in digital form, the computer lacks the faculties of discernment needed to allow it to "comprehend" the nature of a part. Classification and coding take the place of those faculties, and they present the part to the system in "digestible" form.

3.6 CAPP, CADD/CAM, and CIM

A functioning CAPP system is a boon to the designer, because its very functioning provides feedback regarding the producibility of designs. It helps the manufacturer by making better use of capital equipment and reducing manufacturing costs and lead times. In a full computer-aided manufacturing (CAM) environment, feedback from the automated manufacturing processes can be routed directly to the CAPP system to further enhance its performance. CAPP is thus an essential go-between for CADD and CAM, taken in conjunction with GT. It is also a necessary and integral part of computer-integrated manufacturing (CIM), in which all design and manufacturing functions are tied together by means of computer and communication systems.

3.7 The Future

An apt cliché for the penetration of computers into factories today is "islands of automation." As with other islands, such as CADD, CAPP's benefits on its own are interesting but not revolutionary. In concert with group technology, solid modeling, and full CAM, however, CAPP will figure centrally in the elimination of that most expensive and inconvenient of factory artifacts—the employee.

Elimination of the expert process planner is not far off; it awaits the application of principles of artificial intelligence and expert systems that are already well known. Coding for GT may eventually be made obsolete by more sophisticated modeling systems than are common today. Process planning, before 2000, will probably lose its identity as both an activity and a profession. It will blend smoothly into the sequence of events leading from concept to complete product, and it will be effected, in many instances, with no human intervention whatever.

3.8 Summary

CAPP offers manufacturers potential for savings in process planning itself, in direct shop labor, in materials, in scrap and rework, in tooling and in WIP inventory. Even without a complete CIM system, CAPP's benefits can be enormous. No U.S. manufacturer can afford to ignore them.

CHAPTER

4

Numerical Control Systems

Ann E. Meister

4.1 Programmable Automation and Computer-Integrated Manufacturing

4.1.1 The information perspective

The objective of a manufacturing enterprise is to produce a quality product that can be sold at a profit. To do so, marketing first identifies a need and engineering then designs a solution, documented as product definition data. To make the design a reality, manufacturing interprets the product definition data to plan the orderly sequence of events that must occur to convert raw materials into the physical objects described in the product definition data documents. The documents may be physical or electronic, for example, blueprints, specifications, or a computer-aided design (CAD) database. The automated factory requires an orderly flow of both physical goods and information.

Computer-integrated manufacturing (CIM) requires the convergence of three viewpoints of the automation process: a user view, a hardware/software view, and a neutral view of the enterprise. A framework for information resource management required to support the CIM environment is shown in Fig. 4.1. It is based on the three schema architecture proposed by the American National Standards Institute ANSI/X3/SPARC study group on database management systems in 1977. The information architecture supports the user view; the computer systems architecture supports the hardware/software view; and the control architecture supports the enterprise view with the organization, standards, and procedures required to create and maintain the integrated environment. With this approach, CIM can achieve its purpose, which is to fulfill the objective of the manufacturing enterprise: producing the quality product which can be sold at a profit.

Programmable automation, in the form of numerical control (NC) technology, is frequently used to address production process requirements in the CIM environment. This technology is represented by the physical implementation of the computer systems architecture. Production optimization requires that many complex and interrelated functions be integrated. Therefore, the focus of the components of the CIM system is to make the production process work smoothly and without bottlenecks while maintaining cost and quality. The components of the CIM system, such as CAD, group technology (GT), process planning, material requirements planning (MRP), manufacturing resources planning (MRP-II), and production and inventory control, are covered in other chapters of this handbook. If there is no production process, there is no need for these components to support it. These applications are represented in the user view or information architecture.

In the CIM environment, programmable automation systems provide the

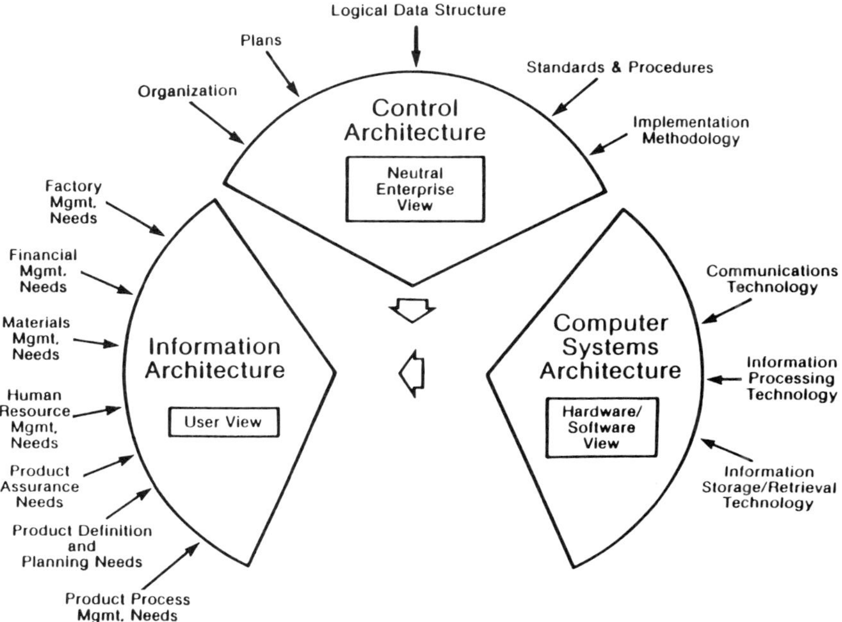

Figure 4.1 The three-schema architecture approach to CIM information resource management. *(Courtesy of D. Appleton Company, Inc.)*

greatest means for leveraging both the information assets and the fixed assets of the manufacturing enterprise. The information assets are the data contained in the integrated database. This integrated database is the heart of CIM, and it is necessary for truly flexible, data-driven automation in the manufacture of discrete parts. The integrated database is derived from the conceptual schema or logical view of the enterprise which resides in the control architecture. The control architecture provides for application of a consistent set of operational or business rules which govern the enterprise. The fixed assets, represented by programmable automation hardware, are useless without the application of the information assets through the control architecture.

The ability to understand the changes taking place in manufacturing today is required of all levels of management. Management must control the introduction of new technology and determine what of the old technology to keep. Addressing production needs without regard to current problems in the support functions and systems, or future business or operating conditions, may result in the purchase of unnecessary new equipment. Indeed, significant productivity improvements can be realized without expensive capital improvements. To do so requires the availability of timely and accurate information to current manufacturing processes and the application of industry and corporate standards for data and technology. Both standards and information (its quality and accessibility) are key issues in successful NC systems.

Technology for the sake of technology is more likely to cause problems, not solve them. This chapter will emphasize planning strategies, the leveraging of information, and the requirements for integrated applications, rather than the variety and constantly changing aspects of automation hardware.

4.1.2 Numerical control defined

One of the primary problems in dealing with new and evolving technology is the lack of commonly accepted terms and definitions. In addition, many ambiguous abbreviations and acronyms are used. These may represent different words, depending on context or usage, which changes the meanings of the terms significantly. Glossaries which address this problem for CIM and factory automation exist, but until a commonly accepted terminology is established, the best solution is to include definitions in any discussion of the technology.

The *numerical control concept* addresses the control of processes by numbers and symbols. It provides for the marriage of the brains of a computer to the brawn of industrial production equipment. It is important to realize that NC provides the means of machine or process control; it is not the production process or method itself. By its basic definition, an NC system puts the concept of NC to work; it is a system in which the direct insertion of programmed symbolic and numerical values, prerecorded and stored on some form of input medium, is automatically read and decoded to cause and control a corresponding action in a machine or process.

In more physical terms, a *numerical control system* is made up of a number of components. The basic component is a numerically controlled machine or group of machines chosen according to a logical arrangement of parts or operations. Material transport and handling may be accomplished by using humans, robots, or controlled vehicles. Sensors, inspection, and/or test devices provide control and feedback which might otherwise be provided by human interaction. In addition, the system may include other programmable devices and/or human operators. A means for hierarchical communications is implied. The enterprise-specific information, which is the lifeblood of the system, is provided by critical support functions including NC programming, process planning, tool management, and material, inventory, and production control. In addition, engineering must provide a product designed with appropriate considerations for quality and production automation. Some or all of these functions may be verified through computer simulation of the system or process.

4.1.3 The development of numerical control

Efforts to eliminate human intervention from the production process became a goal of management with the introduction of the concepts of interchangeable parts and mass production. Low-volume, batch-oriented manufacturers could not take advantage of high-volume production-automation techniques owing to cost and inflexibility. Numerical control could provide (1) flexibility to increase low-volume productivity and (2) stored instructions to decrease direct labor.

Numerical control technology was the first application of computer-aided manufacturing (CAM), the application of information processing technology to industrial automation technology. The original NC machine tool was developed under Air Force contract by the Massachusetts Institute of Technology (MIT) Servomechanisms Laboratory and demonstrated in 1952. The need at that time was for the military aircraft industry to produce highly complex and frequently modified parts on an emergency basis. The complexity was due in part to the discovery by design engineers of integrally stiffened skins used to lessen the structural weight of military aircraft. Traditional manufacturing processes had difficulty creating those parts.

The first commercial installation of NC equipment was in 1957. Except for the highly sophisticated multiaxis profilers in aircraft plants, the original NC machines were standard, general-purpose machine tools to which a control unit and some form of servomechanism, such as step motors, had been retrofitted. Most were two-axis point-to-point milling and drilling machines, since the ability to coordinate axis movements for angular, circular, or sculptured surface cuts required more sophisticated and expensive controls.

Control technology developed in parallel with digital computers, from vacuum tubes through transistors and integrated circuits to the more capable and reliable minicomputer- and microprocessor-based control units, which are referred to as computer numerical control (CNC). The older control units were hardwired, which means that the logic circuits were physically interconnected by wire on a backplane to give a fixed pattern of events which is essentially unalterable. It was this hardwired logic which made the controller a point-to-point circuit card wire-wrap machine or a five-axis contouring mill. The hardwire control was less flexible in its ability to read and respond to part program instructions, because there were usually no more than one or two blocks of buffer storage. The part programs were usually stored on punched tape, which had to be read through the tape reader on the controller each time a part was to be made.

In CNCs, the hardwired logic is replaced by a software executive program which gives the controller its identity. In addition, part program storage is provided. Many controllers now accept computer logic operations such as variables, branching, and subroutines in the part program instructions. CNCs may also allow the executive logic to be reprogrammed to provide, as an example, the interpretation of instructions to operate an inspection probe. The terms NC and CNC are now frequently used interchangeably.

This new technology required the creation of two new interchangeable job titles: NC programmer and part programmer. It is the job of the NC programmer to prepare the specific and complete set of data and instructions for the purpose of manufacturing a part on NC equipment. The data and instructions are called the part program, and our programmer is of the new breed of information workers in the new Industrial Information Revolution. As development of machine and control technology progressed, a need was recognized for a programming method to manipulate and translate engineering and manufacturing information to create control media for complicated three-dimensional parts. Again under Air Force sponsorship at MIT, a project was initiated to develop an English-like language which could be used by manufacturing engineers. The result was the computer-assisted system called APT, for automatically programmed tools. Whereas machines have amplified the capability of human muscle, the computer is now amplifying the capability of the human brain.

There has been a dramatic increase in the number of conventional machines available under numerical control. That has been due in part to new control and servo capabilities, as well as a better understanding of machine design requirements. The ability to coordinate movements by computer control using sophisticated servo drives has led to the creation of affordable equipment which can accomplish tasks that earlier systems could undertake only at great expense and with great difficulty. As more human intervention is removed from equipment operation, human controls, such as handwheels, have been removed and the operator accessibility requirement minimized. Other operator functions, such as loading a tool carousel or mounting workpieces on a pallet, are now made more accessible. It is interesting to note, however, that an attempt has been made to make the control units more user-friendly, even though the machines are being designed for decreased operator intervention.

As machine designers realize the power, versatility, and reliability of computer numerical control, more machines are being designed to accomplish specialized tasks more productively. Gears, impellers, and turbine blades are easier to manufacture by using NC technology. Processes such as traveling-wire electrical discharge machining (TWEDM), water jet and laser cutting, and high-speed machining would be impossible without numerical control.

The Air Force has continued to support the development of computer-integrated manufacturing; it has sponsored such projects as the factory of the future (FoF), the integrated sheet metal center (ISMC), the integrated information support system (IISS), and the integrated electronics factory (IEF). The industrial modernization incentives program (IMIP), sponsored by the Department of Defense, is an effort to promote productivity improvement through technology modernization for all tiers of defense contractors. In addition, the National Bureau of Standards has established the Automated Manufacturing Research Facility (AMRF). This will provide a test bed where integrated manufacturing system measurement research, as well as research on interface standards, can be performed.

4.2 Structuring an NC System

Numerical control applications should not be viewed as the simple and direct automation of previously manual processes. The embodiment of this view has been a shortcoming of computerized systems of many types. The move to NC provides an opportunity to reevaluate manufacturing operations, tools, and fixtures for adaptability to programmable and flexible automation. If appropriate, the basic product designs can be revised to improve producibility, both generally and specifically for the automated environment. The hardware and applications software of an NC system do not work unless they are fueled with information. Planning is the essence of NC.

4.2.1 The human element in the computerized system

Creating an automated system is not an automatic process. Though computerization implies removing the human element from the job to be done, in reality the primary human input has been relocated to a position earlier in the process. In the case of NC, the brain-to-hand neural information pathway becomes a brain–to–machine control media pathway. Rather than a machinist cranking handles to initiate the appropriate machine motions, a part programmer must define each movement of the machine, along with all auxiliary functions, and record this data on a reusable medium which will allow the operation to be repeated whenever it is required by the computerized equipment. This should allow the process to be duplicated indefinitely at a level of consistency which the human machinist cannot equal. In theory, the knowledge and experience of an expert machinist is being captured for the future, at least for that part or family of parts.

As the computerized production system becomes more complex, a higher level of expert information must be gathered in order to define system requirements, prototype a functional design, and build the system. Here is where the conceptual schema of the control architecture provides the source for data-driven system development and supplies the method for mapping the external or user view of the data requirements to the internal or physical implementation requirements, as shown in Fig. 4.2. Indeed, the relations or rules documented in the conceptual schema can become the basis for an expert system. The equipment on the shop floor under numerical control is a user of data and has a user view.

A computer is an obedient servant; it does what you tell it to do. If you want it to do something complex, such as operate a group of unattended machines, all the tasks to be performed must be completely and accurately

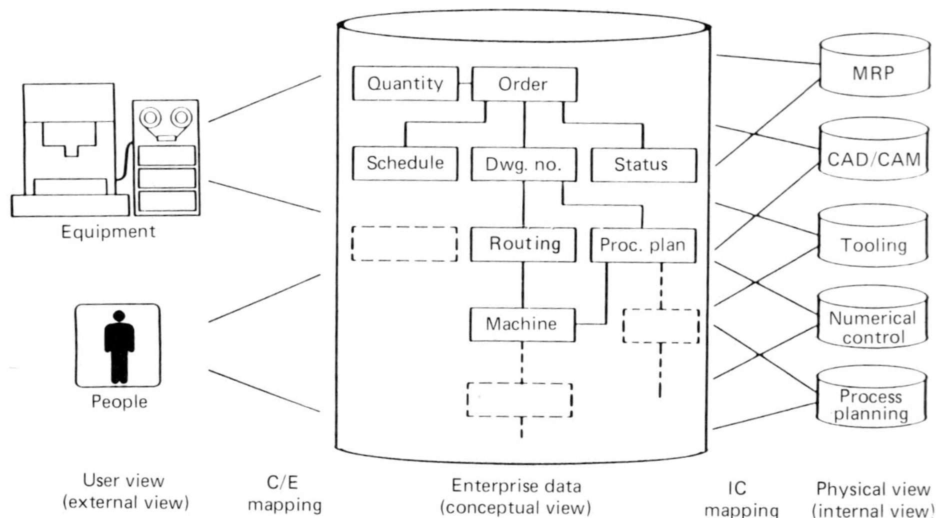

Figure 4.2 Architecture for an integrated database. *(Courtesy of D. Appleton Company, Inc.)*

described in their logical sequence, in a manner comprehensible to the computer. Any decisions previously made by a human operator must now be structured for the computer. If, for any reason, there is a deviation from the prescribed sequence of events, the computer program must provide the means for making decisions to prevent the destruction of the system. To facilitate responsiveness, decisions in both human and computer systems should be made at the lowest appropriate level.

4.2.2 System hierarchy

The necessary levels of responsive decision making help to define the structure (or architecture) of programmable automation for the evolving factory of the future. One of the Air Force integrated computer-aided manufacturing (ICAM) projects defined the levels of an implementation hierarchy to provide a mechanism for top-down planning and bottom-up implementation in an orderly, modular fashion. Those hierarchical levels, shown in Fig. 4.3, are as follows:

1. *Process.* A manufacturing process is a single operation or set of operations carried out by a person or machine not aided by an external hierarchy of program-driven circuitry or computerized software. Processes are primarily controlled by a person or a station controller. The highest level of control for a process is a cell. Deburring is an example of a process.

2. *Station.* The station is the lowest level of automated control. It is composed of sets of manufacturing processes under the control of software resident in or under the direction of the respective station. Stations control processes. Stations do not control other stations, and stations are controlled by cell controller software. A CNC lathe is an example of a station.

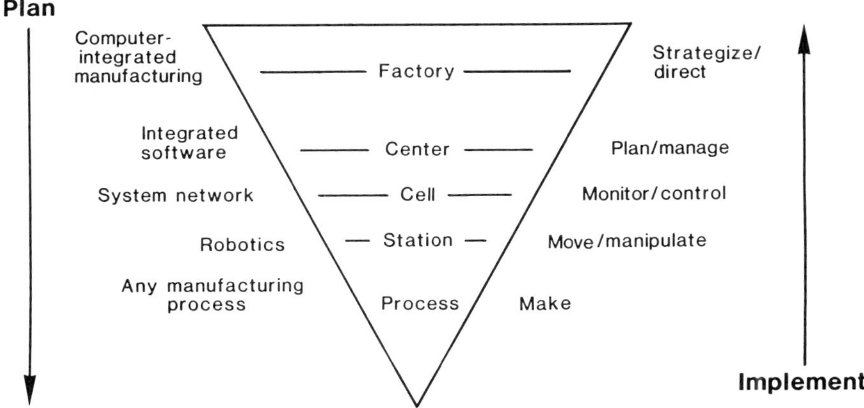

Figure 4.3 Factory-of-the-future architecture. *(Courtesy of GCA Corporation, Information Systems Group)*

3. *Cell.* A cell is the automated control of two or more stations. It may include a single process external to any station control in the respective cell. A single station under cell control must be accompanied by a process not under station control. Cells are controlled by center software. A CNC lathe and a CNC mill with an intermediate deburr operation and a robot controlling the movement of workpieces between operations are an example of a cell.

4. *Center.* A center is the automated control of two or more cells. It may include a single station external to any inclusive cell. A single cell under cell control must be accompanied by a single external station. Centers do not control other centers. Centers are controlled by factory controller software with emphasis on scheduling, production, and management information.

5. *Factory.* The factory is the integration of marketing, engineering, production, and support functions working in concert to produce the product of the enterprise. Factories are controlled by management personnel applying the policies and procedures of the enterprise.

Unfortunately, this terminology is not precisely applied, especially among the hardware vendors. A "machining center" or a "turning center" does not correspond to the definition of "center" stated above. Each is a stand-alone CNC machine with more functionality than a conventional mill or lathe. In addition, the term "flexible system" is used; it may correspond to "cell" or "center." Machining or turning centers may be incorporated in a flexible system.

4.2.3 The basic stand-alone CNC system

The basic module of the NC system is the stand-alone numerically controlled machine and its dependent support systems. The hardware component consists of the process equipment itself, the machine control unit (MCU), the interface between the machine and the control, and job-specific accessories such as tools and tooling. The software component consists of the executive program, diagnostic programs, and part programs. The critical component is the information specific to the enterprise. The technology of the hardware and application software provides the means to accomplish a task. The system, however, functions in response to the information provided; if the information is incomplete, inconsistent, or incorrect, so too will be the results. This follows the first law of computers: garbage in, garbage out (GIGO).

Between the machine and the control is the interface which functions as an input/output (I/O) clearinghouse and provides the means for transmitting the signals between the MCU and the appropriate machine control devices addressed by the part program. The signals transmitted, for example, contain information to turn on the spindle motor to 435 rpm clockwise, rotate the tool turret to select tool 2, turn on the flood coolant, and engage the Z-axis servo motor at 0.01 IPR (inches per revolution) for 1 in of linear motion. In a closed-loop system, a resolver or other device will provide the feedback to the controller to inform it when the programmed stopping point has been reached. The information processing required of this interface becomes more important as sensors and other devices are added to the original equipment functions. The interface may be a programmable logic controller (PLC) and/or electrical relays.

The MCU provides the interface to the outside world. Here resides the executive program which gives the system its identity, as well as diagnostic programs to help locate and solve problems when the system is acting improperly. Input to the MCU may be in the form of punched tape, magnetic tape cassette, floppy disk, or a remote link with another computer (which will be discussed under DNC). If a machine is purchased for current stand-alone use, but with the intention of future integration into a cell, center, or DNC system, the capability of the controller to communicate with the outside world is of prime importance. Other features, such as the amount of memory, the ability to store multiple programs, leadscrew error compensation, and terminal emulation capability, may be needed. Many features of the MCU provide part programming and editing capabilities on the production floor. Management policies and procedures should dictate how extensively these capabilities should be used.

Stand-alone NC systems have met with less success than anticipated. The *1983* American Machinist *Inventory of Metal Working Equipment* showed that only 4.7 percent of the machine tool population in the United States is numerically controlled. This figure is somewhat misleading when related to machine tool usage; new CNC machines are more productive, and older, manual machines may still be on the books but not in daily use. Though problems of cost justification lead the list of reasons for not using NC, one not so obvious reason why stand-alone NC has met with less success than expected is that the enterprise-specific integrated database required for maximum productive benefit did not exist. This integrated database is the physical manifestation of the communication, coordination, and cooperation required between engineering and manufacturing. As automation projects are undertaken, the critical nature of this data as a requirement to ensure the success of the project becomes apparent. Indeed, the (almost) paperless factory is the result of an integrated database.

4.2.4 DNC—communications for the paperless factory

DNC is the abbreviation for either direct or distributed numerical control. One of the principal reasons for DNC was to eliminate punched tape and the associated tape punch and tape reader problems. In addition, configuration management could be applied to the machine control data stored in the host computer. The basic elements of a DNC system include:

1. The host computer
2. Communications hardware such as modems and cables
3. A machine interface unit (MIU), terminal emulation capability on the MCU or other DNC adapter
4. A behind-the-tape-reader (BTR) interface, serial communications port (for example, EIA RS-232), or other appropriate input/output capability

Both direct and distributed NC provide data communications networks which link one or more NC or CNC machines to a host computer, where part program data is stored and then sent electronically (downloaded) to the MCU. In more sophisticated DNC systems, information from the MCU may also be sent back (uploaded) to the host computer and other information, such as inspection, setup, and tooling data, may be available for viewing on the shop floor.

In a *direct numerical control system,* the part program data is sent block by block to the MCU as if it were on a punched tape. This method is appropriate to hardwired controls, which have no capability for storing part programs, or to CNCs which do not have sufficient memory to store the entire part program for the operation to be performed. The primary shortcoming of the direct system is that, if the host computer fails, all machines receiving direct instructions also fail.

In a *distributed numerical control system,* whole part programs are sent from the host computer for storage in the memory of a CNC unit. The operation of a machine tool is therefore no longer directly dependent on the host computer.

Either system may require a terminal between the host computer and the machine control unit. This machine interface unit (MIU) is used to communicate with the host, since many MCUs do not have a terminal emulation capability. Another possibility is that the MCU may have a direct communication link with the host via a hardwired adapter. The terminal capability serves two primary purposes: data integrity checking and system access security. The communications hardware and software must provide error detection and correction on all downloaded information to ensure that the data sent by the host computer is indeed the data received. A seemingly small data error, such as a missing minus sign or decimal point, can cause expensive damage to the machine tool, fixture, or workpiece. System access security procedures also ensure data integrity and file protection by appropriately limiting authorized transactions and preventing unauthorized access.

In some cases, the machine interface unit may provide intermediate storage of part program data. It is then essentially an intermediate host, and it requires a means of managing and protecting the information in its memory. The MIU may also be able to distribute information to more than one MCU or provide a direct link to the MCU. If the ability to edit machine control data is allowed at either the MIU or CNC level, a means to provide feedback to the original machine control data file must be available. The reason for and nature of the edit, and also whether the edit is a temporary or a permanent change in the operation instructions and the source part program, should be reported.

If the MIU has graphics display capability, pictorial instructions concerning setup, tools, tooling, operation sequences, and inspection requirements may be viewed. Here we see the application of DNC terminals to the paperless factory. Remember, however, that the elimination of paper is predicated on the availability of information, which may exist in a heterogeneous networked environment. In order to provide communication between the various systems, networking standards such as manufacturing automation protocol (MAP) are being developed.

DNC systems also provide a means for shop floor collection of management, production, quality, and maintenance data. This can be done by operator entry or by automatic input, with appropriate means for feedback and reporting. In order to provide automatic feedback, the system must be equipped with the means to obtain appropriate sensory input and pass that information to the appropriate control level. For example, if the MCU receives a signal from the lubrication system of the machine that the system has reached the upper limit of operating temperature, it initiates a shutdown cycle. If appropriate communications are installed, maintenance can be informed of the problem. In addition, production control can be notified that the machine has been taken out of production.

4.2.5 Sensors—the key to automated control

Sensors provide the system with the capability to respond to its environment. This is a necessity for automated operation. A sensor is a transducer or other device whose input is a physical phenomenon and whose output is a quantitative measure of that physical phenomenon. For our purposes, the output should be intelligible for computer input. In the manufacturing environment, sensors provide input on workpiece recognition and orientation, worn or broken tool detection, cycle counting, and equipment status feedback. In an unattended operation, sensors provide in-process quality control checkpoints. As is appropriate for any inspection system, this monitoring is most advantageously applied to detect reject states at the point of lowest value added. Sensors are of two types: contact and noncontact. Contact sensing includes force, torque, touch, and weight sensors; noncontact sensing includes vision, acoustic, laser, and proximity sensors.

Adaptive control systems in chip-making operations provide the ability to automatically optimize feeds and speeds based on machining conditions. This allows for maximum cutter life and/or lower machining cost. These systems monitor torque, horsepower, spindle deflection, or vibration. They provide a means of feedback to the MCU for adjusting feeds and speeds in a prescribed manner based on limits established in the part program. Some of these systems also detect torsional overload indicating imminent catastrophic tool failure. With programmed torque expectancy, they can initiate an automatic interrupt if no torque is experienced, which indicates a broken tool or missing part surface.

Touch probes are the "tools" of NC coordinate-measuring machines, and they can be used on machine tools. They are useful for part or feature recognition and orientation, in-cycle inspection or gaging, and automatic tool setting or gaging. Probe systems consist of five basic components: the probe body, a replaceable stylus, a means for signal transmission, a control and interface printed circuit board in the CNC, and appropriate software probing routines (to establish hole centers and diameters, Z-axis surface location, etc.). Cycles are usually run at slow feed rates to accommodate mechanical conditions such as following error and to provide accuracy and repeatability. Therefore the trade-off of probe time vs. potential for error detection and correction must be analyzed. Feedback to the MCU to regrid the machine to accommodate variations in workpieces (castings) or fixture setup, or to adjust tool offset or tool length compensation, provide the means for automatic operation of equipment. In addition, statistical trends can be analyzed to provide input to the decision algorithms concerning tool replacement, or dimensional inspection reports can be output for quality control.

Machine vision systems can be used for gaging, verification, cosmetic evaluation, recognition, identification, and location or position analysis. Visual inspection is currently a major, though largely undocumented, task of all production workers. Building that versatility into a machine vision system is beyond the current state of the art, although this is changing rapidly. The essential components of a vision system are the means of image acquisition (its eye), which requires adequate and appropriate illumination of the object, and the means of image processing (its brain), which requires explicit information about what it is looking for. Current systems can be categorized as specific-task-oriented (such as offline dimensional measurement), multifunctional systems (which are configurable to a variety of applications but perform a limited number of tasks), or dedicated specialized inspection stations. Areas of technological investigation to enhance functionality include sensor resolution, imaging time, evaluation of a three-dimensional scene, color processing, and the time to process an image to make a decision.

Acoustic sensing can be used for object recognition, position and orientation verification, missed operations or component detection, sorting, and measuring the three-dimensional shape of an object. The system measures the interference pattern of continuous ultrasonic waves detected by an array of transducers mounted near the parts to be verified. It is taught the acoustic signature of an object, and multiple signatures can be stored. In operation, the objects to be inspected are acoustically measured, and the system either identifies the learned object or signals that an unknown object has been detected.

Basically, a sensor must recognize what is happening within its realm of perception and transmit that information to its controller. The signal is then sent to the machine control unit or system controller for information processing. An appropriate response is initiated, depending on what type of response mechanism or alert has been incorporated into the system and how the devices have been programmed to operate. This is one of the most critical and difficult design problems in building a system, especially since the system interfaces frequently require the interfacing of vendors as well as technology.

4.2.6 Flexible automation

The basic premise of flexible automation is to make discrete parts manufacturing function in the manner of a continuous process environment. Flexibility can be achieved through the use of NC systems. The most frequently stated objectives for flexible automated manufacturing systems are:

1. Reduced operating costs and capital investments
   - Reduced labor costs
   - Reduced work-in-process inventory
   - Improved equipment utilization
   - Reduced tooling and associated costs
2. Increased design freedom and part volume flexibility
   - Adaptability to new workpiece introduction
   - Ability to react quickly to most engineering changes
   - Reduced lot sizes
3. Improved capability to respond to market demand
   - Ability to optimize lot size based on demand
   - Ability to minimize leadtime

Though NC technology is used at all volume levels of production (including high-volume, high-speed transfer lines), the needs of the batch manufacturer in the mid-volume, mid-variety range are receiving particular attention. This covers yearly production volumes of 15 to 15,000 per part number combined with part varieties ranging from 2 to 800, as shown in Fig. 4.4. The term being used to describe this type of system is the abbreviation FMS, where the M can represent manufacturing or machining. A flexible manufacturing system (FMS) is a highly automated, complex multimachine system which may perform any type of manufacturing operation, including milling, turning, grinding, fabrication operations such as punching, bending, and shearing, or assembly operations. A flexible machining system (FMS) may be a flexible manufacturing system which is specifically concerned with chip-making operations, or it may consist of a single CNC machining center equipped with automatic tool changer and pallet shuttle for unattended operation. This may also be called a flexible cell or module.

Flexibility is the result of planning, not an excuse for a lack of planning. The system must be designed for a specific set of workpieces, which require a specific set of operations, at a specific production rate per month or year, based on normal daily production hours. Simulation modeling can assist in the design as well as in the operation of a system. Indeed, the experience of designing a computer simulation model may suggest real-world changes in the actual system.

The concept of a flexible manufacturing system is to have palletized workpieces of different types randomly and simultaneously transported between workstations and processed at those workstations according to processing and production requirements for each piece, all under automatic computer control. Workstations usually consist of standard general-purpose machine tools and equipment such as coordinate-measuring machines. Information is the lifeblood of the system, and the system control

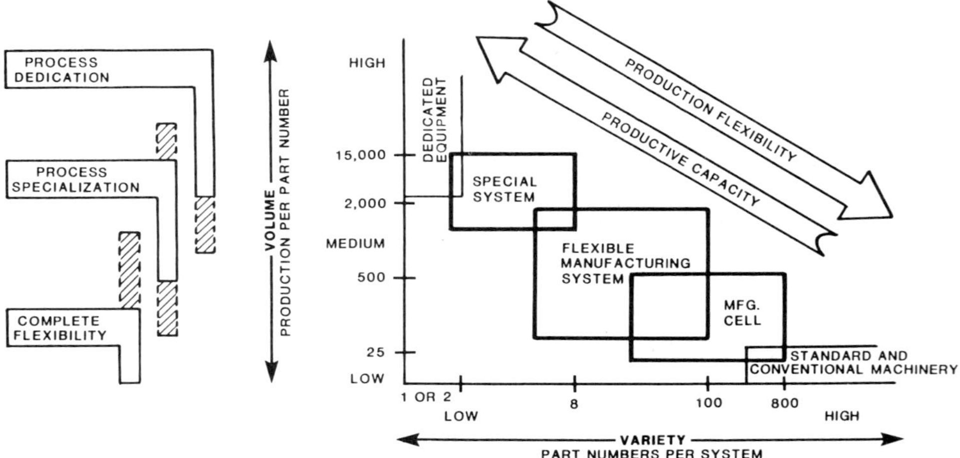

Figure 4.4 Manufacturing concepts based on volume and variety mix. (*Courtesy of Kearney and Trecker Corporation*)

center is the information clearinghouse for system operation. It communicates with production control for workpiece-scheduling information and routings and with MCUs and other device controllers to download NC programs and operational instructions and upload feedback signals. In addition, the system controller acts as the traffic manager for the automated materials handling system to assure that the correct palletized workpieces arrive at machines having the correct tools for the required operations. It handles machine status information, and it can dynamically reroute workpieces when machines are taken offline for maintenance or other reasons.

Flexibility does not necessarily require a totally automated facility which can randomly process all operations on all parts. A special system may be dedicated to a few part numbers produced at relatively high volume, frequently a flow-line type of process which may require specialized process equipment. This may apply to an automated, sequential-station printed circuit board assembly line which can stuff a variety of boards. Here the flexibility is expressed in terms of variety/volume mix within specified constraints, rather than random routing.

Manufacturing cells are usually dedicated to a single part family or a group of related part families defined by similar shapes and manufacturing processes. A cell will usually include all varieties of equipment required to process the family members completely. The exception would be very specialized processing such as heat treating, which could not be handled economically in the cell. The cell may be physical or virtual. In a physical cell, all the equipment is grouped on the shop floor according to the needs of the cell. In a virtual cell, specific machines from the functional lines of lathes, mills, grinders, and drill presses are assigned to specific cells and scheduled accordingly without physically changing the location of the equipment. A physical cell may be automated and serviced by a robot, whereas a virtual cell, though containing NC equipment, would not have that benefit. The flexibility of a cell can be expressed in terms of reduced planning, tooling, and setup for each member of the part family and the ability to economically handle small lots.

Prime considerations in planning for flexible automation are the selection of vendors, the time-phasing of equipment purchase and installation, and full system integration and implementation. All are dependent on financial considerations, as well as requirements to integrate existing equipment. Some companies will decide on a turnkey, single-vendor system in an attempt to minimize system integration and vendor relation problems; others will purchase equipment over an extended time frame and then integrate their own systems. A lack of hardware and software standards, e.g., in pallets, communication protocols, tools, and other devices, can cause unanticipated problems.

As the specifications for a flexible automation system are being defined, it is important to remember that all the tasks to be automated must provide for all activities previously performed by human beings, such as workpiece unloading and loading (including compensation for variance, orientation, and positive location), detection of incomplete prior operations, burr removal, thread gaging, chip removal, application of special lubricants for drilling and tapping, recognition and replacement of worn and broken tools, and so on. The system must have the ability to recognize when a problem has arisen, select the appropriate corrective action, verify that the corrective action has indeed solved the problem, provide feedback to resume operation or alert a human, and report on the sequence of events to the system control center.

Many current systems are designed to run with minimal human intervention, though more attention is being given to designing unattended systems. In that case, efforts should be made to prevent known problems from occurring in the automated system. For example, jamming of chip conveyors is a known problem. The human solution is to first reverse the conveyor when a jam has occurred and then untangle the mess. In the automated system, rather than wait for the jam to happen, it may be easier to reverse the conveyor every so often to prevent the tangles, though a fail-safe jam detector should also be provided. Or better yet, new tool materials, tool configurations, and/or processes may prevent chip tangling altogether. Flexible systems are frequently run more conservatively than conventional equipment to protect the machines and tools and to minimize the potential for catastrophic failure. Every element in the system must be extremely reliable and exhibit long service life.

4.3 NC Programming Requirements

The information provided to the NC machine by way of the part program determines the basic efficiency and effectiveness of an NC investment. Buying a machine without considering programming requirements is like buying a Cadillac and forgetting it needs gas.

4.3.1 The NC programming function

The primary benefit of CIM is the leveraging of information. Since NC programming requires direct access to product definition data, the ability of engineering to provide data to manufacturing in a form which is usable without modification or alteration and requires little or no interpretation is of critical interest. That is true whether the production of NC programs involves a human programmer or has been automated. Until the generation of machine control data is fully automated, the effective and efficient human NC programmer must follow the three C's of CNC: communication, coordination, and cooperation.

The basic instructions for the coordination of NC operations are provided by the process plan, routing sheet, or operations list. Tool design, the tool crib, the part programmer, and the setup person must communicate with each other so that all know which tools and fixtures are to be used on a job, including the appropriate documentation to assure that the entire operation can be duplicated in the future. Production control must coordinate the scheduling of the operation so that materials, tools, and NC programs arrive at the right machine at the right time. The NC programmer, working with the machine tool operator, should know how to optimize the operating parameters of each piece of NC equipment for which he or she is responsible. To close the information loop, any changes made at any level must provide feedback to the group which initiated the information. Numerical control productivity is a team effort.

4.3.2 NC programming methods

The method of NC programming must provide for the description of part geometry, the means of relating manufacturing operations to the geometry, and a means for specifying and formatting machine-specific information. There are four basic methods of NC programming.

1. Manual (by numbers)
2. Digitizing (by templates)
3. Language-based computer assist (by words)
4. Graphics-based computer assist (by pictures)

Manual part programming is known as programming by the numbers. It requires working with the blueprint to compute the precise numerical values for the sequence of locations which the tool will traverse for the operations to be performed. The data is then formatted by using the appropriate codes for the machine tool–control unit combination; the input media (e.g., punched tape) is prepared; and the program is proofed by dry-running it on the machine or by back-plotting from the machine control data. Manual programming is done at the machine language level. This method of programming is basically incompatible with the CIM operating philosophy, though it may be appropriate for the small NC shop.

Manual data input (MDI) numerical controls are frequently stand-alone units unable to communicate electronically with the outside world. Some offer sophisticated NC programming capabilities, including graphics, with programming workstations similar to the machine control unit. Some of these systems may integrate horizontally for shop floor automation but not vertically with the product definition database.

Digitizing is appropriate if a nonmathematical or free-form scaled drawing or a pattern of the part is all that is available for creating the NC data. Digitizing means obtaining a digital (or numerical) representation of an object or drawing. This can be done by using a mechanical mechanism and position encoder (e.g., using a stylus connected to an arm to follow a model like a tracer template or mold pattern). An alternative method uses an electrooptical or electromagnetic mechanism to follow a scaled drawing placed on a special digitizer bed. The data obtained is processed by a computer in order to generate NC data. Digitizing provides a means for inputting physically represented data to a digital database. This method has been used to generate aircraft tubing on NC tube-bending equipment and to create the digital database which allowed tube masters to be made obsolete.

Language-based computer-assisted part programming uses a high-level language, such as Automatically Programmed Tools (APT) or Compact II, for writing a manuscript, or source program. This program contains definitions of the workpiece and cutting tool geometry, a description of the tool motions, and the sequence of machine functions. The source program is then processed by the computer. The output of APT is a cutter location (CL) file or CL data. This can be plotted for toolpath verification. The data is then postprocessed to generate the machine control data. Though language-based programming is widely used, its basic incompatibility with CIM is the requirement to recreate, redefine, and reinterpret geometry, rather than use design data stored in a digital database. Some CAD systems do provide output of named or labeled geometry, which allows the programmer to use the digital database for the geometric description of the workpiece. But the programmer must still write a source program to describe the sequence of manufacturing operations in relation to the geometry. However, the NC processor is then able to integrate the digital geometric description with the tool motion statements of the source program. A problem arises if the NC programmer is able to modify the geometry definitions, because that will introduce inconsistency between design and manufacturing data. Based on the shortcomings of graphics-based processors in some applications and the legacy of existing NC source programs, language-based part programming is viable and will continue to be in use for many years.

Graphics-based computer-assisted part programming allows the digital database created on a CAD system to be used directly to create toolpath data. This allows for the maximum leveraging of information. The term "assisted" rather than "automated" is still appropriate because the NC programmer must still follow the process plan to indicate what operations are to take place on which pieces of geometry. As the CAD database and its associated applications programs and interfaces become more intelligent, the automatic creation of manufacturing sequence information becomes feasible. This includes integration with automated process planning, automated tool selection, and machinability data tables. Indeed, some special application programs have been written to automate the design-through-manufacture process. These are usually based on families of parts or other special applications.

One of the prime benefits of graphics NC programming is the ability to overlay the toolpath on the workpiece geometry to provide process simulation and visual verification of the correctness of the toolpath and its relations with tools and fixtures, which should also be designed by using the product definition database. This provides a tremendous economic advantage over the on-the-machine tape tryout procedure required for other programming methods.

The geometric descriptions used by most graphics NC processors are based on wireframe and surfaced models. These can be ambiguous to the computer. Solid geometric modelers provide an unambiguous description of geometric shape. Their use for NC processing is increasing.

The geometric model created on the CAD system is a mathematical description of the part to be produced, and it must be complete and accurate. Since this data is to be shared, engineers and drafters who create the models must become more sensitive to the needs of manufacturing. Previous drafting practices, such as standard notes and callouts for fillets, radii, and chamfers "unless otherwise specified," must be graphically incorporated into the geometric model. All information should be entered at full scale. This ensures a dimensionally accurate database that requires no interpretation by either a human operator or an automated program.

Tolerancing practices can create some serious problems requiring the redrawing of the model, thereby endangering the integrity of the design. Since manufacturing frequently splits the difference and works to the mean, nonsymmetrical or unilateral tolerancing is inappropriate for data to be used by numerical control. For example, if a feature is dimensioned and toleranced as 2.000 +.010/−.000, the NC programmer will define it as 2.005 +/−.005 to conform to normal machining practices. In the CAD database, the computer has calculated and stored the distance between these two pieces of geometry as 2.000; the tolerance is essentially ignored. When manual NC programming or language-based computer assist is used, this poses no problem; the programmer is in control of the defined geometry and adjusts it accordingly. When a CAD database is the source of the geometry, the NC programmer must either redraw the part to the mean (2.005) or create auxiliary geometry to compensate for the tolerance offset. A similar problem arises with max/min dimensioning. These procedures can cause a serious update problem when an engineering change order must be incorporated. In order to share data, the engineering standards and practices of the enterprise may need to be rethought and revised to accommodate digital technology.

The software legacy has hindered the acceptance of graphics by some companies that have developed either a formal or informal group technology system with a set of part family NC applications programs which minimize NC programming lead time but are not necessarily integrated with the engineering world. The programs are frequently written in APT or Compact II subroutines which are not necessarily conveniently translatable to graphics programming languages. Though this may be a gap in the CIM system, the productivity advantages of proven programs may outweigh a perceived need for purity of concept. For the long term, however, a migration path must be included in the overall integration strategy.

The preceding discussion has been concerned with programming the normal machine operations. As the manufacturing system becomes more complex, other devices, such as sensors and robots, may be controlled from or interfaced with the MCU. There is little standardization in how this is accomplished. Some devices can be programmed offline. Some robots must be "taught" their instructions by leading them through the appropriate movements. Some sensors are taught by allowing them to sense a correct pattern. For example, a machine vision or acoustic system may be "shown" what it needs to recognize. The integration of these operations into the basic machine control data or part program may be cumbersome, manual procedures which must be duplicated if the part program is revised to accommodate an engineering change or if the requirements for the device operations change.

4.3.3 Postprocessors and NC data format requirements

The output of language, graphics, or digitized programs must ultimately be processed into the data format appropriate to the MCU. This is done by a postprocessor, which is a computer program that takes the generalized cutter location (CL) data and adapts it to the specific machine control unit–machine tool combination that will be used for the operation. In general, one postprocessor is required for each control unit–machine tool combination. In an attempt to simplify the postprocessor problem, a standard CNC programming format, the EIA Standard RS-494 or binary cutter location data exchange format for numerically controlled machine tools (BCL), was developed. This standard defines a specific format for CL data to be used as input to the CNC. Essentially, the postprocessing function has been moved to the executive program in the MCU, with "postprocessing" done at the machine during program execution.

The data format accepted by most CNC units is called word address, where the location to which a signal must go is identified by a single-letter "word" code. A block, or complete instruction, to the MCU may look like this:

N5G01X35000Y-5000Z-7500F100T1M3

The N word, for example, identifies the sequence number of the block. Unfortunately, there is a lack of standardization in many aspects of the data required. The numeric values may require leading or trailing zeros or optional decimal points. Though some word codes have generally accepted meanings, such as the X, Y, and Z that are designations for cartesian coordinate axis references, other letters of the alphabet are variously employed by the control builders.

The G codes, referred to as preparatory codes, may indicate how an X, Y, or Z value is to be interpreted by the MCU (absolute or incremental, inch or millimeter) or what type of operation is to be performed (cut at linear feed, cut a clockwise arc, or move at rapid traverse). A G code may initiate a canned cycle such as threading on a lathe or pocketing on a mill. The M codes are referred to as miscellaneous, machine, or mechanical codes. They would be used, for example, to turn the spindle on clockwise or initiate a tool change cycle. Though G and M codes may have commonly accepted functions, there is a lack of standardization in values, which will cause problems as more devices, such as in-cycle inspection probes, are interfaced with the MCU.

4.3.4 Guidelines for analyzing programming needs

A basic needs analysis for selecting a programming method would include answers to the following questions:

1. What types of parts are to be programmed? Are the machined parts to be programmed simple (two-dimensional turning or drilling) or complex (three-dimensional sculptured molds)? Are special applications such as coordinate-measuring machines, nesting and punching for sheet metal parts, or printed circuit board drilling and component insertion required? Will the programming of special devices, such as sensors or robots, be incorporated into the part program?
2. On what medium is the product definition data available?
   - A physical pattern, model, or template.
   - A traditional blueprint.
   - The digital counterpart of the blueprint on a two-dimensional computer-aided drafting system.
   - A three-dimensional digital database on a computer-aided design (CAD) system.
3. If digital data is to be transferred between dissimilar systems, is initial graphics exchange specification (IGES) neutral data format processing available?
4. How many programs will be needed in a week or month? How often will existing programs need modification (e.g., to incorporate engineering change orders)?
5. Are a variety of machine tool–controller combinations with different input code requirements to be programmed? Do any of the controllers accept BCL data?

6. Will the NC system be part of a larger production system in the future? Is there an existing software legacy which must be accommodated?

4.4 Tool Management

Tools and tooling address critical issues of time and cost in manufacturing, especially as they affect a programmable automation investment. The economic order quantity (EOQ) of traditional batch manufacturing included in its cost equation the amortization of the cost of setup over the quantity of pieces in the lot. If the EOQ of the flexible system is to be reduced to 1, as is suggested by the just-in-time operating philosophy, tools and tooling must be provided to support the system economically.

Tool management is made up of tool design and usage control and tool inventory control. Tool design and usage control promotes standards, prevents the proliferation of unnecessary tools, and devises means to simplify work holding. Tool inventory control, with the addition of tool requirements planning (TRP), is the traditional tool crib function. An example of a flow diagram for such a tool management system is shown in Fig. 4.5.

4.4.1 Tool design and usage control

When NC equipment first came into use, one of the high-ranking benefits was the ability to use simple workholding devices such as vise jaws. When the workpiece was clamped in a known position, the NC machine would run its cycle and produce a good part, provided the correct tools were used and set at the appropriate lengths and so on. With the new flexible systems, the same thing holds true. At the National Bureau of Standards Automated Manufacturing Research Facility, a flexible, programmable vise has been developed; it is able to modify itself under directions given by a special fixturing controller in the workstation. This enables the vise to accommodate a variety of parts, as well as robot loading. In lathe applications, the ability to program the opening and closing of chuck jaws along with the ability to change chuck jaws from a tooling carousel on the headstock automatically allows for quick changeover of setups, as well as unattended operation. Modular or "erector set" tooling allows minimization of hard tooling cost by providing for the reuse of the components or modules. The ability to design the fixture on a CAD system is also an aid to fabrication. A complete bill of materials, as well as a detailed and accurate layout showing both the workpiece and the fixture can be provided.

A more unique solution, phase-change fixturing, is being researched at the Industrial Technology Institute (Ann Arbor, Michigan). Phase-change fixturing is the ultimate conformable workholding device: it is able to accommodate all varieties of geometric configuration. It exploits the ability of certain classes of materials to change from a fluid to a solid and back to a fluid again. In the solid state, the material envelops the workpiece and provides uniform distributed support; in the fluid state, loading and unloading can be easily accomplished.

The application of group technology concepts to the selection of parts in the flexible system allows for a significant decrease in both setup time and tooling costs by using the same setup for all members of the family

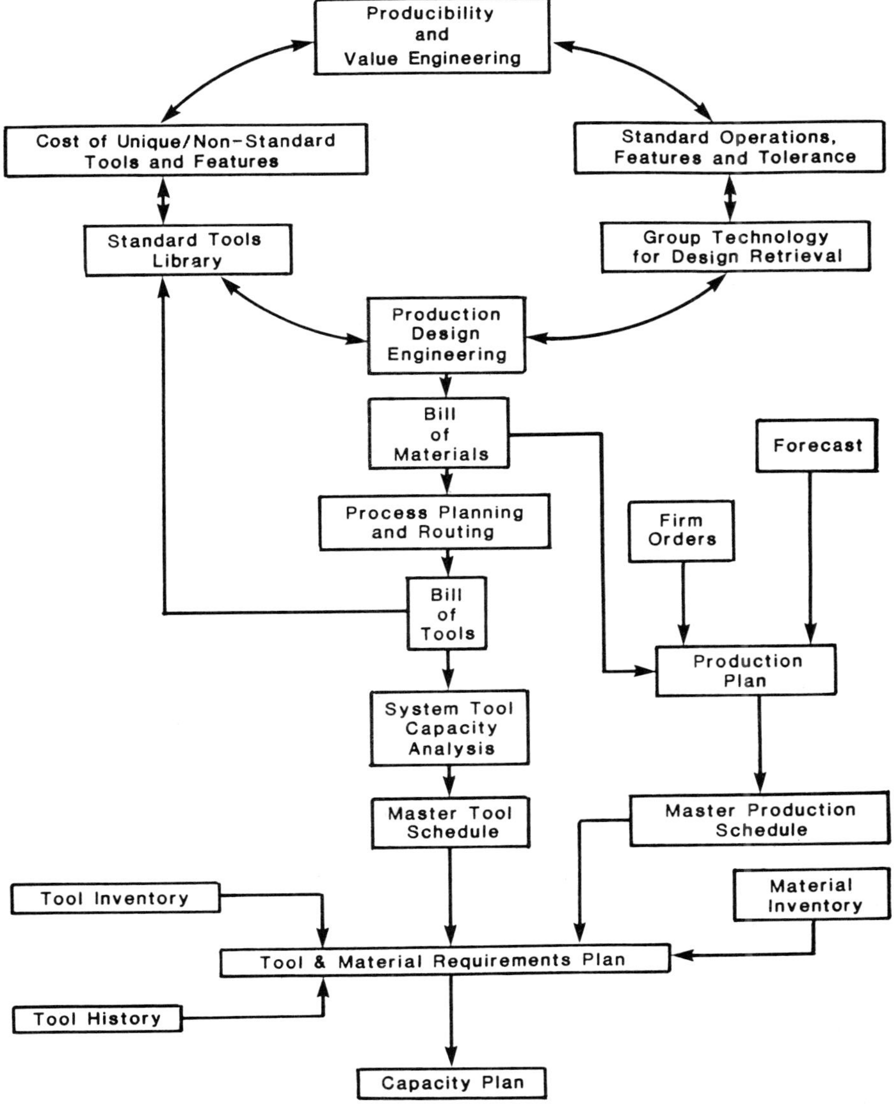

Figure 4.5 Tool management system.

of parts. This can include preset tool turrets for quick changeover on lathes. The fixturing is designed to accommodate all family members, and families based on common operations and workholding requirements can be created. All of this is facilitated if the fixturing can be designed on a CAD system. That provides the opportunity for all family members to be placed in the fixture before the fixture is built to discover unanticipated interferences or other problems.

The institution of standards goes hand-in-hand with preventing the proliferation of unnecessary tools, and a two-pronged attack is required. First, standard tools and tooling must be defined and identified. This may be done in conjunction with creating standards for process planning. One company noted that 431 different drills under ½ in diameter were called for in the process plans for parts routed through the FMS. Second, the standards must be applied to the product design during the original design process. Costs associated with the use of unique or nonstandard tools must be consistently applied.

Tool proliferation is not a new problem. In the automated manufacturing environment, it was first encountered when NC machining centers and turning centers were introduced. The solution was to increase the capacity of the tool carousel or magazine, add tool turrets, or use gang tooling. The need for each tool was not questioned. In the FMS environment, the solution has been to add the ability to change tool magazines, not just tools. However, system tool capacity is not infinite. A better solution might be to institute a tool review board to analyze tool requirements and implement tool management concepts. Also, a computer model for simulated operation of the FMS can include tool capacity as one of the system constraints.

The tools and tooling which will be considered standard in your shop must be identified. A standard tool list contains the items which are commercially available and/or are in normal use in your operation. One method for organizing this data is through the application of group technology classification and coding systems. The system must allow for flexible access for a variety of users from design, manufacturing, material control, and purchasing. Applying the standards to the design process is a delicate situation; it can cause accusations of infringement on the creativity of the design engineers. However, the use of standard tool parameters, such as the width of a commercially available grooving tool as the standard for a nonfunctional thread relief groove, and standard operations, such as how to create a 1/4-20 tapped hole, can effectively limit the number of tools required to do a job.

4.4.2 Tool inventory control and tool requirements planning

A wide variety of items is covered in the term "tool," including cutting tools, fixtures, gages, NC tapes, and robot end effectors. These are critical items;

without them, work cannot be accomplished. In the age of automation, the scheduling of tools and tooling has become increasingly important as the investment in the machines which use the tools has increased. Tool shortages that cause production bottlenecks must be eliminated. To facilitate the elimination, tool requirements planning requires a bill of tools for each operation for each part to be created and maintained. This will require enforced configuration control over process plans and other manufacturing documents which identify tool requirements. In essence, TRP is MRP specifically for tools. Its use identifies shortages and contentions for special tools.

A basic requirement to make the tool management system work is a formal inventory control system in the tool crib which provides for the orderly storage and retrieval of all classes of tools and includes an accurate count of items on hand. Tool tracking is a convenient way of locating tools in use, and it may be an important part of the tool maintenance, refurbishment, and calibration schedule. A tool history file can be of value for reporting tool life. It can also be used to track problems with machines, maintain quality checks on vendors, and provide information to purchasing.

4.5 Planning for Shop Floor Automation

Planning for shop floor automation is not simply a manufacturing management function. When the top-down strategic planning efforts underway in many manufacturing enterprises at this time are considered, it becomes evident that shop floor automation must fit in with the overall strategy of the company. If the engineers of the company, or the company's prime customer, have redesigned the product to use new continuously variable transmissions rather than gearboxes, there is no need to build better gearboxes. Perhaps a decision has been made to make a new product out of composites or ceramics or that near net shape forgings have better material properties than the old style forgings. These decisions will have a dramatic effect on production processes.

The production philosophy must be compatible with the enterprise philosophy. A management decision to go just-in-time will require not only a change in manufacturing methods but also the reeducation of the entire work force. No longer will bins of parts be stacked by each machine as a symbol of job security to the worker. Even a change to group technology–oriented cells requires the reeducation of those who are used to the traditional manufacturing flow. The elimination of direct labor by the installation of unattended systems requires the full attention of management. Investing in new automation technology requires dedicated planning.

It is unfortunate that the planning, analysis, and system simulation which are being undertaken for flexible manufacturing systems are not being applied to stand-alone NC machines. If it were, it would become apparent that the latter also could be used more productively. In addition, the cooperative problem solving efforts required to realize the unattended flexible system environment are just as appropriate to productive use of the attended stand-alone NC machine environment.

The flexibility that programmable shop floor automation seeks to achieve is the result of strategic planning. Crisis management, the norm in many manufacturing facilities, must be replaced with knowledgeable decision making based on timely and accurate information if automated systems are to perform to their full potential. Data is the fuel of the automated factory. The source is an integrated database which allows for the integration of data-sharing applications such as group technology, process planning, CAD/CAM, MRP, tooling, shop floor control, and production scheduling. The critical factor, however, is not the machines or the data, it is the dedicated group of people who have the full support of top management. People make the system work.

CHAPTER

5

Robotics

Peter Walsh

5.1 Introduction

This chapter has to do with industrial robot systems and their role in computer-integrated manufacturing (CIM). The approach taken is to examine the control, programming, and integration of industrial robots with particular focus on the evolution of robot control technology and the current directions in that evolution. There are unique features of industrial robot technology which are important in understanding the issues related to the integration of robot systems into the manufacturing environment. This chapter presents those distinguishing features through a discussion of the evolution of controller technology.

5.1.1 Flexibility

A key contribution made by industrial robots to manufacturing automation is the flexibility which they give the manufacturing automation environment. Industrial robot systems provide the integrated combination of general-purpose manipulating hardware and a computer-based controller. Reprogrammability enables such a system to be applied with a high degree of flexibility. This flexibility is found in the capability of the robot system to handle part batches of varying size and mix as well as its capability to adapt to process variations and uncertainty in the working environment. For example, a robot performing tasks such as arc welding and spray painting can process any of a family of parts presented to the robot system which have similar processing requirements if the processing is within the limits of the work envelope and performance characteristics of the robot.

In an automated robot system, the identity of the incoming parts may be determined automatically by a device such as a bar-code reader or vision system. The identity of the part can be used to load the appropriate robot program into memory and initiate program execution. Just as the proper program can be loaded into the robot controller, so the proper robot end-of-arm tooling can be attached automatically through the use of interchangeable tooling. The attachment of the appropriate tooling for each part may be controlled by the program which has been automatically loaded into the robot controller.

The flexibility found in the ability of a robot system to adapt to variability in the work environment is achieved through the use of sensors for determining the state of the work environment and feeding that information into the robot controller. The controller uses the information to adapt its behavior. This is accomplished in an arc welding system, for example, by sensing the location of a seam to be welded and applying corrective actions to the path of the welding torch as the robot moves the torch along the seam.

Many manufacturing tasks cannot be automated, in terms of either technical feasibility or economic justification, unless the associated equipment is adaptable to changes in the manufactured product or is able to handle the uncertain and unstructured nature of the manufacturing environment. Conventional hard automation solutions, which are not adaptable in this regard, are comparatively more risky investments because of their single-purpose nature. The characteristics and the current evolution of robot technology are, to a great extent, motivated by the required flexibility of robot system performance. As will be discussed in the subsequent sections, flexibility is a motivating force behind the current developments in areas such as programming methodology, functionality supported by programming languages, and the application of sensors for the control of robot activities.

5.2 Programming Methodology

The methodology for programming industrial robots is of primary importance in viewing robot systems from the perspective of CIM. The evolution of robot controller technology and the complementary technology of offline programming systems are progressing rapidly. The methodology of programming through teaching served early applications of industrial robots, but there is a strong need to augment teaching with significantly more powerful techniques which are becoming available in the form of offline and interactive programming systems.

5.2.1 Teaching

The traditional method for programming an industrial robot is a lead-through or teaching system in which instructions are programmed sequentially by push-button selection of instruction types and their arguments by using a teach pendant. There is no standard for its design, but the teach pendant typically contains one button for each instruction type and a numeric keypad accompanied by a limited display. There is also some means for controlling the positioning of the robot axes by either push buttons or a joystick. Some controllers augment the teach pendant by a conventional computer terminal or some form of keyboard and display at the robot controller.

During the programming of motion instructions, the robot is positioned at each of the locations to be programmed by the use of the teach pendant. Once positioned, a positioning instruction which causes the current locations of the axes to be stored in memory is selected. Similarly, all input and output functions, branching, and other functions are taught by the use of the teach pendant. While this programming methodology is expedient for simple applications, it becomes cumbersome as the tasks become more complicated and is prohibitive for the more advanced applications which require either extensive programming or intricate positioning. Applications which are to be programmed with the teach method can be infeasible based solely on the programming requirements if they have a large number of positioning instructions for a single part, must be programmed for many different parts, or require an extensive amount of logic programming.

5.2.2 Interactive terminal-based systems

Teach pendant programming has been augmented on some robot controllers by the use of more conventional terminal-based programming. With such a system, the teach pendant is still used to manipulate the robot and can be used as a simple operator interface. However, the programming of positioning instructions can also be accomplished at the terminal by the specification of the coordinates of a position. Previously programmed positions can be altered or the actual current robot position can be manipulated by commands from the keyboard. Most systems of this type cannot be used for programming without interrupting the operation of the robot system, and they are characterized as being online programming systems.

The ability to edit programs at a computer terminal is a significant advantage over systems which employ only the teach pendant. Programs can be written and edited at the terminal by using a traditional textually oriented programming environment. The programs can be entered, displayed, and manipulated without the cumbersome use of a teach pendant. However, the teach pendant is still generally used for the manual manipulation of the robot axes. The positions of the axes can then be read directly for use as arguments in positioning instructions.

5.2.3 System start-up

The online nature of such programming systems is very powerful for the interactive operation and debugging of robot systems. The debugging and initial running of a robot system are extremely important but often initially overlooked aspects of the application of industrial robots to manufacturing automation.

In the initial operation of a newly installed robot system, one must deal with:

- Proving out the logic of the robot programs
- Fine-tuning the precise locations to which robot motions are programmed
- Verifying and correcting the operation and performance of associated special tooling or fixtures
- Verifying and correcting the operation of and interfacing with other components in the system
- Compensating for the impact of the actual vs. anticipated variability in the parts which are being processed by the robot system

It is not just the robot program which must be debugged, but the entire integrated robot system and all that it entails. The capability to operate interactively and modify the operation of the system is, therefore, very powerful, and it can significantly enhance the rapid and smooth introduction of the system into the manufacturing environment. Although the need for some of the online debugging can be mitigated by the use of simulation in offline programming systems, the initial system debugging will remain an important aspect of the industrial application of robots.

5.2.4 Offline programming

The alternative to the traditional method of programming through teaching is the use of offline programming with either a text-based programming language or a graphics-based programming system. The motivation for offline programming comes in many forms. The programming of tasks which involve the processing of a part with a large number of positioning instructions will generally benefit from the use of offline programming if part geometry data is available and usable by the offline system for the generation of robot-positioning instructions. A robot system which processes very simple parts may also benefit from offline programming when a large number of parts are to be programmed.

Extensive programming of positioning instructions is not the only motivation for using offline programming. Tasks with relatively few positioning instructions may also be unwieldy for the conventional teaching method if there is significant complexity in the logic programming of the task. That is, a task may require extensive use of input and output functions and corresponding testing and branching in order to control program execution as a function of the state of the integrated system. Programming such a task by using the teach method is cumbersome and inefficient.

Regardless of the specific nature of its application, offline programming provides the capability to program a robot system without taking the system out of production. Not interrupting production can, by itself, be sufficient motivation for using offline programming. As the manufacturing process becomes more integrated, this becomes particularly important. Programming the system offline also permits the programming activity to be conducted prior to the actual installation of the system. This reduces the number of sequential tasks required for the introduction of the system into the manufacturing process, since system hardware and software development can occur simultaneously. The shop floor environment is not particularly conducive to the programming of difficult tasks, and so performing the programming task at an offline programming station can be beneficial because of the resultant efficiency.

5.2.5 Textual and graphics-based programming systems

Systems for offline programming can be divided into two primary types: text-based and graphics-based. Text-based programming systems provide the means for creating and editing robot programs by working with a textual representation of the program on a computer system which may be external to the robot controller. Such systems are primarily of value for programming tasks which require a considerable amount of logic programming. They provide the means for working with the entire program on a display terminal rather than dealing with single instructions through the restricted teach pendant methodology for programming and editing. These systems are, by themselves, not as well suited as graphics-based systems for the offline programming of motion instructions. The specification of coordinates for motion instructions from a keyboard may in some cases be preferable to teaching, but it is generally not an effective means for entering motion instructions offline.

If, however, an interface to a CAD database is incorporated in the system, coordinate data may be imported from the external database. The part geometry data from the CAD system must be converted into tool position data. This can generally be achieved by converting the part geometry data to the tool position data on the CAD system and then transferring the positioning data to the offline programming system. The primary advantage of text-based programming systems is the capability to deal with difficult programming tasks without using the more sophisticated and expensive graphics-based systems. The interactive terminal-based robot controllers discussed previously have many of the characteristics of the text-based offline programming systems.

In comparison to text-based systems, graphics-based programming systems are better suited for dealing with geometric data for describing the parts, other equipment in the work cell, and associated fixtures and tooling.

The graphics-based systems have facilities for modeling the robot mechanism, the tooling, and the surrounding equipment. The cell layout for the robot system can be designed and analyzed. Simulations can be performed to estimate cycle time and to study the timing effects due to changes in the layout and in the robot program. Comparative analyses can be performed for different robots in the same cell.

The incorporation of robot kinematic models into the graphics-based systems permits the programmer to deal directly with tool position rather than the positions of the robot axes. Graphic analysis can be performed to determine the limitations of the work envelope of the robot as well as the detection of potential collisions between the robot arm and surrounding equipment. Robot position instructions are programmed by specifying the position and orientation of the robot end-of-arm tooling in a fixed reference frame. Once robot positions are defined relative to the location of a part in the work cell, the part can be repositioned elsewhere in the cell and the appropriate transformation of the robot positions can be applied to maintain the same relative position and orientation between the part and the robot end-of-arm tooling.

5.2.6 Generation of executable robot programs

Once the design and analysis of the robot cell have been completed, it is possible to generate executable robot programs which can be loaded into the robot controller. The representation of robot activity in offline programming systems must be converted into the internal formats used by the robot controllers. Many of the robot vendors have developed offline programming systems in support of their own products. The more generic offline programming systems which support robots from a number of manufacturers must deal with the lack of standards in robot controllers, programming languages, communication protocols, and formats for the transfer of programs. Much of the required format conversion is of a straightforward and mechanical nature. However, the disparity in functionality of robot programming languages inhibits the generic offline programming systems from supporting and taking advantage of the more advanced features of the various programming languages. Therefore, these generic systems tend to be limited to supporting a level of functionality which is the "common denominator" of the functionality of the individual robot programming languages.

The offline generation of executable robot programs solves many problems but creates a new one. The motion instructions generated will not, in general, be as accurate as the robots are repeatable. That is, there are many sources of error which affect the accuracy of offline programmed motion instructions but are not significant when teach mode programming is used. Such factors as manufacturing tolerances for the robot mechanism, the specification of zero reference points for the axes, and the effects of gravity loading all contribute to the error in positioning for locations which are specified offline. In fact, taught programs are not generally usable when interchanged between different robots of the same model and manufacture because of the uniqueness in construction of each mechanism. The models of the robot kinematics used in the offline programming system and in the robot controller are idealizations of the actual robot and do not capture all of the important phenomena. For further discussion of this point see Sec. 5.4.

5.3 Programming Languages and Controller Technology

Although the concept of a robot programming language may tend to be associated with robot systems which have conventional keyboard terminals for programming, the teach pendant controllers have equally well defined languages which happen to be accessible only through the teach pendant system. The underlying features and functionality of the corresponding languages can be examined independently of the methodology used for entering the instructions.

5.3.1 Explicit vs. implicit implementations

In general terms, the functionality of a robot controller is implemented by providing the capability with either explicit language functions or implicit support through a sufficiently general-purpose programming language. Explicit implementations provide predefined functions which will, in general, be easier to use but, because they are predefined, may be limited in their use. Controllers cannot generally be categorized as being strictly of one type or another. There are controllers which provide a general-purpose language as well as have explicit support of certain functions.

General-purpose programming languages have been implemented on many industrial robot controllers. They support the definition and use of variables, support a computational capability for computing values for those variables as functions of other variables and inputs from external sources, and support the use of variable arguments in the program instructions. Further generality may also be provided by the implementation of user-definable subroutines or procedures, various structures for control of program execution, and high-level functions for coordinate transformations.

As an illustration of the difference between the explicit and implicit approaches, consider the task of palletizing parts in a regular, rectangular pattern. In a possible explicit implementation of the operation the programming could require only three corner positions to be specified, along with the number of rows and columns in the pattern. A significant advantage is achieved by eliminating the need to program the positions which correspond to the elements in the pattern. The restrictive nature of explicit implementations can also be seen by the inability of such a palletizing routine to handle variations such as a skewed pattern.

5.3.2 Extensibility

The two methods for implementing the controller functionality exhibit a trade-off between extensibility and ease of use. "Extensibility" refers to the ability of a user to implement new functionality which was not explicitly provided in the original system. The implicit implementations are inherently extensible in that the user is able to implement new functionality by using the general-purpose programmability of the system. Such programming requires a higher level of competence of the programmer as compared with that which is required for purely explicit implementations. This is not necessarily a drawback, since the more complex programming tasks can be handled by one individual while the relatively simple task of programming for each member of a family of parts can still be handled by a second individual with a lower level of expertise. Another aspect of the implementations which is related to extensibility is the complexity of the applications which can be handled. The implicit systems are better able to handle more complex applications because their extensibility lets them be adapted to meet the requirements of advanced applications. The explicit implementations maintain an advantage of being relatively easy to program by individuals with little programming expertise.

5.3.3 Motion control

The closed-loop control of robot axes is typically achieved by feedback of measurements of the speed and position of the actuators. This direct feedback control is generally based on actuator position, e.g., motor shaft encoders or resolvers, rather than the actual end-of-arm-tooling location. This is relatively easy to implement, and it facilitates the achievement of good stability and performance characteristics for the control loops on an individual basis. However, it leads to inaccuracies in the static positioning of the robot end-of-arm tooling as well as shortcomings in the dynamic performance of the robot system as a whole. In general, the design of the loop closures does not take into account the dynamic interaction of axes or the variability of the loop characteristics due to changes in the payload or the effects of varying moments of inertia due to changes in the positions of the robot axes. Better dynamic performance and higher payload-to-manipulator weight ratios will be achieved with the development of more intelligent axis control, greater use of external sensors, direct drive motors, and reduction in the weight of the mechanisms.

Initial implementations of closed-loop controllers provided independent and uncoordinated control of the motion of the individual axes of the robot. Programmed motions were executed by driving each axis to its target position without consideration for synchronization with the other axes and with no consideration for control of the path of the tooling at the end of the robot arm. Although that was adequate for early systems, there was a great need for robot controllers to provide a real-time axis coordination capability for control of the path of the end-of-arm tooling during program execution as well as for use as a valuable aid in positioning the robot during the teaching process. Without coordinated control, positioning of the end-of-arm tooling during teaching requires an operator to manipulate all of the robot axes directly even for simple straight motions of the tooling position.

5.3.4 Coordinate transformations

The availability of greater computing power has enabled robot manufacturers to develop new versions of their controllers which incorporate coordinate transformations in the control of the robot arms. The coordinate transformation, which is based on a geometric model of the robot mechanism, converts between actuator positions and the corresponding tool locations. These tool locations are specified in a convenient external frame of reference which is generally referred to as the world frame of reference. Transformations are made in both the forward and backward directions, i.e., from actuator to world coordinates and from world to actuator coordinates. When a motion instruction is programmed by using the teach method, the robot is positioned at a desired location, the actuator positions are read and converted into equivalent world coordinates, and the equivalent world coordinates are stored in the controller memory.

During execution of the motion instruction, a world coordinate interpolation is made in real time to compute the path to be followed for moving to the desired position from the current position. This interpolation is usually a straight line in the world frame of reference, though other trajectories such as circular interpolation may also be supported. These transformations and the associated interpolations generally support both orientation and position of the robot tooling. The interpolation provides a sequence of world coordinate locations along the trajectory which are converted to actuator coordinates for use as reference trajectories for the axis control loops.

The use of coordinate transformations allows for the direct control of the tool center point in a world frame of reference. The tool center point is generally definable where it is most useful, such as at the center of a gripper, the tip of a welding arc or welding gun, or the point of contact with a part for processing applications such as plasma spraying and gluing. This enables a programmer to consider only the tool motions without the need to deal with the individual axes of the robot. This is particularly important in connection with the various jointed arm configurations of robots.

A further benefit of coordinate transformations is the ability to control the tool position and the tool orientation independently. This applies during both manual teaching and automatic run-time motion control. Because the tool position is being controlled directly, the velocity of tool motion is controlled accurately. That is critical for processing applications, such as arc welding, in which the feed speed is a primary process parameter. Multiple tool center points are typically definable, which allows multiple tools to be handled. Because the location of the tool center point is usually definable, changes in the tooling can be compensated for without a complete reprogramming of an existing program by redefining the location of the tool center point.

5.4 Positioning Performance Characteristics

Both robot controllers and offline programming systems use kinematic models for the computation of robot actuator positions. As a result, positioning error sources which were not evident with the teach method of programming become important. A number of distinct performance measures are needed to describe the positioning characteristics of a robot system fully. The key characteristics are resolution, repeatability, and accuracy.

Each of the performance measures can be defined at a point in space, but an actual value will depend on the particular point in space at which it is being examined. That is particularly true of, but not restricted to, the articulated robot arm configurations. The position dependence of these characteristics can be readily illustrated with an articulated arm example. A robot which rotates about its base will exhibit side-to-side positioning resolution which is dependent on the distance of the tool center point from the centerline of the base rotation. If the angular resolution of the base rotation is the limiting factor in the side-to-side positioning resolution, then an increase in the radial distance to the tool center point will result in a corresponding increase in the value of the linear positioning resolution. From the applications point of view, the worst-case values over the entire work envelope for each of these parameters must be considered unless it can be assured that the task will always be performed in a known subset of the work envelope.

5.4.1 Resolution

The resolution is the smallest displacement that the robot can move; it is measured at the end of arm tooling. The controller may be able to represent smaller motions in the motion commands but this is the smallest motion which can actually be performed. The resolution is primarily determined by the measurement resolution of the actuator sensors, which are typically incremental encoders or resolvers mounted to the motor shafts. In practice, the sensor resolution is generally the determining factor in the control loop resolution, although mechanical backlash and friction may be significant. The positioning resolution of a robot system is an important performance measure because it can indicate the tolerance of fine motions which the robot can make during such tasks as the mating of components in an assembly.

5.4.2 Repeatability

The repeatability is the error in returning to a point at which the robot had previously been positioned. It is the measure of system positioning stability; it reflects any variability in the control loop functioning as well as the characteristics of the various components which make up the drivetrain from the actuators to the tooling position such as hysteresis, backlash, and friction. Repeatability is the most important positioning characteristic when the teach method of programming is being used. During teaching, the actual values of the actuator sensors are read. At subsequent positioning at the same point, the axis control loops are commanded with the same set of actuator positions. Although the forward and reverse coordinate transformation may be applied, the repeatability is unaffected by kinematic modeling errors because the original position commands are given by the taught actuator positions rather than by specification of the world coordinates.

5.4.3 Accuracy

Accuracy is the characterization of the positioning error when the position commands are specified in a world coordinate reference frame. The world coordinate data is converted into actuator coordinates by using coordinate transformations which are based on a kinematic model. This model is an idealized representation which does not accurately reflect the actual robot mechanism. The components of the robot have manufacturing tolerances which deviate from the nominal values which are used in the kinematic model of the robot mechanism. They include such factors as the link lengths, the squareness of bearings, and the gear ratios. The point of reference for each axis, i.e., the zero reference point, will have some error when it is established for each robot, whereas the model assumes that it is set without error.

The positioning accuracy of the robot will also be affected by gravity loading. Errors due to gravity loading will occur as a result of deflections in the gearboxes and structure of the mechanism and because of inadequate disturbance rejection in the servo loops. Positioning variations will occur as the load at the end of the robot arm changes and as the self-load on each axis changes as the robot changes position. All the preceding are sources of error for static positioning of the robot. When the robot is in motion, dynamic loading occurs and causes further deviations from the desired trajectory. Accuracy is the important positioning characteristic for offline programming. Whenever the positioning coordinates are being calculated rather than being read directly from the position sensors, as in teach pendant programming, accuracy is important because the position calculations are based on the idealized model of the mechanism. This is not restricted to offline programming since any frame shifting, trajectory generation and sensor control, which are all performed online by the robot controller, will also calculate position coordinates by using the kinematic model.

5.4.4 Accuracy enhancement

The effective accuracy of the robot can be enhanced either by offline compensation, in which a robot calibration procedure is used, or by online compensation, in which sensors are used to adapt the robot position actively.

With offline compensation, the kinematic model is more closely matched to each individual robot mechanism. This requires that a set of parameters which characterize the robot mechanism be measured for each robot through a calibration procedure. The parameters, which characterize the specific robot, are then used by the coordinate transformations which generate the positioning instructions. More complex representations of the robot mechanisms are required in order to fully represent the robot with such calibration procedures. Any changes in the mechanism, such as those from service or wear, will degrade the performance of the accuracy enhancement procedure. Such calibration procedures are not widely used because they are difficult to implement and also because use of offline programming in practice has only recently begun.

The offline technique for accuracy enhancement is an attempt to isolate, characterize, and compensate for the various sources of positioning error systematically. In contrast, the online techniques do not explicitly identify the sources of error. Instead, sensors are used to actively determine the position of the robot arm relative to a part or a reference point. The net positioning error is measured directly, and an offset is applied to the robot position in an attempt to cancel the net effect of the various sources of error. The implementation is considerably less complex than the offline compensation procedure. It does, however, have an effect on the in-process operation of the robot system, and it may be limited by the capacity of the sensors being used to determine critical locations with sufficient accuracy and speed of operation. The online technique is the approach most commonly applied for accuracy enhancement.

5.4.5 Sensor integration

With servo-controlled robots, sensors which are internal to the robot system typically provide measurement of axis velocity and position for feedback control. Integration of external sensors for more advanced control of the robot provides information to the robot controller about the environment. The information is typically in the form of absolute or relative part locations or process variables. The integration of sensors into robot systems is required by the need for flexibility in the robot system for handling uncertainty or variability in the environment. Active sensor-based control of the robot system reduces the requirements for high levels of precision in fixturing and part presentation which are generally needed in systems without sensors. With sensors, variability of part dimensions can be handled and process requirements for such tasks as monitoring, error detection, and in-process testing can also be met.

The application of sensors may be divided into three categories: (1) control of program execution, (2) static position adaptation, and (3) dynamic position adaptation.

Control of program execution Program execution is controlled as a function of the state of the work environment. Sensor information is typically presented to the robot controller in a simple form, such as discrete input bits representing pass/fail test results, interlock signals, and so on. Branching in programming execution is a function of the input values as well as internal control variables such as counters or flags.

Static position adaptation Static corrections provide offsets to the programmed positions; they allow translation and possibly rotation of the frame of reference in which the positions are defined. This is used for local calibration, in which a translation of motion instruction arguments is based on the sensed position of the end-of-arm tooling relative to a reference location. This form of calibration is referred to as local because it can provide enhanced accuracy local to the sensed reference location. The correction will not usually be valid for positions which are far from the calibration position. Local calibration is typically used to compensate for small part-to-part positioning variations.

The sensors used for static corrections may be as simple as limit switches for contact detection. The robot can be programmed to search in a particular direction while the binary input from the sensor is used to interrupt the motion of the robot when a target is encountered. This use of robot motion allows the determination of the target location without any true position-sensing capability in the sensor. A number of such simple searches can be used to build up a three-dimensional offset which is then applied to all subsequent motions.

More complex sensing can also be applied; in it range information is determined by a sensor and fed into the robot controller. The controller can then move the robot to a sequence of positions at which the sensor signal is read to determine the relative location of a part or fixture. Vision systems can be applied to provide multidimensional part location information which is typically input to the robot controller through a communications link. This information may be either relative position, as when the camera is mounted at the end of the robot arm, or absolute information, as when the camera is mounted in a stationary location independent of the robot motion.

Once a set of positioning instructions is programmed, the location arguments for the instructions can be altered through a change in the frame of reference in which the position instructions are defined. This change may occur during program execution or as an offline operation. The ability to relocate previously programmed positions allows compensation for changes in the position of a part being processed or changes in fixturing after the robot programming has been completed. This relocation is a static operation performed on a set of positioning instructions.

The specification of the transformation can be a simple offset for translation or a combined translation and rotation of the frame of reference in which the points are defined. The transformation may be generated by repositioning the robot at a set of new locations for one or more predefined reference points. The new and the old robot coordinates corresponding to each of the reference points are then used to automatically determine a transformation which is applied to subsequently executed motion instructions. As the distance between the old and the new points increases, the accuracy of the new points degenerates. That is because the new points are being calculated rather than taught, so the key performance specification for positioning the robot becomes accuracy rather than repeatability.

Dynamic position adaptation Dynamic position corrections are characterized by the real-time nature of their functioning. With each incremental motion along a trajectory, the robot controller reads information from an external sensor, computes a correction to the nominal path, and commands an updated signal to the axis servo loops. This closed-loop trajectory control occurs continuously as the robot travels along its programmed path.

Position tracking is the most common use of dynamic adaptation in which the sensor signal is used to modify a nominal trajectory that the robot is following. A typical use for this is tracking the location of a seam during arc welding in which the cross-seam position of the torch is being controlled by the sensor. Variations in part fabrication and fixturing can result in random changes in seam location along the entire length of a seam. These variations cannot be compensated for by the offset technique used for static position adaptation. The sensor for such an application provides cross-seam position error information to the controller. This might be obtained by using vision or by processing signals from the welding arc to estimate the cross-seam position error.

Another form of position tracking is the control of standoff distance between a tool and the part being processed. The standoff distance is a critical process parameter in applications such as plasma spraying and arc welding. As with cross-seam position control, a sensor signal is input to the controller to provide the instantaneous standoff distance. The signal is produced by a sensor which determines absolute range information.

In addition to position corrections, the velocity at which a tool travels along a trajectory can be controlled in real time. A signal input to the controller is used to set the instantaneous speed of the tool. This is typically used for tasks, such as grinding and deburring, in which tool forces or deflections are used to determine the feed speed. With such feedback applied, the robot system can compensate in-process for random variations in the amount of material to be removed.

For most of the applications of dynamic position adaptation, the robot is closely involved with the processing being performed, and so the adaptations directly control important parameters of the process. The adaptive performance of the closed-loop sensor-based controls must, therefore, be measured against the requirements of the process being controlled.

5.4.6 Sensor interfacing

Active control of the robot by external sensors allows the robot to be adaptively controlled during program execution. This advanced functionality requires the robot controller to support both the associated sensor interfacing and the appropriate programming capabilities. The sensor interfacing capability must allow sensor signals to be input to the robot controller. The method for accomplishing this includes analog or digital signal interfaces or, in some cases, communication from a sensor system by means of a communications link.

A corresponding functionality must be provided in the language of the robot controller. It must include ability to access the sensor data and use it for performing the desired function, i.e., controlling program execution, offsetting position instructions, or modifying the trajectory or velocity of the tool. The functionality of the required language support has been implemented in widely differing ways. In some cases, predefined features of a language allow straightforward programming of the instructions and the required specification of the sensor interfaces, control gains, and correction direction information. Other implementations are not as straightforward in their programming, but they provide more general-purpose programming features which are less constraining for the user.

5.4.7 Communications and cell control

Industrial robot systems are seldom applied in isolation without requiring coordination with other pieces of automated equipment. The coordination with equipment in the immediate work environment is generally referred to as cell control. For many robot applications, the robot controller has sufficient capability to perform as a cell controller. Equipment in the robot cell can be interfaced to the robot controller through a combination of discrete signals and higher-level communications interfaces. The sequencing and control of the cell can then be accomplished by the program of the robot controller. In some cases, the robot controller must be interfaced to a separate controller which may be acting as a more powerful cell controller or as a supervisory controller. In that case, the robot controller will typically be accepting its programs from the higher-level controller, receiving commands and data for controlling the execution of the robot program, and sending data to the external controller that provides information about the current status of the robot.

5.4.8 Evolution of current technology

Advances in robot technology are occurring in the areas of controller functionality, programming systems, sensor capabilities, and robot mechanism and drive technology. The functionality of robot controllers will be enhanced by the continuing evolution of robot programming languages and the features they incorporate. Eventually, the procedure for programming robot systems will be more task-oriented, which will lessen the need for low-level specification of every step required to perform a task.

Offline programming systems for industrial robots will provide the facilities for simulating more than just the robot motion. The interaction and communication with other equipment will be modeled and simulated, and the effects of sensor control on the robot system will also be accounted for. The generation of executable programs will be supported for all of the components of the integrated system instead of the components having to be programmed independently. Advances in sensor technology will bring about significantly greater flexibility in the robot systems which use the sensor systems. Vision and tactile sensor systems will provide greater capabilities for use in recognition, locating, and decision-making functions. Positioning performance will be improved through the use of lighter-weight mechanisms, more intelligent axis control, and the use of direct-drive motors on the axes.

Standardization, which has been almost completely absent to date, is becoming a more pressing priority in such areas as electrical and mechanical interfacing, communications, and programming languages. Standardization will significantly contribute to an increase in the application of robot technology to improve manufacturing productivity.

CHAPTER 6

Material Requirements Planning and Inventory Control

Hal Mather

Computers have been used extensively for planning and scheduling since the late 1950s. Early attempts were failures, mainly because the techniques selected were improperly applied. A breakthrough occurred when the independent/dependent demand principle was established. Quick progress followed, and it led to a further breakthrough when all elements of a manufacturing control system were placed in their correct orientation. Data linkages between techniques were then obvious.

Current manufacturing control systems are too simplistic and need huge amounts of human interface to be successful. The future will see computer-aided design (CAD) systems providing bills of materials, computer-aided process planning (CAPP) providing routings, and scheduling-optimization routines replacing human judgments. Artificial intelligence will complete the move to an automated factory.

6.1 History

Improved planning, scheduling, and inventory control have been objectives ever since the invention of computers. The beginnings were in the late

1950s and early 1960s. But real progress did not come until IBM introduced the 360 series of computers with significant (for those days) disk storage and retrieval capability as well as a reasonably fast processor.

Mathematics was thought to be the solution to all of the problems of manufacturing. Many statistical forecasting formulas were developed, the most famous being exponential smoothing. And this did give a real lift to the forecasting of make-to-stock finished goods and spare parts.

Statistical forecasting was combined with the order point–order quantity inventory replenishment system. Again, this was a real help to the make-to-stock product. But the same concept was also applied to components and raw materials used to manufacture finished goods, a serious mistake by the practitioners.

Operations research then became the vogue. Queueing theory, linear programming, and other mathematical tools were advanced to solve manufacturing's problems. The overall results were fair at best. Many of the simulation approaches were too simplistic. Others were too concerned with perfection; hence, they became too complex for practical use.

6.1.1 Material requirements planning (MRP)

Another development was occurring in parallel with the mathematical approaches at a number of companies, most notably J. I. Case. There Dr. Joe Orlicky developed the independent/dependent demand principle, a real breakthrough in current thinking [Ref. 1]. The principle is to forecast end product sales and use order point methodology to control finished goods replenishment now. The calculation of needs for components and raw materials should be based on the plans to manufacture finished goods. Netting logic will manage these component and raw material inventories.

The independent/dependent demand principle cleared up confusion about the application of techniques. It didn't negate the value of statistical forecasting, but it did reduce the scope of forecasting to independent demand items.

In reality, MRP was not a new concept. It was used by cavemen and the builders of the pyramids; it was and still is used by cooks preparing meals. Manufacturers also used it through the early 1930s. They manually calculated the needs for parts and raw materials to support their order books of end items. Increased product variety, shrinking order books, and a more dynamic marketplace prevented manual MRP from continuing—people couldn't keep up with all the calculations. So the order point methodology replaced MRP.

Order points were an easier and faster technique to handle manually, but the technique was actually invalid in that application. Computers allowed companies to come full circle back to MRP with their ability to handle the calculations, even with a dynamic marketplace and a wide product variety.

Many inventory control practitioners rejected the move to MRP, and most upper-level managers were not aware of its existence. The American Production and Inventory Control Society (APICS) launched the MRP crusade, which was spearheaded by the three fathers of MRP (Joe Orlicky, George Plossl [Ref. 2], and Oliver Wight [Ref. 3]), and supported by many others. Seminars and courses were held and writings were published on the value and logic of MRP. Companies using MRP, most notably Black & Decker, opened their doors to others to come and see their approach.

The result was oversell. Now MRP was being touted as the panacea for all manufacturing's ills. That also wasn't valid, but it was a start, if somewhat naive. Forecasts for end items were automatically plugged into the bill of material explosion run. There was no concept of capacity checking before calculating detailed schedules for the factory. Master production scheduling was yet to come.

6.1.2 Manufacturing resource planning

The term MRP was modified to mean manufacturing resource planning, an expansion to all elements of a complete manufacturing control system. The confusion this generated was not helped by referring to material requirements planning as small mrp and manufacturing resource planning as big MRP, or closed-loop MRP. Then MRP II came along: manufacturing resource planning plus the financial system, an additional source of confusion.

But these efforts created another real breakthrough regardless of the confusion in terminology. The complete spectrum of manufacturing planning and scheduling techniques was organized in a cohesive systematic way. Now we could see how all the techniques fitted together and what their inputs, outputs, and prerequisites were. The role of each management department and level of management also became clear.

6.2 Current State of the Art

A complete operations planning and control system has now been defined to manage the flow of product from vendors through the factory to the customer. This flow of product is a common focus for all departments and all management levels. Flowing the right material quickly and efficiently from vendors through the factory and into the hands of customers is the mission of all industrial companies.

6.2.1 The manufacturing control system

A representation of a complete manufacturing control system is shown in Fig. 6.1. Each element will be discussed in turn, and its relations to all other elements will be clearly defined.

6.2.2 Strategic business plan

The strategic business plan is long-range, typically 5 to 10 years out. The impact of technology, market position, economic and demographic influences, and competitors, as well as a company's strengths and weaknesses (low-cost producer, technology leader, quality reputation, and so on) are some of the items considered in this plan. The dashed box in Fig. 6.1 represents the plan as not being totally a part of the manufacturing control system. Other activities are driven by the plan, e.g. research and development, new product introductions, marketing strategies, and financing needs.

Sensitivity analysis, often by using spreadsheet programs, can show the influence of volume, price, purchased material costs, and direct labor charges on return on investment, cash flow, and market share. This analysis is usually performed with large aggregate data, e.g., sales volumes by large product groupings and time periods of years. The major objective is direction, not detailed product plans.

Subplans are derived from the strategic business plan, still in high levels

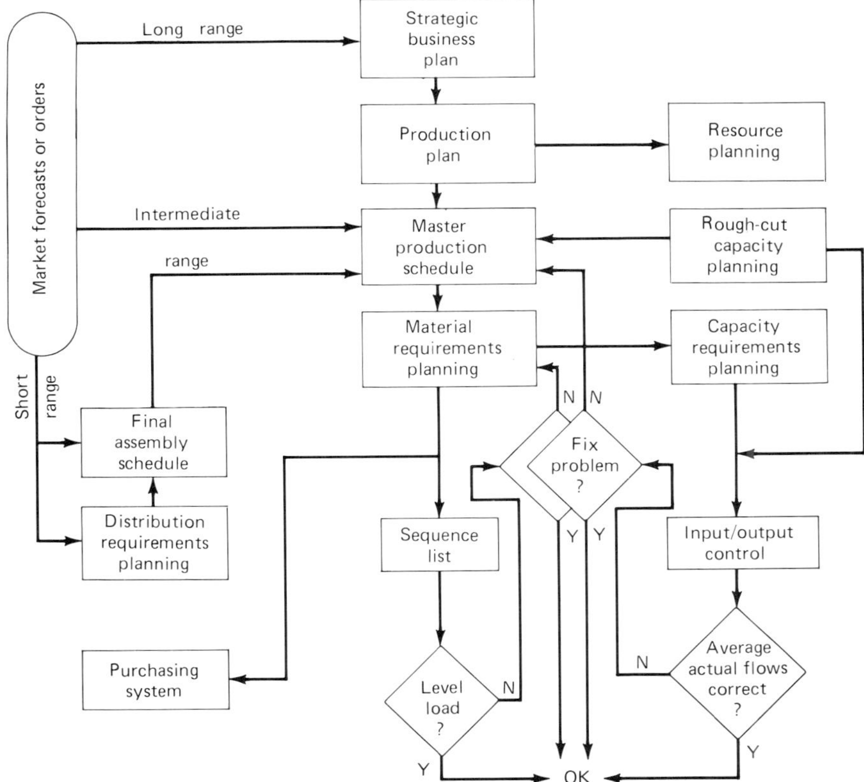

Figure 6.1 Manufacturing control system.

of aggregation. These subplans are for marketing, R&D, finance, and production; the latter is of greatest importance for the manufacturing control system.

6.2.3 Production plan

The production plan is a long-range plan expressed in aggregate product groups suitable for resource planning. Time buckets are large, e.g., quarters or years. A fundamental requirement is for the level of definition of product in the production plan to be convertible, through some historical or planned relationship, into the resources necessary to execute the plan. An example of a production plan is shown in Fig. 6.2.

6.2.4 Resource planning

Lead times to change resources are long. New plants take up to 3 years to build and perhaps 20 years to amortize. Machinery can also take up to 3 years to justify, buy, install, and debug; amortization can typically take 5 years. Defining the need for and adding other resources, e.g., vendors, direct labor, and technical or managerial skills, also takes time. Hence, a planning system is needed to predict when resources will be required well ahead of their actual need dates and to show a continuing need for the resources to amortize their purchase or hiring and training costs.

A resource-to-product profile is shown in Fig. 6.3. It includes key production areas. It could also include technical skills, which are especially necessary for an engineer-to-order company. The example in Fig. 6.3 also includes inventory as a turn figure; hence, inventory levels by product family can be predicted and the impact on cash requirements can be sent to the financial planning group.

Product	Jan.	Feb.	Mar.	Apr.	Yr 5, qtr 2	Yr 5, qtr 3	Yr 5, qtr 4
Family 1, $	10,000	10,000	10,000	10,000	50,000	50,000	50,000
Family 2, units	412	490	380	408	2,000	2,000	2,100
Family 3, tons	15	40	10	25	80	85	90
•							
•							
•							
Family 12, equiv. units	3,000	3,000	3,100	3,200	12,000	11,000	10,000
Repair parts, $	25,000	25,000	25,000	25,000	100,000	100,000	100,000

Figure 6.2 Production plan.

Sales value	$1000
Cost of goods sold (COGS) (55% of sales)	$550
Purchased material (60% of COGS)	$330
Direct labor + burden (40% of COGS)	$220
Assembly hours	10
Fabrication hours	12
Inventory turns	4

Figure 6.3 Planning ratios—product family 1.

Spreadsheet analysis programs are often used in production planning to simulate the effects on resources of changing a variable. Specific projects that provide most benefits, e.g., improve productivity, reduce purchased material costs, and increase inventory turns, can now be selected. See Fig. 6.4 for a resource plan for a given product family.

6.2.5 Market forecasting

All manufacturers use forecasts in their planning. The degree of detail and use of forecasts vs. orders differs with the type of manufacturer: make-to-stock, assemble-to-order, or engineer-to-order.

Long Range "Long range" typically means longer than one year. Its use is for strategy and resource planning, as defined earlier. Because of the horizon, most forecasting is based on economics, demographics, or technology trends. Long-range economic models, Delphi forecasts, and the like are used to predict the direction of a given technology or market.

Predicting a given product's sales that far out, or more important, your share of a product's market, is not the objective. The idea is to look at long-term trends and potential opportunities in the marketplace. The strategy you and your competitors implement, e.g., being the low-cost producer, high-technology manufacturer, or quality leader, will to a major degree decide the portion of the sales volumes you will enjoy.

Intermediate range "Intermediate range" is typically 6 to 9 months. Its primary role is to help build a master production schedule; hence, its level of detail must be applicable for that role. Time units are usually months or in some cases weeks. Items that are being forecast are product families or groups; they may be families of end items or families of options. (The latter is especially true of assemble-to-order products.)

Computerized statistical analysis can be of assistance in forming a foundation for a forecast. It is especially useful for companies with huge catalogs of end products or a great variety of options or variants. A large amount of human judgment also is required in this time frame to introduce factors not included in the statistics, e.g., new product introductions, promotions, and the economy.

The key is to accumulate data by product group and forecast the group directly. It is not valid to forecast each individual end item based on the item's actual historical sales and sum all resulting end item forecasts in the group to get to a group total. Aggregate to detail is the correct direction, not detail to aggregate.

Short range "Short range" is typically a few weeks. Its role is to define exactly which end item to assemble out of the group forecast or to help define how many of what products to ship to a specific warehouse. Actual booked orders, not forecasts, will drive the final assembly schedule (FAS) for assemble-to-order companies. Warehouse replenishment orders will drive the FAS for make-to-stock companies. Extensive computerized statistical analysis, combined with the finished goods inventory replenishment system, is very effective here. Decisions of what products to assemble or ship are routinely based on the short-range mathematical predictions.

Significant deviations in sales from the norm should be highlighted for human review. Tracking signals can identify an unusually large order or selling period as well as detect drift or bias in sales below or above a product's historical sales pattern. Exceptions must be analyzed, and decisions about their inclusion or rejection from the future planning system must be made.

Period	Jan.	Feb.	Mar.			Yr 5, qtr 3	Yr 5, qtr 4
Sales forecast, $	10,000	10,000	10,000			50,000	50,000
Cost of goods sold, $	5,500	5,500	5,500			27,500	27,500
Purchased matl. , $	3,300	3,300	3,300			16,500	16,500
Direct labor and burden, $	2,200	2,200	2,200			11,000	11,000
Inventories, $	16,500	16,500	16,500			27,500	27,500
Assembly hours	100	100	100			500	500
Fabrication hours	120	120	120			600	600

Figure 6.4 Production plan—family 1.

6.2.6 Order entry

Customer orders for stock items and custom-built products need to be entered in the inventory and/or production system. The capabilities of order entry systems vary with the type of product. For make-to-stock products, it is important to have access to inventories by item, by location, by existing commitments against those inventories, and by pending additional production or deliveries by item. Delivery promising becomes timely and accurate under these conditions.

For assemble- or make-to-order products, order entry also creates the product build specifications. A good example is an automobile; the customer-selected options are aggregated together into the finished car bill of material and routing. Figure 6.5 shows the order entry form for a hoist, an assemble-to-order product. This order entry form, labeled a "menu" because of its similarity to a restaurant menu, is completed with the customer-requested options. The bottom line, listing the specific mix of options for this order, develops the bill of material and routing for the order, prices and costs the product, generates pick lists, and so on. This process also checks inventories and planned production of the options to develop delivery promise dates (see Sec. 6.2.7).

For engineer-to-order companies, the specifications provided through order entry can also be used to develop the end product build details. This is a more advanced version of the assemble-to-order logic, and it approaches computer-aided engineering. Good examples are medium-duty transformers, highway trucks, and public telephone exchanges. Here, customer specifications are used to actually engineer a unique product. This differs from the assemble-to-order logic in that nonstandard assemblies or parts also can be defined.

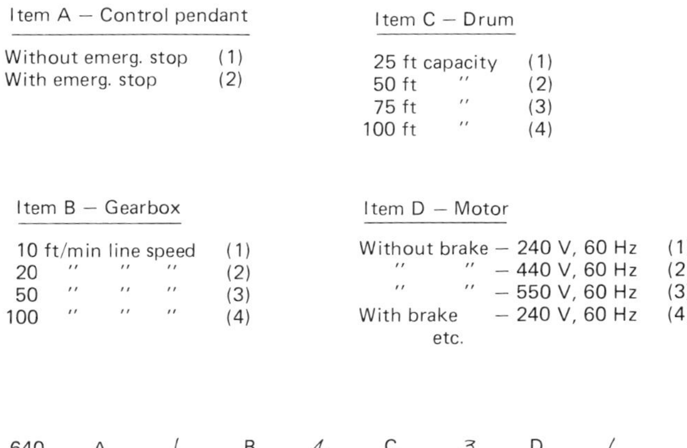

Item A — Control pendant

Without emerg. stop (1)
With emerg. stop (2)

Item C — Drum

25 ft capacity (1)
50 ft " (2)
75 ft " (3)
100 ft " (4)

Item B — Gearbox

10 ft/min line speed (1)
20 " " " (2)
50 " " " (3)
100 " " " (4)

Item D — Motor

Without brake — 240 V, 60 Hz (1)
" " — 440 V, 60 Hz (2)
" " — 550 V, 60 Hz (3)
With brake — 240 V, 60 Hz (4)
etc.

640 A __1__ B __4__ C __3__ D __1__

Figure 6.5 Hoist menu—model 640.

Online access of order status is a critical element of order entry. Customer inquiries can easily be answered with this facility, and customer changes can also be introduced quickly and accurately. Many companies are moving their online order entry processes further and further upstream, and some companies are giving that capability to their customers. The trend will continue and will accelerate. The key now is for a company to decide how far to allow its customers access to its planning, scheduling, and inventory system. In the short term, the pattern is to greatly restrict customers' capability, leaving free access only to authorized company employees. In the longer term, the trend will be to allow customers much freer access, so that eventually customers can schedule actual production of their vendors within certain limits.

6.2.7 Master production schedule

The production plan sets overall rates of production for large product families. In the short term, these rates of production must be broken down into enough detail, in both product and time, that specific raw materials and parts can be purchased and production schedules issued. As a minimum, the horizon of the master production schedule (MPS) is the stacked lead time for a product (Fig. 6.6). A longer horizon than the minimum can be useful for capacity planning and showing vendors planned future ordering. The trick is not to confuse accuracy and precision. The further out the MPS is driven, the more inaccurate it becomes; the calculation from this inaccurate data of needs to support these numbers is very precise.

The role of the MPS is to drive the detailed planning and scheduling system. All specific procurement and manufacturing decisions are tied mathematically to the MPS. That is the reason for the first word, "master."

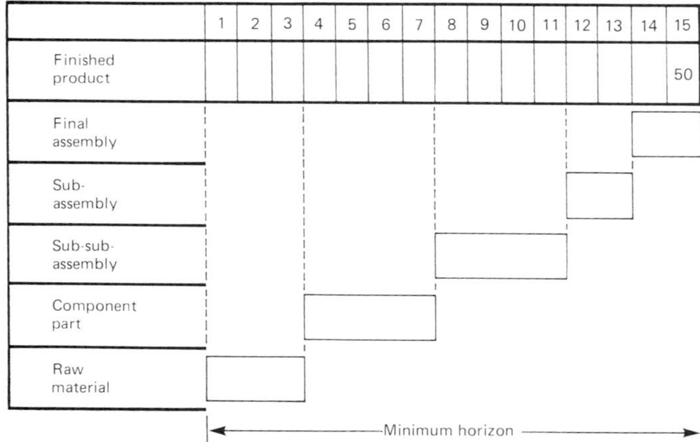

Figure 6.6 Master production schedule.

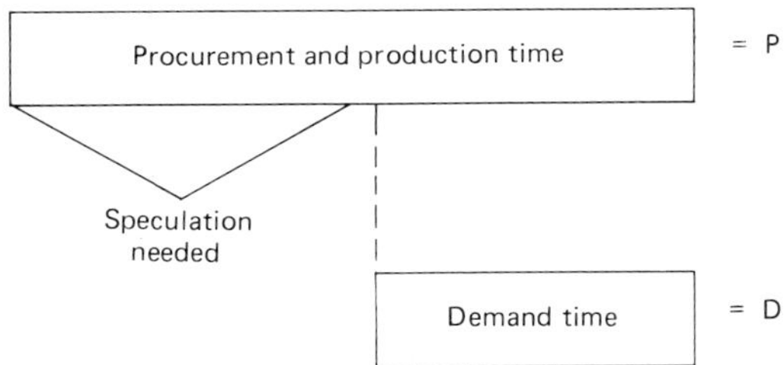

Figure 6.7 P:D ratio.

Hence, the MPS must receive maximum management attention to ensure its validity and guarantee correct detailed actions at lower levels.

The planning dilemma Figure 6.7 shows the planning problem to be addressed by the MPS. As shown in Fig. 6.6, P is the stacked lead time of procurement and production of a product and D is the customer lead time. D can be a competitive length of time or simply the horizon at which customers can predict their needs with reasonable accuracy. Utopia is obviously a $P{:}D$ ratio of 1:1 or $<$1:1. No speculation is required under those conditions. Large engineer-to-order projects, e.g., military products, ships, and special-purpose machine tools, create this scenario by pushing the D time out to meet the design, procurement, and production cycle.

When P exceeds D, speculation by the manufacturer is mandatory and is present in the MPS. Now two schedules are required. One, based on speculation, is the MPS covering the time frame P minus D. This causes specific procurement and production actions for intermediate parts or subassemblies. The other is the finishing or final assembly schedule covering the D time frame. The FAS consumes the items provided by the MPS. Figure 6.8 shows this relationship clearly.

There is no direct linkage between the MPS and FAS. Frequently, the product definitions in each schedule are quite different. However, over time, these schedules must relate to each other in both volume and mix. What is being provided by the MPS must be consumed by the FAS to avoid excess inventories and parts shortages.

Make to stock The MPS for a make-to-stock product is largely forecast-driven. However, it would be a serious mistake to conclude that a forecast is the MPS. An MPS defines a product, quantity, and time when something is to be produced. This is where the middle word "production" comes from. And what is to be produced rarely will be what is forecast to be sold.

Take a seasonal product, for example, or a product subject to marketing promotions. In either case, the sales patterns have spikes of demand and low spots. But factories are rarely flexible enough to be able to adjust their production rates to match seasonal sales peaks and valleys. Inventories decouple the sales outflow from the production inflow.

Because the MPS drives all detailed procurement and production actions, it's obvious it will be level-loaded compared to sales. Hence, the forecast is an input to create an MPS, but it is not one and the same. Capacity constraints and production hatches are reasons other than seasonality or promotions to decouple forecasts and production.

An inventory technique called a time-phased order point (TPOP) can be used to generate a preliminary MPS for a make-to-stock item. This technique is shown in Fig. 6.9. The forecast of 20 per week is shown as the requirement for this item. In the warehouse there are 80 completed units, 55 of which have been designated the safety stock quantity to take care of forecast error. We assemble these items in batches of 100 with a 2-week final assembly lead time. There is an assembly order currently in process for delivery in week 2.

The netting logic, identical to that used by material requirements planning, creates a planned assembly order for receipt in week 7 to cover the negative available inventory. Offsetting this by the lead time of 2 weeks pushes the planned assembly order start date to week 5. This technique differs from material requirements planning simply by the source of the requirements. In this case the requirements come from a forecast, so the item is an independent demand item. Hence, the technique is called a TPOP. If the requirements come from a bill-of-material explosion or from the MPS, then the item is a dependent demand item and the technique becomes MRP.

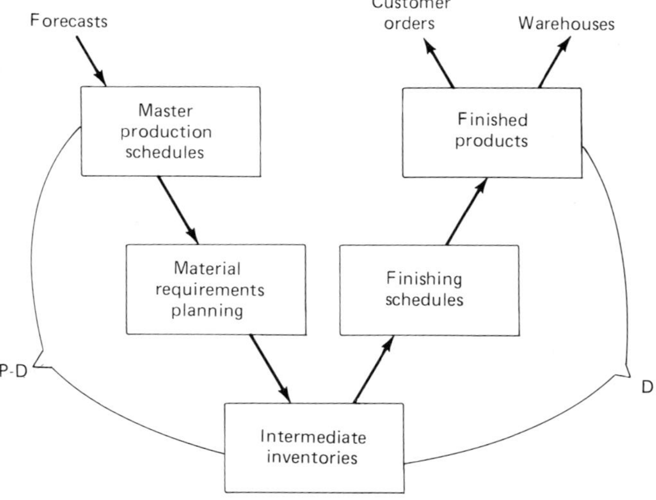

Figure 6.8 The planning dilemma.

Forecast = 20/week On hand = 80
LT = 2 OQ = 100 SS = 55

		Weeks							
		1	2	3	4	5	6	7	8
Requirements		20	20	20	20	20	20	20	20
Scheduled receipts			100						
Available	25	5	85	65	45	25	5	−15	−35
Planned order due								100	
Planned order release						100			

Figure 6.9 Time-phased order point.

The question is, "Which line of information (requirements, scheduled receipts, available, planned order due, or planned order release) is the MPS?" It cannot be the requirements line because that is what we hope to sell; it is not what we intend to make. The available line can also be eliminated, because that is a picture of resulting available inventory.

That leaves three lines: scheduled receipts, planned order due, and planned order release. Earlier we defined the minimum horizon of the MPS as the stacked lead time for a product. The scheduled receipts extend only the final assembly lead time, not the stacked lead time. Hence, this eliminates that line, which is in fact the finishing schedule (Fig. 6.8). The scheduled receipts line is consuming the inventories the MPS provided by making actual finished goods.

That leaves lines 4 and 5. Line 4 is a prediction of the next final assembly order completion. Continuing with the separation of the MPS from the FAS, line 4 is therefore a predicted FAS, not the MPS. Line 5 provides the inventory for the FAS to consume. Hence, it is the desired MPS for a make-to-stock product.

A TPOP is a very nervous technique. Actual sales rarely match a forecast each week or each month. A forecast is considered good if it tracks a long-term average. But a TPOP treats each week's sales as absolute (Fig. 6.9). Actual sales in week 1 were 26. At the beginning of week 2, then, the available line drops by 6, creating a negative available condition in week 6. An exception message will be generated to move the MPS 1 week earlier to cover the negative available. Let's assume we did that. If the actual sales in week 2 were 14, giving us a perfect forecast of 20 per week on the average, the message would be to delay the MPS 1 week.

Dampening these trivial messages is necessary to free up time to review how to handle significant forecast errors. For a more detailed discussion, see Hoskin [Ref. 10]. So far we have talked about the desired MPS. Planned orders coming from a TPOP define how you would like to replenish the warehouse. But what about capacity? How about level loading? If the desired plan exceeds capacity or provides peaks and valleys of work load the factory cannot handle, then the desired plan needs modification to suit the factory limitations. Figure 6.10 shows this logic clearly.

Assemble to order The assemble-to-order products are typified by having a wide variety of options that can be selected in various combinations and permutations to make a large number of finished products. An automobile is one of the best examples, but many other products, e.g., machine tools, computers, and trucks, have similar characteristics.

A relatively simple example of an assemble-to-order product, a hoist, is shown in Fig. 6.11. The hoist is assembled from five option groups: hook, motor, drum, gearbox, and pendant. Each option group, apart from the hook, offers several choices. Multiplying these choices together gives you the possible number of end items, assuming no conflicts among choices. In this case, it's $30 \times 4 \times 10 \times 2 = 2400$. If the P time for this product is 15 weeks and the production rate is 50 per week, as in Fig. 6.6, the impossible end product forecasting problem is clear. Which 50 specific hoists, out of a potential 2400, will customers buy 15 weeks from today?

Let's assume the D time for hoists is 4 weeks. That means 11 weeks $(P - D)$ of procurement and production must occur, on speculation, before

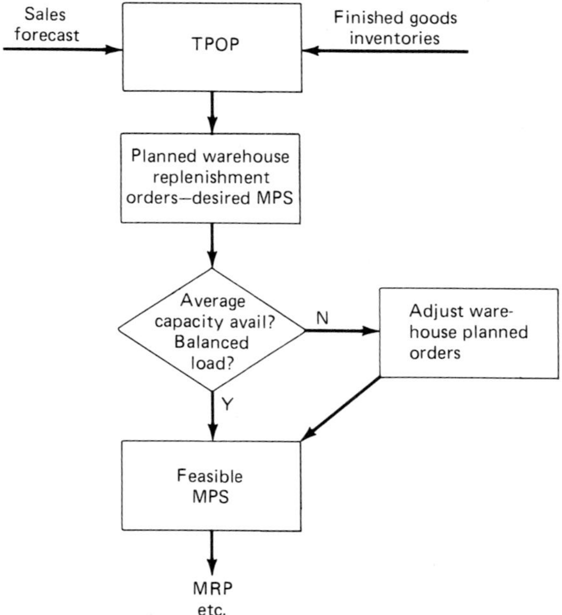

Figure 6.10 Desired vs. feasible MPS.

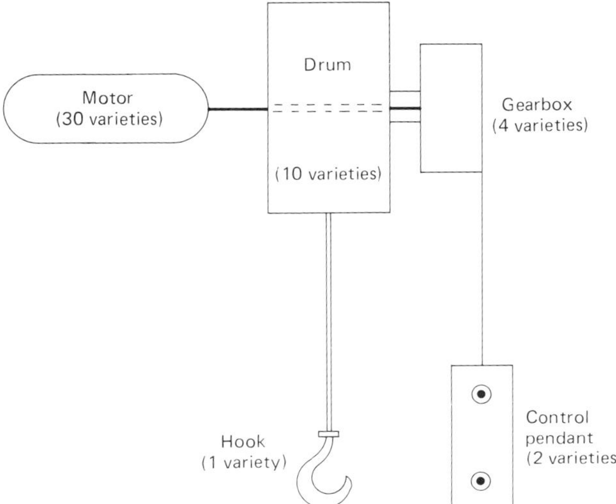

Figure 6.11 Hoist arrangement.

firm customer orders are booked. It's obvious that the forecasting problem is simplified if the various options are forecast as end items, separate from their potential combinations. In this case there are only $30 + 10 + 4 + 2 + 1 = 47$ things to be predicted 11 weeks out. Customers' orders will determine the combination method and will not be forecast.

Of course, the assumption is the final assembly can be done in 4 weeks. What if final assembly cannot be done in the *D* time, if, for example, it's a make-to-stock product with many combinations of options such as a washing machine or refrigerator? It still makes more sense to predict the options long range as separate items, get them moving through the factory, and leave the forecasting of specific end items until the last moment, in this case the final assembly lead time. These examples should make Fig. 6.8 much clearer.

Returning to the hoist example, let's look at the forecasting problem in more detail. Predicting the needs for hook assemblies is relatively easy. The total production of hoists is set at 50 per week, so obviously that is the MPS for hooks. Moving on to the control pendants, there is a mix problem. The total is 50 per week, but how many of control pendant 1 vs. control pendant 2? Past historical sales or a forecast of the future will be needed to develop the mix.

A key function of the order entry process is to gather past history of sales to assist in the master scheduling exercise. Most sales statistics are concerned with sales by territory, salesman, product family, and so on. This data is rarely constituted to show the mix of options within the product; add this feature to sales reports for businesses with option variability.

Let's assume the past history of pendant sales showed a 60/40 split between pendants 1 and 2; let's also assume no forecast input to change that mix. It's clear that the MPS for pendant 1 is 30 (60% of 50) and for pendant 2 is 20 (40% of 50). The same logic could apply to the other option groups, i.e., gearboxes, drums, and motors. The larger the number of choices within an option group, the more questionable this technique becomes. The percentages become small and highly variable.

Breaking these options, e.g., the motors, into subgroups, as we did with the hoist, could be a solution. Making frequently sold options to forecast with short delivery cycles and infrequent sellers to order with long lead times is another choice. Holding risk inventory for the infrequent sellers and replenishing on a min/max basis could also be done.

The actual sales of hoists in a given period will rarely demand the options in the exact split as the past historical average or a market forecast. If the mix varies even a small amount for the motors, drums, gearboxes and pendants, only a few finished hoists can be assembled. One solution is to have safety stock for these options; another is to use the technique called "overplanning," a modified and improved method of adding contingency. Instead of using the average historical mix to plan the options, use an inflated number so the total of options planned per group exceeds 100 percent. Figure 6.12 shows this idea for the control pendant.

The extra numbers in the MPS can create safety stock at differing levels in the bill of material depending on the stacked lead time of the product and the week or weeks in the MPS chosen for the overplan quantity. If the extra is put in week 1, safety stock of pendants is created. In week 3 (assuming a 2-week assembly lead time) safety stocks of components are created. In week 8 (assuming 2 weeks assembly and 5 weeks fabrication time) there are safety stocks of raw materials. Now the inventory investment can be balanced against the speed of response to variable marketplace demands.

This discussion of the MPS and the problem of predicting the future has a direct impact on the structure of bills of materials and influences product design. The MPS drives detailed materials planning and factory scheduling

Option	Historical %	MPS	Overplan %	Overplan MPS
Hook	100	50		
Control pendant 1	60	30	70	35
Control pendant 2	40	20	50	25
Control pendant tot.	100	50	120	60
Gearbox 1				
Etc.				

Figure 6.12 Planning production.

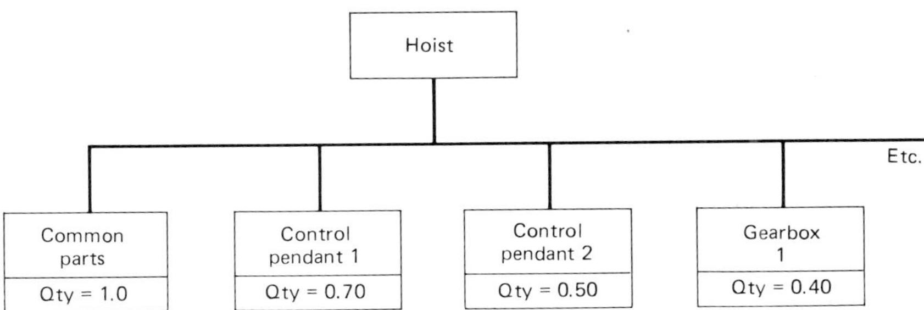

Figure 6.13 Hoist percentage bill of materials.

activities through a bill-of-materials explosion. Hence, individual bills of materials must be structured to contain the parts and raw materials necessary to produce a given option. They are called modular bills-of-materials, quite a deviation from many bill-of-materials arrangements.

The design of products must also consider the MPS/FAS relationship. Uniqueness at the last moment must become the key phrase for designers and manufacturing engineers. CAD and CAM systems must rate various designs and processes against their forecastability and resulting impact on customer service and inventories, as well as their traditional rating of cost and function.

Another type of bill-of-materials structure is called a planning bill, a poor term because it covers a variety of bill-of-materials structures. The most common is a percentage bill of materials; an example for the hoist is shown in Fig. 6.13. Instead of individual modular bills, each of which is driven by its own unique MPS, with unique overplanning by module, all options are placed in a comprehensive bill of materials for the hoist. Each bill is linked to the hoist by a percentage figure as the quantity used. An MPS for the hoist of 50 per week explodes into the requirements by option.

This is a poor substitute for modular bills for many reasons. For example, additional product structures are needed for planning, and they are separate from those needed by manufacturing, finance, or engineering. Close control of each option is not possible without additional computer logic. Use this technique only when it is necessary to reduce the number of items in the MPS; that is the only real value of the percentage bill-of-materials approach.

The MPS is often considered to be one homogeneous set of data stretching into the future. Nothing could be further from the truth. We have already talked about separation of the MPS and FAS. How about further separation of data, one that depends on the time horizon?

Figure 6.14 shows a moving scroll idea, first advanced by Orlicky. Time is continually being wound onto the left scroll to become history and is being unwound from the right scroll to keep a constant view of the future.

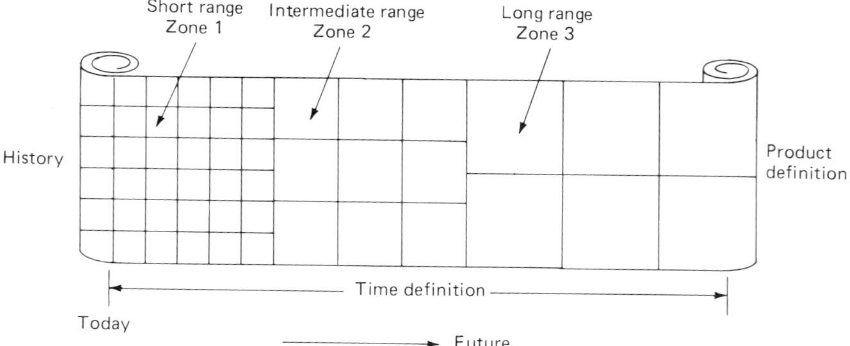

Figure 6.14 Moving scroll of time.

In the short range, detailed finished product schedules are needed with short time increments, e.g. days. That is because, during this time period, you will produce warehouse needs or the items customers have ordered. This is the FAS in reality. In the intermediate range, the level of product definition decreases to families or options and the time increments get wider, maybe to weeks. The job of these numbers is to produce the items the FAS will consume. Contingency planning between zones 1 and 2 allows great flexibility in the specific mix of items scheduled in the FAS.

In the long range, the level of product definition drops to large aggregate product groups and wide time increments, maybe months. The job of these numbers is to procure common, long-lead-time raw materials that zone 2 of the MPS will consume. Contingency planning between zones 2 and 3 will allow a flexibility similar to that between zones 1 and 2.

Along with the material side of planning, direct labor needs, machine capacity, tooling, and perhaps vendor capacity must be considered. The various zones of the MPS must also provide sufficient data for these resources to be evaluated. It's obvious, therefore, that the bills of materials and process definitions used in the three zones will vary with the degree of detail needed. Creating the various relations and controlling designs and processes to assist with the zones idea is a critical activity to generate a successful MPS.

Time fences are often used to separate the various zones of the MPS as well as to control the overplanning quantity. A time fence is a specific technique that controls the automatic computer manipulation of data, and it is always set at a point in time from today. As time goes by, the time fence rolls forward at the same rate to always stay in the same place relative to today. Overplanning quantities, when they reach the fence, are automatically rolled into the future as time elapses. It takes a human input to jump the fence and convert the overplan into actual production.

Different logic can be applied to numbers within the various zones. For example, planned master schedule quantities, when they reach the final assembly (zone 1) time fence, can be held and flagged for conversion into final assembly orders. Sometimes the data driving MRP differs with the MPS time zone. For example, in zone 1, perhaps only customer orders drive the system. In zones 2 and 3 the greater of forecast or booked orders in a period drive the system. The master production schedule should be linked to order entry to quote delivery dates to customers. This is called available-to-promise logic. The menu order entry process, shown in Fig. 6.5, assists this linkage.

A simplified example of an MPS for a company making an assemble-to-order product is shown in Fig. 6.15. This is a machine tool with a basic configuration plus options. The company has modular bills of materials, one for the basic machine (i.e., all common items) plus two more containing the parts unique to options A and B. The real example has up to 100 options that can be ordered in various combinations, but only 2 are shown here for simplicity and understanding.

Option A's forecast usage is 60 percent of the basic machine's production, option B's is 10 percent. The MPS for the basic machine is based on order book, forecast, capacity constraints, financial goals, and so on. The MPSs for the options are based on their percent usage plus extra (overplan) for contingency.

As orders are entered, the basic machine and options are considered to be separate entities. Only the sales orders link them together. The promise date to ship customer orders is based on the availability of the gating item,

Basic Machine											
Weeks	1	2	3	4	5	6	7	8	9	10	11
Master schedule	—	50	—	—	50	—	—	60	—	—	60
Booked orders	—	50	—	—	35	—	—	10	—	—	0
Uncommitted	—	—	—	—	15	—	—	50	—	—	60
Option A(60%)											
Weeks	1	2	3	4	5	6	7	8	9	10	11
Master schedule	—	35	—	—	35	—	—	40	—	—	40
Booked orders	—	28	—	—	21	—	—	5	—	—	0
Uncommitted	—	7	—	—	14	—	—	35	—	—	40
Option B(10%)											
Weeks	1	2	3	4	5	6	7	8	9	10	11
Master schedule	—	6	—	—	6	—	—	8	—	—	8
Booked orders	—	5	—	—	6	—	—	0	—	—	0
Uncommitted	—	1	—	—	—	—	—	8	—	—	8

Figure 6.15 Customer delivery promises.

easily seen for all items. In this example, all basic machines have been sold for period 2 but options are still available. If they are not needed for field retrofits or spare parts, they are superfluous. They must be removed from period 2. The extra availability comes from unsold overplan or undersold forecast.

In the case of option A, because period 5 (the next production lot period) still has plenty of overplan left, the excess 7 in period 2 should be removed from the plan completely. For option B, though, period 5's sales have sold out all available options including overplan. Hence the extra unit in the MPS in week 2 needs rolling out to period 5 to provide availability then and to ensure correct priorities in the plant.

Defining the logic to promise deliveries to customers is fairly straightforward, and many companies have computerized the activity. Handling unsold options is more complex. Many companies leave it to human judgment because of the complexities involved.

Engineer to order An engineer-to-order company is really selling capacity; hence, its MPS is defined in units that relate to capacity. As an example, a foundry could define its MPS in molds or tons of product. These numbers long range can be used to procure pig iron, alloying elements, chemicals, sand, and so on, and to set operating levels. Capacity and manning plans can also be generated from this data. Actual orders for castings consume this plan and convert it from an MPS into the finishing schedule. Evaluating available capacity against incoming orders can set delivery promises and help direct the salesforce to sell capacity that is not being consumed at an acceptable rate.

More complex products can follow the same pattern. Figure 6.16 shows a company making special, medium-duty, variable-speed, ac motors. The

Family 1										
Week	1	2	3	4	5	6	7	8	9	10
MPS	30	30	30	30	30	30	35	35	35	35
Booked orders	28	27	29	34	29	36	36	25	10	2
Overplan							35			
Uncommitted	−2	−3	−1	+4	−1	+6	+1	−10	−25	−33
Family 2										
MPS	100	100	100	100	100	100	100	100	110	110
Booked orders	103	105	101	97	103	92	90	81	28	3
Overplan							100			
Uncommitted	+3	+5	+1	−3	+3	−8	−10	−19	−82	−107

Figure 6.16 Family MPS.

P time is 18 weeks: 12 weeks to procure rough castings, steel bar and sheet, wire, and so on, and 6 weeks to fabricate, assemble, and test the end product. The *D* time is 6 weeks.

The MPS is for a family of motors. Because each motor is custom-engineered, the company can't produce end items or components to forecast. The MPS sets an operating rate that defines capacity needs by large groups of resources. It also generates raw material requirements based on a family bill of materials. This is a percentage bill of materials that defines, on the average, how many end bracket castings of various types are used per 100 motors sold. The family bill of materials also contains average usages for steel sheet for the laminations, steel bar for the shaft, wire, bearings, and so on, by using percentages where there are multiple choices. The averages are maintained automatically, and they can be calculated on the basis of various periods of history, say, the past 3, 6, and 12 months. They can also be calculated on the basis of the current order book or a mixture of order book and history.

Contingency is added in period 7 (6 weeks fabrication, assembly, and test time) to guarantee extra raw materials in stock (safety stock). The time fence at the end of period 6 signified by the double line keeps the overplan always 7 weeks from today. All booked orders are sent to engineering, where specific designs are created to suit the customer applications. These specific orders replace the family forecast and become the detailed fabrication and assembly schedule (zone 1).

The overplan quantities allow mix variability, e.g., produce more of family 1 and less of family 2, as well as volume variability, e.g., produce more of both families. In the latter case, additional capacity in the form of overtime or temporary hires also is necessary.

In zone 1, up to and including week 6, only booked orders drive the system. They use specific bills of materials and routings for unique end items. After week 6, booked orders plus the overplan quantity plus uncommitted (negatives are ignored) drive the system. Booked orders use specific bills of materials and routings; the overplan plus uncommitted use the family bill of materials plus average routings.

6.2.8 Rough cut capacity planning

Rough cut capacity planning is a refined version of resource planning discussed earlier. It is based on an average relation of work needed by key resource for each product family. The resource profile is sometimes time-phased to show the demand on resources relative to the completion of the end product. A typical resource profile is shown in Fig. 6.17. It defines the amount of work in some meaningful unit of measure by resource per unit of output. In this example it is per 1000 units of output. Also included in the example are vendor resources, a frequently overlooked resource constraint. The example is not time-phased. Time phasing is appropriate only if you make a widely differing mix of products each period with significant differences in demands on resources or if the rates of output frequently rise or fall significantly.

A resource profile is best developed by using a weighted average of sales approach. The total work content by major resource per end product can be calculated by using each end item's bills of materials and routings. Extending these by the share of total family sales that each end item takes provides a weighted average applicable to the family. Another way to develop a resource profile is to use historical data of work expended by resource by product family and calculate an average per unit. Estimates by floor supervisors or industrial engineers are often close enough. Cost accounting can sometimes convert dollars to labor-hours per product.

Extension of resource profiles by planned master schedules can simulate capacity needs by major resource. Spreadsheet programs are often used for this purpose because of their ease of use and speed of results. Capacity needs must be checked against average available capacity and level loads. For labor-intensive production, available capacity is usually defined as past average actual output. Needs that differ from that must be resolved by management decisions relative to overtime, hiring, layoffs, subcontracting, and so on. Companies with machine-intensive production usually use engineered standards to calculate available capacity. An assumed mix based on past history or forecasted demand is necessary.

It cannot be stated too strongly that this step, ensuring that the capacity needed to execute the MPS matches the capacity available, is one of the activities most vital to a successful manufacturing control system. Overloaded master schedules create false priorities and a disillusionment with the system. Unbalanced master schedules with peaks and valleys of demand cause excessive queues, late deliveries, and large amounts of expe-

Product Family 1

Work center	Std. hours per 1000 units
110	10
108	8
105	14
104	28
103	9
101	17
Vendors—Castings	250 molds
Chemicals	400 gallons
Fasteners	4500 pieces
Steel	520 lb

Figure 6.17 Resource profile.

diting around the system. A well-designed rough cut capacity planning system with easy simulation capability can go a long way toward avoiding these problems.

6.2.9 Material requirements planning

The material requirements planning technique is the original one with the acronym MRP. Now it is sometimes referred to as mrp (little MRP) or MRP I.

Basics The logic of material requirements planning in its simplest form is shown in Fig. 6.18. Supplemental techniques based on this logic will be discussed later. The process starts with several necessary inputs: a master production schedule, bills of materials for all items in the MPS, on-hand inventories for all raw materials, components, and subassemblies (intermediates), summary status of open work orders for manufactured items, and open purchase order status for purchased items and raw materials. Basic operating parameters, e.g., lot sizing rules or quantities, lead times, and safety stocks, need definition. Various codes also are necessary depending on the specific software for each item, e.g., purchased or manufactured, planner identification, and ordering policies. Many optional pieces of data can also be present for informational reasons, e.g., ABC classification, cost, and past usage.

The MRP process starts with the master production schedule, which is exploded through the bills of materials. At each stocking level in the bill of materials, inventory netting takes place. Also, needs to replenish inventory are calculated and decisions are made, automatically on the basis of parameter inputs, for lot sizes and lead time offsetting. Replenishment decisions at one level in the bill of materials are exploded into requirements at the next lower stocking level. This can be seen clearly in Fig. 6.18. The MPS quantities for the projector are exploded through the bill of material to the head assembly, causing requirements that match the MPS quantities. Obviously, the bill of materials quantity for this major assembly per MPS item is 1. This same condition will be assumed for all other bill of materials quantities for simplicity's sake. In real life, of course, bill of materials quantities can be any number.

The head assembly requirements are demands that must be satisfied. The first place to look for this satisfaction is on-hand inventories. We currently have 10 in stock, insufficient to cover the first requirement of 20. The next place to look is work in process, shown as an open work order (scheduled receipt) for 10 pieces due week 2. The on-hand plus on-order satisfy the first requirement, leaving zero inventories available. Hence all future requirements must be covered by future planned production.

The lot-sizing rule in this case is lot-for-lot, meaning you should order exactly the same replenishment quantities as the needs in a given period. The head assembly lead time is 1 week. Hence planned orders, signifying future production, are automatically generated: the completion date to suit the requirement date, the start date simply one lead time earlier. The start dates and quantities of future planned production of the head assembly are exploded through the bill of materials to the next stocking level, in this case the wiring assembly. These start dates and quantities create requirements for the wiring assembly. The same process of netting on-hand and

```
PART NO 1140   DESC=PROJECTOR
WEEK NO    P/D    1    2    3    4    5    6    7    8    9   10   11   12   13   14  TOT
MPS                     20        30        25        35        20        30        25  185


PART NO 1401   DESC=HEAD ASSY      ON HAND=10   U/M=EACH         SFTY STOCK= 0
ORD QTY=L4L LEAD TIME= 1   ALLOC= 0   P/M=M   ACTION= 0
WEEK NO    P/D    1    2    3    4    5    6    7    8    9   10   11   12   13   14  TOT
REQUIRE                 20        30        25        35        20        30        25  185
SCH REC                 10
AVAIL       10   10    0    0  30-       25-       35-       20-       30-       25-
PLN DUE                           30        25        35        20        30        25  165
PLN REL                      30        25        35        20        30        25       165
PEGGED:1140-PROJECTOR


PART NO 4110   DESC=WIRING ASSY    ON HAND= 0   U/M=EACH         SFTY STOCK= 0
ORD QTY=L4L LEAD TIME= 2   ALLOC= 0   P/M=M   ACTION= 0
WEEK NO    P/D    1    2    3    4    5    6    7    8    9   10   11   12   13   14  TOT
REQUIRE                      30        25        35        20        30        25       165
SCH REC                      30
AVAIL        0    0    0    0    0  25-       35-       20-       30-       25-
PLN DUE                                25        35        20        30        25       135
PLN REL                      25        35        20        30        25                 135
PEGGED:1401-HEAD ASSY


PART NO 1014 DESC=BULB             ON HAND=100 U/M=EACH         SFTY STOCK=10
ORD QTY=100 LEAD TIME= 4   ALLOC= 0   P/M=P   ACTION=PLN REL   FCST REQT =10
WEEK NO    P/D    1    2    3    4    5    6    7    8    9   10   11   12   13   14  TOT
REQUIRE           10   10   35   10   45   10   30   10   40   10   35   10   10   10  275
SCH REC
AVAIL       90   80   70   35   25  20-  10-  30-  10-  40-  10-  35-  10-  10-  10-
PLN DUE                                100                 100                          200
PLN REL          100                   100                                              200
PEGGED:4110-WIRING ASSY
```

Figure 6.18 Typical MRP display.

on-order inventory, calculating uncovered requirements, and automatically generating future planned orders to cover the needs occurs now for this subassembly. The same process is repeated for the purchased component. But in this case spare parts requirements of 10 per week are added to the calculated needs to get the total requirements per period. The example follows only one chain in the bill of material. Obviously, similar calculations are occurring for all items at all levels.

The netting process cannot be done for items common to several higher-level products until the requirements from all the higher-level products are posted to the common item. That is controlled through a technique called low-level coding, which determines when all requirements have been posted and then triggers the netting process.

Outputs Two basic types of conditions are flagged, by exception, by the MRP process. The first condition requires action by the pertinent planner, and for that reason the exceptions are expressed as action messages. These messages are generated by a planned order release date maturing into the close-in time frame or by master schedule revisions, scrap, engineering changes, customer returns, or correction of faulty records. Three basic types of actions can be recommended by the system:

1. Order release. The planned order start date for an item has arrived in the close-in time frame. This time frame could be defined as the past due plus current week or could be past due plus several future weeks. The latter choice simply provides more short-term information for planning, grouping items, and so on. The planner's job is to review the order recommendation, consider information external to the system, e.g., pending engineering changes, price breaks, and tooling constraints, and convert the planned order to a released order. This is usually done in conjunction with scheduling for factory orders and purchasing for vendor orders.

2. Expedite. A released order, either work order or purchase order, is needed sooner than currently scheduled. It is identified simply by having a requirement for an item uncovered by on-hand inventory earlier than the due date of the order for the item. The planner must evaluate if the order can be moved up, bearing in mind capacity, tooling, and vendor limitations. If it can be, the due date of the order is changed to satisfy the open requirement. If it cannot be, changes to the higher-level assemblies demanding this item, maybe as far up as the master production schedule, must be made to move the uncovered requirement to a period where it can be covered.

3. Delay. A released order, purchased or manufactured, is scheduled earlier than necessary for current requirements. It is identified by having on-hand available inventory at the end of the period in which the order is currently scheduled. Here the planner needs to evaluate the cost of the

inventory if the order comes in early, the impact the order has on factory priorities, and how much disturbance to the factory or vendor delaying this order will cause. If the order should be delayed, changing the order due date will reposition the order correctly.

The second condition flagged by the system is mainly informational in nature, and the messages are normally called exception notices. Typical exception notices warn of requirements in the past-due time period, items with inventory but no requirements in the planning horizon, and so on. The action taken on these exceptions varies greatly with the exception conditions. Generally, exception notices are not as time-sensitive as the three action messages described earlier.

Supplemental techniques Many other techniques can be built around or in the basic logic of MRP. Brief descriptions of the more common ones follow.

1. Lot sizing. MRP is a technique for timing when to reorder an item. How much to order is a separate and distinct activity. Many mathematical and judgmental approaches are available. Especially dangerous within MRP are the dynamic lot-size calculations, in which batches are mathematically optimized during each MRP run. As a general rule, stay with fixed quantities based on judgment or lot-for-lot as the lot-sizing decision.
2. Safety stock. Safety stock can be included in the MRP process as a quantity, as a period of time, or by using overplanning as explained in the master scheduling section. Safety stock is a dangerous concept because it always disturbs priorities and demands capacity earlier. Make sure the real benefits of safety stock outweigh the disadvantages. Include safety stock only on parts with truly predictable events that safety stock can cover.
3. Pegging. Pegging is a technique that records where an MRP requirement comes from. It is similar to a where-used list, but it is very specific to a particular item's MRP requirements.
4. Firm planned orders. Firm planned orders provide a method of human control over computer-controlled planned orders. The timing of the order and the quantity ordered are the controlled elements.
5. Allocations. Allocations are a method of reserving components or raw materials to make their parent product; they are usually used to keep the system in demand-supply equilibrium after a shop order is issued and before the parts are picked from the stockroom. They are also used to test component availability before issuing a shop order; sometimes called staging on paper.
6. Dampening rules. Also called nervousness filters, dampening rules are computer logic that inhibits expediting and delaying messages. This is usually done on a time basis, e.g., don't report expedites of less than 1 week or don't report delays less than 2 weeks. It can also be done on a quantity basis, e.g., don't plan new orders or request expediting when sup-

ply in a period is 95 percent or more of the required quantity. These are dangerous ideas. Nervousness is caused by a problem somewhere, and the only value of these rules is to inhibit the number of action messages. Attack the causes of change, and the number of messages will drop automatically. Use these rules only if the number of messages is still excessive.

7. Bucketless system. A data processing term and a user term; for data processing, a bucketless system means only that data is stored. Early MRP systems created a series of time buckets for the full planning horizon for various data elements, e.g., requirements and schedule receipts. Most of these buckets were blank. For the user, "bucketless system" is really a misnomer; it means the time bucket used for planning is 1 day. Hence, MRP can tell if the demand-supply balance is correct to the day of the week instead of balanced somewhere within the bucket, traditionally 1 week long.

Many companies using bucketless systems use dampening rules to get the equivalent of a weekly bucketed system because of the extreme precision of bucketless systems. Make sure your data and master production schedule prediction capability can support a bucketless system for the user before adopting it.

Supporting schedules MRP is mainly for controlling materials and components, but its look-ahead feature can also be used in other ways. For example, consumable tooling can be built into the bill of materials by using mortality rates as the quantity per. Consumption of tooling will now be calculated along with materials and components.

Toolroom kitting of fixtures and nonconsumable tools can also be a by-product of MRP. Planned orders to make parts can be listed in start date sequence, and tool numbers can be tied to the part numbers they make. Hence, the list of parts to be built in date sequence can inform the toolroom which tools to preset, kit, and have ready for the production start date.

6.2.10 Capacity requirements planning

The process of capacity requirements planning is an extension of the routing standards by released and planned shop orders. Summarizing these by resource and date can show the upcoming standard hours by machine, department, tool, skill group, and supervisory control. The routing process and standards can be manually entered into the system or generated by using a CAM or CAPP system. In either case, in the selection of machines to perform work one must consider the current and planned work loads in relation to available capacity.

Inputs Material requirements planning provides released shop order information of part number, quantity due, and latest due date. Open shop order status records show the condition of this order by operation, e.g., complete, partially done, quantity good, quantity scrapped, balance due, and the start and stop dates for each operation.

It is imperative that the action messages from MRP be resolved before capacity requirements are calculated. A failure to expedite or delay orders needed to keep synchronization within the overall plans means the calculated capacity requirements will be invalid. Material requirements planning also provides planned order data on part number, quantity, and start and finish dates for all manufactured parts. Because these orders have not been released yet, no order status information is necessary or available for them.

Routings provide the list of machines, work centers, tools, and labor grades used in the process of making a part plus time standards for producing it. Sometimes setup times also are provided. Actual setup times, even for the same item, are highly variable in many industries; they depend on the preceding job in the machine. For that reason, setup time is often ignored in the capacity requirements calculation and is added simply as a historical percentage of the calculated run times. Other companies treat setup time just as definitively as standard run times.

Scheduling rules define the average flow of a product through a plant. They usually contain the following elements:

1. *Run time.* For each operation, run time is simply an extension of the standard hours per piece by the number of pieces in the lot divided by the number of standard hours generated by the operation's resource each day. It is usually rounded to increments of one day.
2. *Setup time.* The setup time is optionally included; it is added to the calculated run time for each operation before the elapsed time is calculated.
3. *Move and queue time.* Usually expressed in increments of days, move and queue time often requires a matrix of resource from and to, especially for outside subcontracting operations. Different time allowances can then be based on an individual part's process through the factory.
4. *Unmeasured activities.* Some operations do not contain standard time estimates per piece, e.g., inspection, plating, and heat treating. In this case, time allowances, usually in increments of days, are used for the total of run, setup, move, and queue times.

Scheduling rules can be much more complex than the simple examples given above. For example, various parts of the factory could be operating at different rates, one shift versus two or three, 5-day weeks versus 6- or 7-day weeks, routine overtime each day or including the weekend, and so on. They can also consider "overlapping," the sending ahead of part of a lot to the next operation instead of waiting for the complete batch before movement occurs. These more complex rules still apply to the predicted average movement of a batch of product through a factory with MRP orders and routings used as the baseline.

An M-day calendar is the device that defines which days in the year are workdays. It is an internal computer mechanism whereby standard calendar dates are cross-referenced to a simple numerical numbering of workdays. Weekends are omitted unless the operation is 6- or 7-day, and holidays and vacation shutdowns are always omitted. Calendar dates for specific actions can be entered into the system and converted internally to the corresponding M-day; scheduling can be performed internally by using M-days; and outputs can be converted back into standard calendar formats.

Detailed scheduling and loading The inputs described above are used by the scheduling and loading system to create start and stop dates for every operation on the routing. Hence, the part number, quantity, and start and stop dates of MRP orders are converted to work center (resource), standard hours, and start and stop dates of each operation.

Figure 6.19 shows a routing for a part scheduled by using simple rules. In this case, the standard hours generated by each resource per day are 16 (a three-shift operation, allowing for efficiency, nonmeasured activities, and indirect work). One-day move and queue is allowed between operations in the same department, 2 days are allowed between operations in different departments. This order is a planned order from MRP, so no work has been done yet. If it were a released order (scheduled receipt), then an open-order status report would be needed to show what has already been done and so should be excluded from future capacity needed.

The order has been backward-scheduled, which means to start at the finish date for the order and work backwards up the routing by using the scheduling rules. The capacity of all resources is assumed to be infinite, and the scheduling rules are assumed to be correct for all orders.

Part No. A935 Qty = 10,000 Start: Wk 5, day 4
Order No. M7531 Due: Wk 10, day 1

		Std. hours				
Operation	Dept.	Setup	Run/pc	Start date	Due date	Load (Std. hr)
10 turn	45	16	.003	Wk 5, day 4	Wk 6, day 2	46
20 mill	107	8	.006	Wk 6, day 4	Wk 7, day 4	68
30 mill	107	4	.002	Wk 7, day 5	Wk 8, day 2	24
40 drill	210	1	.005	Wk 8, day 4	Wk 9, day 3	51
50 inspect	50	—	—	Wk 9, day 5	Wk 10, day 1	—

Figure 6.19 Detail scheduling and loading.

Forward scheduling is another approach. It is rarely used in industry and then usually where there are few resources and they are largely machine- or equipment-paced. Forward scheduling means to start at today and prioritize all orders that go across a given piece of equipment. Orders are loaded in up to the capacity of the resource. Dates based on the hours of work each order contains and the available capacity of the resource are established. Secondary operations schedules are based on the completion dates of the primaries. Obviously, secondary operations cannot be started until the primary operation is complete unless overlapping is used.

The mathematics involved in forward scheduling are immense for most companies. The choices of schedules among which to choose also are huge. Some companies do the job manually, especially when there is one key bottleneck resource and all other resources have available capacity. Few software packages address this complex scheduling problem; the preference is for the simpler logic of backward scheduling and assuming human input can resolve any capacity imbalances that are generated. As a general rule, bottleneck and machine-paced resources are forward-scheduled and nonbottleneck and labor-paced resources are backward-scheduled.

Outputs The standard hours calculated for each routing operation during the scheduling process are summarized by work center by date, which is the required capacity each week to produce the detailed schedule exactly as established. Obviously some leeway is possible in the dates on which hours are needed. The amount of leeway depends on the extra allowances added when standard hours are rounded to the nearest day and when setup, move, and queue times are added. Figure 6.20 shows a capacity requirements plan for a work center. The column definitions are as follows:

- *Capacity.* Average actual output in standard hours of the past 4 weeks.
- *Released hours.* Standard hours of work scheduled to arrive in this work center from orders or schedules already authorized and released to the plant.
- *Planned hours.* Standard hours of work scheduled to arrive in this work center from MRP planned orders.
- *Total hours.* The sum of released and planned hours.
- *Load-to-capacity ratio.* A graphical depiction of the total hours divided by capacity hours. The X's are 10 percent increments of released hours; the O's are 10 percent increments of planned hours.
- *Load %.* The numerical value of total hours divided by capacity hours.
- *Cum. over/under load.* The cumulative difference between total hours and capacity hours.

Work Center 342

Week No.	Capacity, hr	Released hr	Planned hr	Total hr	Load-to-capacity ratio 0% 100% 200%	Load, %	Cum. over/under load
1	65	89	0	89	.xxxxxxxxxxxxxx	137	24
2	65	55	2	57	.xx xxxxxxx	88	16
3	65	60	12	72	.xxxxxxxxxoo	111	23
4	65	40	21	61	.xxxxxxooo	94	19
5	65	25	36	61	.xxxxooooo	94	15
6	65	26	55	81	.xxxxoooooooo	124	31
7	65	0	54	54	.oooooooo	83	20
8	65	13	74	87	.xxoooooooooooo	134	42

Figure 6.20 Capacity requirements plan.

This picture is obviously from an infinite-loading, backward-scheduling system. Human judgment is now required to determine how or if to handle the cumulative overload of 42 hr in 8 weeks. People must also check to see if the weekly peaks and valleys will disrupt secondary work center schedules or if there is enough leeway in the schedules to ignore the variability.

The decision to react to the 42-hr cumulative overload is an attack on the average capacity needed. It is obviously the most critical decision. Without enough capacity on the average, all schedules, including the master production schedule, are in jeopardy. The choices at this stage are to adjust capacity by using overtime, offloading work to another work center, subcontracting, or adding people or to adjust the MPS downward to suit the available capacity.

The decision to smooth out the peaks and valleys is a level-loading decision. Here the system must provide traceability through the scheduling system to the MRP orders causing the peaks and valleys. Changing lot sizes, using safety stocks, and reducing lead times are ways of adjusting peaks and valleys of work. Firm planned orders in MRP can be used to adjust those factors, and the result will be visible in the next capacity planning run. This sounds straightforward, but in practice it isn't. One work center can be improved by the process, but others that were OK are then disrupted. The combinations and permutations of product flow, especially in a fabrication and assembly plant, defy the ability of humans to optimize all work centers at the same time.

A finite-loading, forward-scheduling system would force the total hours to meet the capacity hours. Depending on the process, capacity hours could be a systems variable based on overtime, reassignment of people, and the use of alternative methods of production.

It should be obvious that detailed capacity planning in any form is fine tuning for the short-term-future horizon. It depends on complete bills of materials and routing details for all items to be made. Scrap, rework, unexpected demand, short-range engineering changes, and so on, are ignored by the detailed mathematics and must be added by average factors based on past history. To get to manning levels, even more factors are needed, e.g., absenteeism, efficiency, indirect/direct ratios, and nonstandard work.

The major capacity planning system has to be rough cut capacity planning by using resource profiles that can allow for some of these variables. Detailed capacity planning shows us the impact of short-term schedules, and it allows fine tuning over the short-term horizon.

6.2.11 Input/output control

Capacity plans determine the desired flow rates of product through all resources necessary to execute the MPS. Input/output control monitors the actual flow rates of product, compares them with the desired rates of flow, and highlights significant deviations for quick resolution. It also has the objective of smoothing work flows through a plant and reducing unnecessary queues of work to a minimum. This function therefore attacks the *P* segment of the *P*:*D* ratio.

An example of an input/output chart is shown in Fig. 6.21. A unit of measure of product flow that is useful for capacity planning and is easily measurable as work occurs must first be picked. Most companies use standard hours of work needed and produced, but pieces, tons, equivalent units, and dollars can be just as effective. The planned input rate of work is defined by the capacity planning system. It is averaged out over a period of time to override the peaks and valleys of work traditional MRP systems generate. (This feature of MRP systems—scheduling production by assuming capacity is infinite and using average or fixed parameters, such as lead times, scheduling rules, safety stocks, and lot sizes, when they are in fact specific to a point in time or highly variable—is obviously a serious drawback to its long-term use by manufacturers. Further development of systems will remove this illogical trait of MRP. See Sec. 6.3 for a further discussion of the changes coming.)

The planned output rate of work is enough to meet the planned input and also reduce the current amount of queue (270 hr) to a planned level (150 hr). This excess queue is planned to be reduced over the next 4 weeks by working 10 percent overtime. Actual release of work is constrained to meet the average planned input. Each order contending for release in a given period must have its lot size extended by the standard hours per piece

Work Center 103
All Data in Standard Hours

Input:							
Week No.	9	10	11	12	13	14	15
Planned	270	270	270	270	270	270	270
Actual	275	265	230	255			
Cumul. Dev.	+5	0	−40	−55			
Output:							
Planned	300	300	300	300	270	270	270
Actual	295	270	280	295			
Cumul. dev.	−5	−35	−55	−60			
Queue:							
Planned (150)	240	210	180	150	150	150	150
Actual (270)	250	245	195	155			

Figure 6.21 Input/output control.

to see its impact on each work center. Only enough orders to be a close enough match to the planned rate of input are allowed to be released in a given period. See Sec. 6.2.12, on shop floor control, to understand the mechanics of order release.

Actual output of work is recorded by the labor-reporting system. Most companies have such a reporting system in place to update their cost accounting and inventory systems. The rule of this technique is to never let input be more than output, i.e., never let the queues of work grow. This rule can be seen in week 11, where the input hours are constrained because of the reduced output hours in week 10. Decisions now have to be made about the 40 hr of input being deliberately held back. Should it be subcontracted, released, by using alternative routings, to another work center with available capacity, or held onto and released next week if the actual output recovers? This same decision is needed in week 12 because, cumulatively, the actual output compared to plan is getting worse. If the work was held onto in weeks 11 and 12, overtime must be extended in the work center for 2 weeks more to allow the release of work to increase to catch up with the cumulative deficiency.

Some companies prefer to use Q control, which is simply a measure of actual queue hours in a work center divided by average daily or hourly rate of output. It expresses the queue in time terms. Control limits are put on the maximum and minimum number of days or hours of queue allowable. Changes are made in output rates to keep queues within limits.

The focus is put on queue for many reasons. "Queue" is another name for work-in-process inventories, so reducing queues reduces WIP. "Queue" is another name for lead time, so reducing queues reduces lead times. Lead times add up to the *P* time for a product, so reducing lead times reduces *P* times, making a company more responsive and flexible to the marketplace.

It may not be so obvious that reducing queues solves the "what do I work on next" question. If queues are low relative to capacity, then any job waiting to be processed is as good as any other. Only when queues are long relative to capacity does the choice of what to work on next become important. But long queues relative to capacity mean long lead times, which mean long *P* times, which mean that the numbers that drive the sequence system, the MPS, are far in the future and therefore probably wrong. You can't work on the right things at the right time under a long lead time situation. It's impossible, regardless of the complexity of the system used.

6.2.12 Shop floor control

The shop floor control technique takes information from the scheduling and loading system and uses it for sequencing the jobs to be worked on in a work center. Being warned of future jobs to be run in a work center can allow many important support activities to be scheduled. Tooling and materials handling equipment schedules, order picking, and maintenance needs can be triggered from the short-range schedules in shop floor control. Shift and people preferences, overtime, crewing, and other short-term considerations also can be planned from the shop floor detailed schedules.

Inputs The planned start dates of work orders come directly from MRP. They are sorted by starting work center and date to start. Figure 6.22 shows such a sort for the work center depicted on the input/output control chart shown in Fig. 6.21. The selection of which jobs to release is based on the planned rate of input (270 hr/week in this case), material availability, combined setups, tooling constraints, and the mix of work for downstream work centers.

For secondary operations, the scheduled start and stop dates are needed for each work order operation. They come from the scheduling and loading system. Open work order status reports show what has been completed and which work orders are in which work center. Job and operation completions keep the open order status report current and show the movement of work orders from work center to work center. Hence, current available work in a work center is presented, together with a view of upcoming work based on the schedules. Factory problems, such as machine or tooling breakdowns and scrap, can also be flagged to the system. These problems either put jobs on hold or create high-priority recovery plans.

Priority rules mathematically sequence the list of jobs awaiting processing. Some of the better known rules are:

- Critical ratio
- Q ratio
- Slack time
- First in, first out

Week No. 8 Planned Rate = 270 hr
Work Center 103 Total backlog = 270 hr

Part No.	Planned order release	Order qty.	Order hours	Cumul. sched. hours
52	Week 9	450	37	37
18	9	610	113	150
72	9	315	48	198
36	10	580	77	*275*
41	10	220	53	53
23	10	1100	127	180
45	10	725	85	*265*
68	10	500	23	23
53	11	150	62	85

Figure 6.22 MRP items planned order release.

It is important to realize that these mathematical rules only sequence work to suit the schedule in a given work center. They are oblivious to other factors such as whether a downstream work center is either close to running out of work or is badly backlogged. Management factors are ignored; for example, a job in a work center is to make parts for a customer order vs. parts needed to make an MPS item that is forecast to be sold. Some people have developed management overrides to modify the pure mathematical calculations to consider these other factors, but the overrides are very subjective and must be closely managed to be successful.

Outputs The most frequent output is a dispatch or sequence list showing each supervisor or scheduler the sequence of work to perform. A sample is shown in Fig. 6.23. Details of all orders waiting to be processed are shown with priority rankings. Some systems also show in a separate category all orders coming in and their relative priority. The figure states "daily" as the frequency. Some companies produce a schedule each shift; still others show the actual sequence of work to be performed online. The frequency depends on many factors, e.g., the velocity of work through a plant, the volume of changes, the frequency of system update, and cutoff points for data. Select your frequency while considering all these factors very carefully.

Work center performance is another output; it is not necessarily tied directly to shop floor control. It is often a part of the cost accounting system, but the data used, standard hours, actual time taken, number of good pieces produced, and scrap are common to both systems. Efficiencies, indirect hours, and actual standard hours produced in a time period are part of the performance measurement system.

Work Center 103

Date: 10-3 Capacity: 298

Part		Order			Prior rank	Goes to
Number	Description	No.	Quant.	Hr		
117175	Sleeve	841A	1500	22	1	106-1
276112	Gear	920E	300	13	2	110-2
523153	Housing	663B	120	8	3	106-4
319181	Frame	717C	165	10	26	108-2

Total hr in 103 = 287

Figure 6.23 Daily dispatch or sequence list.

Schedule adherence is a direct result of the shop floor control system. Measures of actual standard hours produced in a time period are key parts of the input/output control discussed earlier. Making parts on or off the scheduled due dates is a measure of sequence performance. Both measures are a direct result of the shop floor control system, and they will be discussed in more detail in Sec. 6.2.20.

It is useful to recap the total system at this point. The shop floor control sequence list is the culmination of planning and managing all prior elements of the system. The list must be responsive to changes in those plans as well as our failure to execute them.

The MPS is the starting point, a prediction of orders to be sold and/or today's current backlog. It is tested for feasibility and balance by using rough cut techniques. It is then exploded through MRP, by using standard parameters, to get start and stop dates for all planned and released orders. It is obvious what ignoring the exception messages of expedite and delay from MRP will do to the capacity planning system and dispatch lists.

MRP start and stop dates for orders are broken down into operation start and stop dates by using standard scheduling rules. Fine tuning of the detailed capacity requirements may be necessary. These same start and stop dates per operation are now used to set priorities on the shop floor. Floor reporting keeps the system aware of the status of orders and unforeseen problems at all times.

The sequence lists are updated for all external causes. A master schedule revision, engineering change, scrap report, late vendor delivery demanding rescheduling of higher-level parts, machine breakdown, and so on, affect the start and stop dates of affected orders. In turn, they impact on the start and stop dates of affected operations, revise the priority of operations in a work center mathematically in tune with the changes, and show revised capacity requirements to execute the new plan. The system is modular, is mathematically linked, with human judgment applied at critical junctures. Shop floor control success is directly related to the quality of upper-level plans and the level of management expertise applied to resolve identified problems.

Data collection Considerable feedback from the factory floor is needed to keep the system aware of order movement. This data can be collected a number of ways:

1. Handwritten documents later entered into the system.
2. Prepunched or preprinted documents requiring minimal human input.
3. Data collection terminals: pieces of hardware on the factory floor that accept badge and card input to identify the operator and job being worked on plus some ability to enter variable data, such as quantity

produced and quantity scrapped. These terminals can be linked to a data storage device for later processing of the data or can be linked directly to the files for instant update.

4. Bar-code readers: optical or magnetic reading devices that automatically update an item's progress through the plant.

The selection of which to use should be based on total costs, needed speed of update, and accuracy of data transmission.

6.2.13 Purchasing

Planning and control of the outside factory (vendors) is very similar to that for the inside factory. Rough cut capacity planning provides a view of the total demands on vendor resources by major and critical commodity. This needed rate of flow across the receiving dock should be contracted for with all major vendors. Additional sourcing, financial commitments to existing vendors, and so on, are needed to solve any flow rate imbalances.

MRP provides visibility of specific needed materials, components, and bought-out products short range. They must be converted to flow rate terms (dollars, pieces, tons, etc.) and compared to the long-range commitment. Level loading of the short-term demands to suit the long-term commitment, by adjusting the MRP parameters of lot sizing, safety stocks, and lead times is now necessary. Vendor call-offs or sequence lists for specific items that match the flow rate contract must now be transmitted to all vendors. Some companies send portions of their MRP outputs to vendors for that purpose. Typical commitments might be as follows: the first 4 weeks are firm and should be produced and delivered; the next 8 weeks authorize the vendor to procure raw materials; beyond this period the data is for planning purposes only.

Many companies are linking themselves to their vendors electronically for a variety of reasons. The flow rate contract and call-offs mentioned above can be transmitted quickly and efficiently this way. CAD information can be transmitted from customer to vendor to get vendor assistance in product design based on specialized knowledge. Status of orders in the vendor plant can be accessed quickly and easily by tapping into the vendor's manufacturing control system. It won't be long before customers actually schedule their vendors' factories within some agreed-to constraints and capabilities.

Actual receipts of material update the open purchase orders or schedules. Dock-to-stock receiving systems track the flow of material from receipt, through receiving inspection, and into stock. Return to vendor, rework at the vendor's expense, and material on hold are tracked through such a system.

Vendor rating systems also are driven from the receiving system. Quality ratings are based on the amount of material or number of lots accepted as good compared to the total amount of material or number of lots received. Schedule adherence should be based both on (1) the total number of lots or amount of material received compared to the flow rate plan from the MPS and (2) actual receipts by item compared to the vendor acknowledged date. See Sec. 6.2.20 for further ideas of measuring vendors.

6.2.14 Distribution requirements planning (DRP)

Companies which make to stock need to know when to replenish warehouse inventories. This is obviously a demand on the supplying center, whether that be another chain in the distribution system, e.g., regional warehouse to distributor, or a demand on the factory. Significant freight economies can be gained by planning deliveries to warehouses in truck loads or by combining several deliveries to warehouses in an efficient routing.

The foundation of DRP is the time-phased order point shown in Fig. 6.9. Assume it is for a stockkeeping unit at the end of the distribution chain. The forecast sales are therefore sales of this item out of this distribution point. The scheduled receipt is a distributor replenishment order coming from the supply center. Planned orders are desired replenishments of this distribution point's inventory.

If you conceptually link warehouses together, much as a bill of materials links parents to components, as shown in Fig. 6.24, it is easy to see that planned order release dates for a distributor place demand on the upstream warehouse. The total requirements for an item in the warehouse to replenish all distributors can be calculated from all the planned order releases for that item at the distributors. If the warehouse also sells this item directly to customers, then its individual forecast can be added to the planned

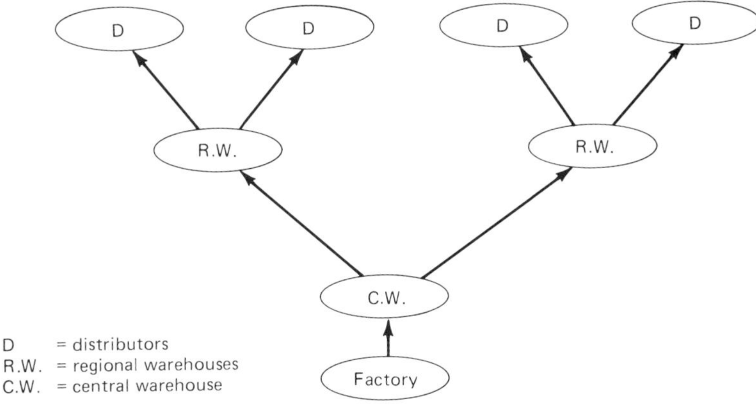

Figure 6.24 Distribution network.

distributor supply orders to get a time-phased picture of the total needs for an item. These needs can be netted against the warehouse inventory, and planned warehouse replenishment orders can be calculated. The process can continue back through all distribution levels to the factory and be an input to the factory's final assembly schedule.

For any one distribution point, the planned orders for all parts show the needs to deliver product in a time-phased manner. Size and weight figures per item can be used to calculate economical deliveries of products by taking into consideration freight, inventory, and possible stock-out costs. This method of replenishment is called a pull system.

Another method of replenishment is called a push or fair shares allocation system. In this scenario, the sales of an item, regardless of location, are aggregated and the total sales are forecast. Planned production is calculated by considering the total inventory in the system, again regardless of the location. This total planned production becomes a major input to the factory's MPS.

As items are produced, the question where to send them remains. The answer is based on the individual forecast of sales at each distribution point and the individual inventories as shown by the time-phased order points. An allocation algorithm distributes the production to warehouses on the basis of need and again on the basis of freight economics.

6.2.15 Final assembly schedule

The final assembly schedule (FAS) is the execution phase of the MPS; it is sometimes called a finishing schedule. Here is where specific configurations of end products are planned and scheduled. Refer to Fig. 6.8 to see the MPS/FAS relationship. A customer order or warehouse replenishment order for a specific item is the starting point. It must be immediately tested against the available inventories provided by the MPS. One of the simplest ways is through available-to-promise logic, shown in Fig. 6.15 and discussed in the relevant section. Another way is through allocation logic, briefly discussed in Sec. 6.2.9, on the MRP.

Allocation logic means that a desired assembly order is entered into the system. The order is immediately exploded through its bill of materials to the first stocking level of product. The needs for the order are tested against the available inventory. This is called a "soft" allocation. If there is adequate availability of all lower-level items, the order is released for production and "hard" allocated, meaning the materials are reserved for the order. If insufficient material is available, the order is not allocated. A report shows the limiting item or items. A decision must now be made to either expedite the missing materials, change the quantity of the order to suit the available inventory, or schedule assembly of the order when materials will be available.

Contingency planning, discussed at length in Sec. 6.2.7, on the master production schedule, can improve the match between the master schedule and final assembly schedule. The final assembly schedule can also point out the need to change the MPS because sales are materializing differently than was expected.

6.2.16 Financial System

Many aspects of the financial system, especially as they are related to cost accounting, are direct derivatives of the data in the operating planning and control system. Product costs come from the bills of materials, routings, purchased material prices, and labor and burden rates. Material receipts update inventory values and create purchase price variances. Labor reports create distribution of costs to overhead, inventory, and variance accounts. Shipments relieve inventories and credit accounts receivable.

The forward-thinking companies tie their physical and financial systems together very closely. Utopia is one system that maintains physical and financial integrity at the same time. Special accounts may be needed to do that to take care of timing differences between physical movement and financial accountability. For example, receipts of material ahead of or behind accounts payable and shipments of material before the receivable is recorded are common timing problems that need resolution.

The physical system can also be used as a financial forecaster. It displays the future in specific physical terms—part numbers, operations, hours, and so on. If costs are part of the same system, ideal inventory budgets, time-phased cash needs for accounts payable and payrolls, and receivables collection can be predicted. These will be ideal numbers needing allowance for nonideal actual conditions, but they can provide a financial forecasting basis built around the plans to operate the business.

6.2.17 Typical database

Each software package has its own unique database to suit the capabilities of its hardware and software. A typical database is shown in Fig. 6.25; the key data files are highlighted. The central focus is the item master file, which contains details of every part number. The product structure file contains the parent component linkages plus effectivity of any changes. Between these two, bills of materials are defined.

The resource profiles are for master schedule feasibility testing. Master scheduled items are specifically identified on the item master file. The work center master, routing detail, and item master together create the complete routing for a part. Sometimes a tool master also is included; it is linked to the routing detail for tool planning.

The requirements file contains all requirements for a part, whether to meet customer orders or to support higher-level production. The order file contains a summary of all open orders, including purchase, shop, and customer orders. A vendor master file and a customer master file contain details specific to those companies.

The operation detail file contains the specifics of all open shop orders and planned orders. As work progresses through the factory, the operation details are maintained by the feedback system. This is a basic set of files needed for all manufacturing control systems. Additional files can be used for sales and vendor history, engineering drawings, product history, employee data, maintenance information, and quality data.

6.2.18 Frequency of update

The frequency of update of information in the system is very different from the frequency of replanning. Don't confuse the two. The choices are:

- *Batch.* A periodic update of data, e.g., inventory records and order billing. It was originally used because of the prohibitive cost of online up-

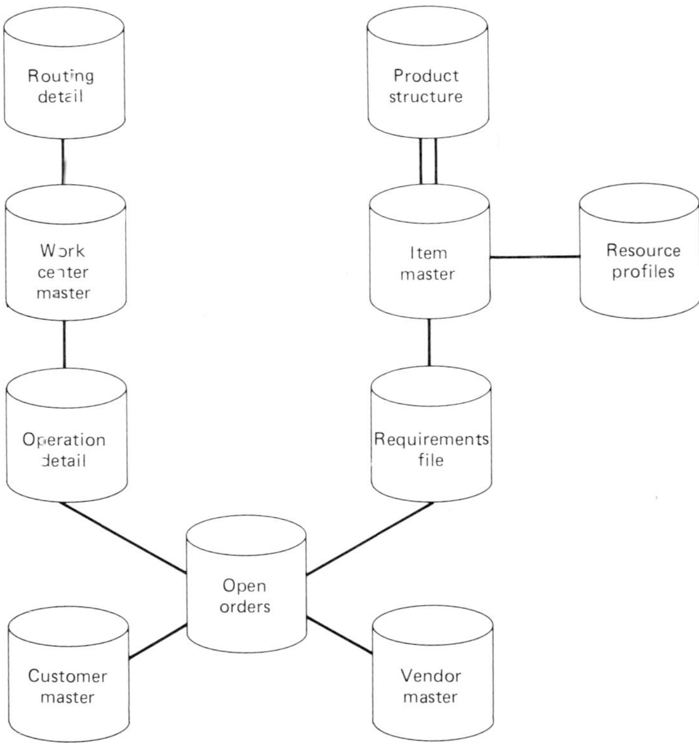

Figure 6.25 Typical database.

dating. Cost is much less a factor today and will become even less of a factor in the future.

- *Online entry, batch update.* Many systems log online data onto a storage device which is periodically polled by the host computer for file update.
- *Online, real time.* Data is entered into the system, and files are updated instantaneously.

As its costs continue to decrease, online, real-time updating will become the preferred method of updating. At the moment this method can be justified when there is a need to know status at all times. An example is a company that produces to stock and makes or loses sales depending on its ability to quote availability of inventory. A day-old or even hours-old report is of little value in this situation. Other data can be effective enough on a batch basis. For example, in a heavy-machine shop, knowing status of all orders on the minute is rarely needed. Day-old data is quite adequate.

6.2.19 Frequency of replanning

Just because the operating data in the system is updated online, real time, doesn't mean the system is completely replanned to stay in step. The details of most manufacturing control systems are replanned with periodic frequency, e.g., weekly, daily, or each shift. There are two reasons for that: the huge amount of computer resource needed to keep all plans in step with any change and the nervousness that is likely to generate in detailed schedules. Refer again to Fig. 6.1 for an explanation of the typical replanning frequency.

Strategic business and production plans need review and possible change on a quarterly basis. Many companies review them annually, but that is not adequate in today's fast-changing world. The MPS should be reviewed monthly. Significant changes detected within the review cycle should be handled by exception. Material requirements planning and capacity requirements planning are usually replanned on a weekly cycle. Input/output control also is managed weekly for most installations, but for some fast-paced assembly operations it is managed daily.

The sequence list is most often created daily. As the pace of material flow in a company increases because of shorter queues of work, this frequency will reduce to shift, hourly, or online, real-time. The replanning being done between MRP runs is to recognize the flow of material through the factory and the changing mix of items in queue. As MRP is usually by week, the sequence list cannot be completely replanned more often than that.

The purchasing system is driven by MRP, so again major replanning is tied to MRP's frequency. Distribution requirements planning is usually by week, and the final assembly schedule's frequency is adjusted to suit the necessary speed of response in the marketplace.

There are two basic methods of computer replanning. One is called regeneration; the other is called net change. Regeneration, as the name implies, means that the total plan is regenerated on a periodic basis. Basic data, e,g., inventory levels, bills of materials, routings, and master schedules, are input to the regeneration run. Material requirements planning and capacity requirements planning are recalculated in total to this latest data.

"Net change" means replanning by exception in the computer, not total replanning. For example, at a point in time, the only change that needs replanning is a scrap report for a major assembly. No other data updates deviate from plan. In this case the total plan, probably regenerated in the past, is left alone. Only the major assembly item that was scrapped and all its lower-level components need replanning. The capacity requirements plan also is correct except for the work to replace the scrapped major assembly. Hence, the affected work centers and time buckets are replanned, not the total plan. This exception replanning process is called net change. Three scenarios are now possible:

Regeneration, periodically The usual regeneration frequency is weekly, but twice per week or twice per month is not unusual. The computer run time for most businesses prohibits regenerating very often. Some companies regenerate monthly with net change processing at intervals between the regeneration runs. The purpose of the regeneration runs is to purge out embedded errors that sometimes creep into net change systems.

Net change, periodically Again the usual frequency is weekly but twice per week or even daily is not unusual. The primary driver is that net change processing is the correct way to update today's interlocking databases. Regeneration is difficult with a database; and if the number of changes is fairly small, net change places far fewer demands on the computer resource.

For the user, the decision is not so easy. Exception output will come from these runs for review and decision making. If the company is changing dynamically, frequent net change would appear to be better than infrequent regeneration. But the volume of output needing human review will overload the planners and schedulers. The informal system will take over immediately. Frequent net change can be effective only for a company under control with few, controlled, changes.

Net change, transaction-driven In this case, as each transaction to the files is processed, the system compares the transaction to what was expected. If anything deviates, e.g., the scrap report mentioned earlier, the item affected, all lower-level parts, and the capacity plans for the affected work centers are replanned instantaneously.

It's obvious that this method of replanning should be reserved for changes needing quick reaction, e.g., a scrap report or a mandatory engineering change. It's not a good idea for a complete master schedule revision, because of the huge number of changes and computer demands it will cause.

6.2.20 Management measures

A manufacturing control system contains millions of bits of active data. This data is used by the system to perform calculations, make logic decisions, and so on. Errors or omissions in the data obviously will result in false plans and action recommendations. A series of "health monitors" is needed to ensure the system is capable of doing the job. It's like checking the tire pressure, oil level, fuel gauge, power steering, and brake fluid levels before going on a journey. Some suggested health monitors are shown in Fig. 6.26. They are meant to trigger ideas more than be the ideas you should use. A concise set of health monitors that are reported on regularly is

Record accuracy:
- On hand
- Shop orders
- Purchase orders
- Bills of material
- Routings
- Customer orders
- Costs

MRP actions volume:
- Expedites
- Delays
- Order releases
- Total

Data omissions and errors:
- Lead times
- Lot sizes
- Manufactured part with no B/M
- Purchased part with B/M
- Order policy codes, others
- Logic errors

Transactions:
- First-pass errors by category
- Aged errors not corrected
- Unplanned issues
- Returns to stores

Capacity plans:
- Past due workload
- Erratic requirements

Dispatch lists:
- Percent high-priority jobs
- Length of queue

Figure 6.26 System health monitors.

needed. Any slippage below minimum acceptable levels needs an action plan to get back on track.

Even if the system is healthy and capable, there is no guarantee it will be used effectively. So another series of measures called operating controls is needed. The purpose of the operating controls is to find out if the system is being driven correctly and if all functions are working in concert with the system's objectives. Some operating controls are suggested in Fig. 6.27; they measure top-management's performance as well as the performance of middle managers and practitioners.

Schedule adherence is a critical concern for vendors and work centers. If fall-downs occur anywhere upstream of the MPS, customer service will suffer, inventories will grow, and costs will increase because of expediting and confusion. An excellent way to measure schedule adherence is shown in Fig. 6.28. The figure was suggested by Frank Gue, Westinghouse of Canada, and is described in detail in his book [Ref. 5]. The horizontal axis is time. The vertical axis is placed at the on-time point; weeks late are marked off to the left; and weeks early are marked off to the right. The

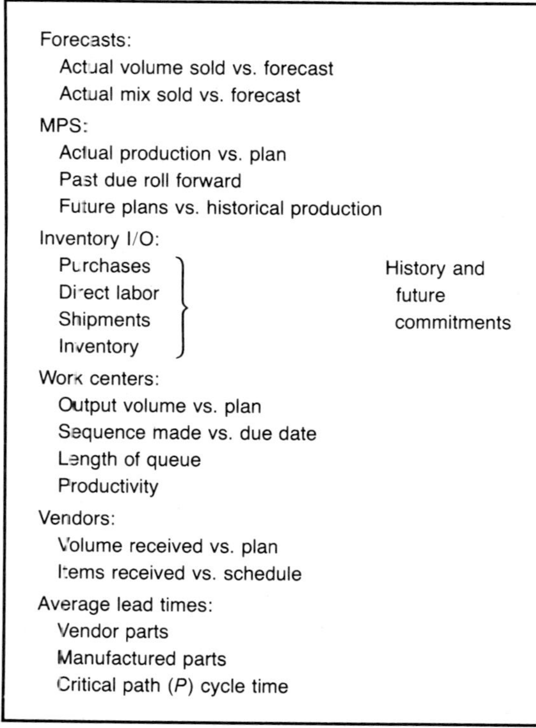

Figure 6.27 Operating controls.

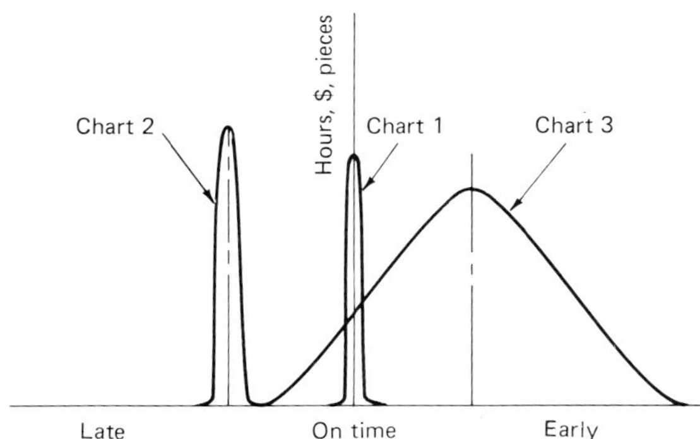

Figure 6.28 Monitoring schedule adherence.

vertical axis is product flow, and it can be measured in dollars, tons, pieces, number of lots, standard hours, and so on.

A labor report is received from a work center or material is received from a vendor. We have a scheduled completion date for this activity in the system and now the actual date. Hence, this receipt can be plotted on the chart: the amount of material or labor expended on the vertical axis and the time relative to on time on the horizontal axis. If all reports from a work center or vendor for a time period are plotted, a histogram will result. This can tell a lot about the resources, whether it is producing enough in total as well as whether it is producing the right things at the right time.

It is obvious that Utopia is chart 1, with the average production all on time and minimal spread between early and late. Chart 2 shows a work center that is on the average late, hence needs its output increased to move the average to on time. The spread indicates that the sequence of production matches the system's need dates well. Chart 3 shows a work center on the average early, but its spread of early and late shows a poor sequence control. This work center has past dues (the tail to the left of the on-time axis), and the temptation could be to add capacity to pay them. But this work center doesn't need more capacity; it needs less! And it also needs better sequence control.

Work centers and vendors can be ranked by their average days late—the worst at the top and those with most average days early at the bottom. Recrewing the work force by taking people away from the resources that are on the average early and putting them in those that are on the average most late will help rebalance the flow.

Ranking work centers by their spread of early and late—the widest at the top and the narrowest at the bottom—will focus attention on sequence problems. Pushing all work centers and vendors to have charts similar to chart 1 becomes the objective.

The system is a forward-planning one, which means it predicts events before they become realities. Summaries of the events can be used to test whether the plans meet business objectives. This was touched on in the earlier discussion of the financial system. Inventory budgets, cash flow for purchases, direct labor payroll, and receivables can be predicted by the system. This data will be ideal, and so it will need modifying to allow for reality. But now ROI and cash flow can be based on the details in the operating system, the details most people are aggressively trying to execute. If the financial prediction of the future is not satisfactory, changes can be made to the plans to get the results desired.

This is a better way to predict financial events than to use a separate financial prediction system isolated from the operating details. The real value is that it allows modification of the plan and a resimulation of the effects before operational plans become reality.

6.3 Where Are We Going?

Today's manufacturing control systems are simplistic and illogical and demand frequent human input. The mathematics used (adding, subtracting, multiplying, and dividing) are understandable by children 8 or 9 years old.

It's obvious that the separated modules with fixed parameters and illogical assumptions—e.g., capacity is infinite as we schedule a plant—need integration and a more logical treatment. It's also obvious that much of the human judgment currently needed to get results is to override the illogical decisions made by the system.

6.3.1 Optimizing routines

A whole new generation of manufacturing control systems that can mathematically optimize a business is on the way. The systems will consider the demands on the business, whether from customer orders or forecasts, the capacity of resources, and the availability of purchased materials and components. An optimized schedule that meets business needs and fits within the constraints will be generated.

The optimized schedule will differ with management inputs. For example, a schedule that maximizes ROI could be quite different from one that maximizes cash flow and again different from one that maximizes customer service. The optimized schedule capability already exists for some process industries. Linear programming can evaluate various scenarios and pick the one that best fits the objective. The question is how to extend that approach to the fabrication and assembly plant, with its huge amount of data and highly variable production parameters, and to the batch processor.

Some systems have attempted this task with variously reported results. Caposs E from IBM, Q control from Wm. Sandman and Associates, and

Opt from Creative Output are the best known to date. Others will come as the mathematical relationships are understood and equations are developed to handle them.

6.3.2 Just-in-time

From the mid-1960s to 1980, our effort was to develop systems to manage a complex factory. Study of many Japanese companies credits their phenomenal success to an attack on the environment. The simpler it is, the easier it can be controlled. An analogy is traffic flow. To move quickly from point A to point B by automobile, you will tell me to use a four-lane highway. Little in the way of systems is needed to control my traffic flow and that of other vehicles around me. But to move quickly from point A to point B in a city is impossible even with all the systems of one-way streets, traffic lights, and police directing traffic. Computerized stop lights can increase traffic flow 20 percent, so from an average of 10 miles/hr we get to 12/hr! We're still a far cry from the speed possible on a highway.

Systems will be developed to help create highways in factories. Group technology is a way to identify similar attributes of parts or materials to assist in creating process flow lines, rather than the traditional functional grouping of machines, to manufacture them. Far more emphasis will be placed on capacity (flow rates) and less on priorities (sequence), because when parts are flowing quickly down a line, priority changes are impossible and are rarely needed.

Better vendor-customer relations and data linkages will be created. Customers scheduling their vendors' factories, with daily deliveries, will become commonplace. Sharing of other common data will help break down the artificial change-of-ownership walls now typical of vendor-customer relations.

6.3.3 Artificial intelligence

Today's manufacturing control systems receive data, perform some calculations within one module, and report out exceptions or summary data. People review this output, make some changes, and reenter the data, and the system performs further calculations in the next module. This start/stop process typifies today's systems.

But many of the decisions being made by humans come from simple reasoning: the data and choices are limited. A reasoning computer could make at least 90 percent of those decisions. That means that data is input only once; the computer plans and schedules after that. Only the real exceptions that require human intelligence will now be reported.

Artificial intelligence in computers is on the horizon today. A major commercial application is in manufacturing: you can't operate an automated factory with today's simplistic computers. Artificial intelligence will dramatically change the roles and numbers of people in traditional planning and control activities.

6.3.4 Automated factories

An automated factory must be viewed like today's automated machine: material goes in and completed parts come out. Tomorrow's automated factory will receive material and produce finished products by automatic planning, scheduling, and operation. Many companies currently making products for stock will be able to convert to make-to-order; faster information flow and processing time will foster the change. But that means electronically linking customer specifications to a design system (CAD), to a process planning system (CAPP), to a scheduling system, to a vendor procurement system, and to the operating systems of the machines and materials handling devices.

There are some serious problems to overcome before all this is possible. Improved logic of planning and scheduling and the need for artificial intelligence for reasoning, both mentioned earlier, are needed. There is a serious problem with converting customer specifications into design and bill of materials data and then into process specifications and routing information. Communication problems involving these various stand-alone, automated systems won't be solved until industry standards are developed.

6.4 Summary

Inventory control was one of the first activities to be computerized. Initially, and naively, techniques were treated as separate, stand-alone, entities. It was quickly realized that the various techniques are in fact linked together. That led to the development of a generic operations planning and control chart that clearly showed the technique and data relations. This was later expanded to include all functions—sales, manufacturing, engineering, and finance. The idea was to have common information, supported by all functions, that all functions can then benefit from.

A huge number of data inputs and outputs are necessary in a typical plant. The actual processing of this data is currently simplistic; human judgment is needed to interpret it and convert it to valid actions. The next moves will be to advanced mathematics to optimize the process, simpler environments (the just-in-time approach), and a linkage with other automation functions, e.g., CAD, CAM, CAPP, robotics, and automated material handling.

Improved data linkages between customers and vendors will blur the distinction between various management groups and improve planning and scheduling even further. Artificial intelligence will complete the picture and give us the fully automated factories of our dreams.

REFERENCES

1. Joseph Orlicky, *Material Requirements Planning*, McGraw-Hill, New York, 1975.
2. George Plossl, *Manufacturing Control: The Last Frontier for Profits*, Reston Publishing, Reston, Va., 1973.
3. Oliver Wight, *MRP II: Unlocking America's Productivity Potential*, CBI Publishing, Boston, 1981.
4. Robert F. Hoskin, "Demand Management by Exception (Not by Exception Report)," *Inventories & Production*, vol. 1, no. 3.
5. Frank Gue, *Increased Profits through Better Control of Work in Process*, Reston Publishing, Reston, Va., 1980.

BIBLIOGRAPHY

Fox, Robert E., "MRP, Kanban or OPT, What's Best?," parts 1 to 4, *Inventories & Production*, vol. 2, nos. 4 and 6; vol. 3, nos. 1 and 2.

Martin, Andre, *Distribution Requirements Planning*, Oliver Wight, Williston, Vt., 1984.

Mather, Hal, and George W. Plossl, *Master Production Schedule—Management's Handle on the Business*, George Plossl Educational Services, Atlanta, Ga., 1975.

———, "Reschedule the Reschedules You Just Rescheduled," *Production & Inventory Management*, first quarter, 1977.

———, *Bills of Materials, Recipes & Formulations*, Wright Publishing, Atlanta, Ga., 1982.

———, "Attack Your *P:D* Ratio," *APICS 27th International Conference Proceedings*, Falls Church, Va., 1984.

———, "Change the System or Change the Environment," *AIIE Fall Conference Proceedings*, Atlanta, Ga., 1984.

Vollmann, Thomas E., William L. Berry, and D. Clay Whybark, *Manufacturing Planning and Control Systems*, Dow Jones-Irwin, Homewood, Ill., 1984.

CHAPTER 7

Production Planning and Control

Darryl V. Landvater

7.1 Introduction

Production planning and control systems evolved from a fundamental problem in manufacturing: managing what's needed and when. In the early days of production and inventory control, order launching and expediting were the primary means of scheduling. We ran our manufacturing operations by using the "informal system" of shortage lists and hot tags. In those days, we used the order point method for ordering what was needed based on the Kardex records. The problem was that the Kardex records were often wrong, and the order point method didn't give us much visibility into the future. The results were material shortages due to the lack of visibility, since the order point projected usage only by looking backward. Most of our scheduling was based on the crisis of the day, and the best visibility into the future was what we could carry in our heads.

Today the tools to help us manage a manufacturing operation properly are available. They can help us run our manufacturing operations better than before and solve the problems of the informal system. This involves all aspects of the company, from production scheduling and material purchases through labor, inventory, distribution, and finance, and it includes support for marketing and engineering while improving product quality and customer service.

Let's take a look at the evolution of production planning and control systems and lead into where we are today.

7.2 Evolution of MRP

7.2.1 Order point and variations

The order point was the first method used to answer the question "when to order" by taking the average usage of an item, projecting it over the normal lead time, and then adding in a safety stock to protect against greater than average demand. The problem was that the order point didn't simulate reality. It looked backwards and didn't account for the constant changes in a manufacturing business.

If we are low on an item, the order point tells us to make more, whether we really need it or not. That's OK for items that we do need, but sometimes we end up making things we don't need; and most companies have neither the capacity nor the capital to make everything they use. Consequently, the order point did not tell us exactly what we needed, when we needed it, and in what quantities.

Other variations of the order point system are the two-bin system, the visual review system, and Kanban. In the two-bin system, there are two bins, or locations, where material is stored. The primary location is used first, and when the second location starts to be used, that is the trigger to reorder material. In the visual review system, a person looks at the inventory on hand and determines what to order by noting the items which seem low. It's an eyeballing of the inventory to determine when to order. The Kanban system is basically a variation of the two-bin system. The difference is that instead of the trigger to order being use of material from a secondary location, the trigger becomes the Kanban card.

There are two types of Kanban cards: a requisition card authorizing withdrawal of material from the feeding operation and a production card authorizing the feeding operation to produce more of what is being withdrawn. As an ordering system, Kanban suffers from all the problems of an order point system. However, it has been used successfully as a material movement system triggering the movement of material in the shop and from nearby vendors.

Unfortunately, none of the variations of the order point system ever answered the real questions on the shop floor: *when do we need the material* and *how much is enough*? That's the job of priority planning.

7.2.2 Priority planning

Priority planning is where material requirements planning (MRP) comes in. MRP uses the fundamental manufacturing equation which says, "What are we going to make? What does it take to make it? What do we have? What do we have to get?" to schedule material. The master production schedule answers the question what are we going to make. The bill of material tells us what it takes to make it, and the inventory records tell us what we already have on hand. The logic of MRP then determines what we need to order (Fig. 7.1).

Let's take a simple example. The master schedule says that next week the family is coming for Thanksgiving dinner. To bake the turkey we need to make stuffing, and the bill of material says that we need bread, butter, seasonings, and onions. We check the inventory on hand and find that we have enough butter and onions, but no bread or seasonings. So our material requirements plan, or in this case our grocery list, says that we need to buy bread and seasonings and have them available for next week when we want to cook the turkey.

The difference between this simple example and a typical factory is volume. We can do MRP manually for Thanksgiving. The typical manufacturing company, though, has hundreds or thousands of end items, 5000 to 50,000 part numbers, hundreds of work centers, and hundreds of vendors. In addition, change is constant. Customers change their minds, manufacturing processes don't work, people are out sick, and so on.

Today we have the tools to solve these problems. We can see months in advance by projecting our needs for every single item and reprojecting them as things change. This is done by using the simple logic of MRP and the massive data manipulation capabilities of the computer.

Even with valid material plans, however, we found that wasn't enough. For example, the material plans for the company could be OK, but capacity could be insufficient. Or the material plans could be OK, but these plans weren't communicated to other departments. In addition to valid material plans, we need a way to measure how well we were doing against the plan. This took us to the next evolutionary step, closed-loop MRP.

7.2.3 Closed-loop MRP

Closed-loop MRP is a set of functions that are needed to represent a valid simulation of reality in a manufacturing company. It ties all the necessary

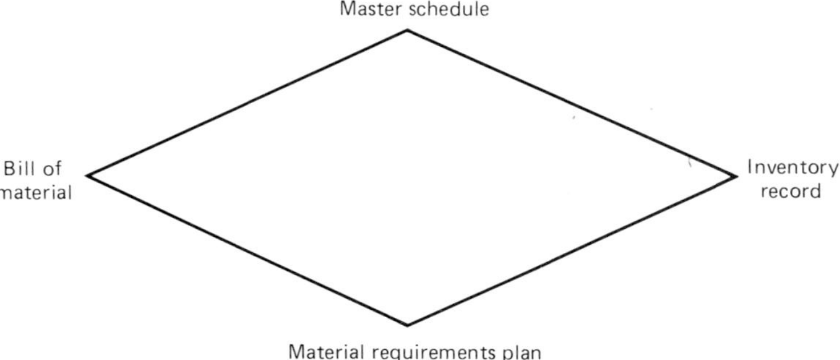

Figure 7.1 The logic of MRP.

functions together as part of one overall operating system. It takes the top-level plans of the company and translates them into specific material and capacity plans. Then people use that information to do their jobs day-to-day and to monitor their performance. (See Fig. 7.2 for a diagram of how this works.) Now let's look at each of these elements.

Production planning The production plan reflects production rates for product families, typically in monthly increments. It is set by the general manager and his staff, and it answers the question, "How many model 30 pumps will be built per week?" It doesn't matter whether you make pumps, widgets, cans, or candy—the logic is the same.

The production plan in a make-to-order business is established by looking at the current backlog of orders and comparing it to the desired backlog. The forecast of what we expect to sell is then added in to determine the overall production rate. For example, let's take an item for which the backlog of orders is currently 500. We want to decrease the backlog to 250 to provide better customer service. The forecast from marketing on this particular family is 500 a month. By taking our desired decrease in the backlog of 250 and adding to it the marketing forecast of 500, we come up with a production plan of 750 this month.

In a make-to-stock business, the production plan is determined by looking at the current inventory, deciding if we want to increase or decrease what we have on the shelf, and adding that to or subtracting it from the sales forecast to determine how many units we want to build. For example,

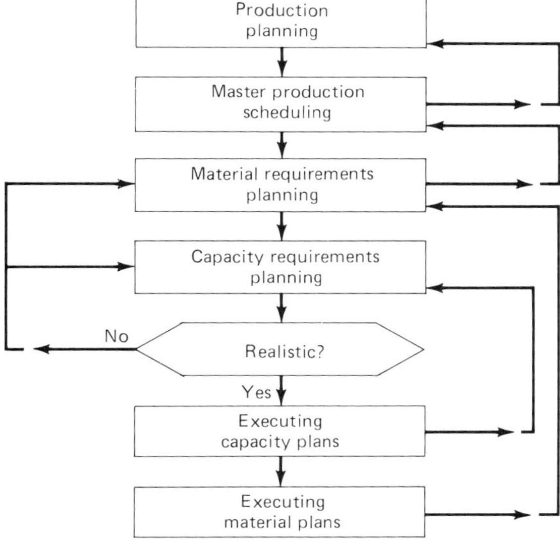

Figure 7.2 The closed-loop diagram.

we might start with 1000 on hand and want to increase that to 2000. The sales forecast for this family is 5000 a month. If we add the desired increase of 1000 to the forecast, we come up with the production plan of 6000 for the month.

The production plan is top management's handle on the business. It is a set of numbers agreed upon by the general manager and his staff which drives the rest of the business. It becomes the input to the next level in the process, the master production schedule, so it's critical that everyone in top management agree to this one unified operating plan.

Master production schedule The master production schedule (MPS) takes the production plan in units for a product family and breaks it down into more detail. It is the next step in taking the plans made by top management and translating them into what can be accomplished in the factory. The MPS is the anticipated build schedule time-phased in weeks or days for each item in a product family. For example, in the make-to-order model 30 pump family mentioned above there may be many different pumps available—the 30-01, 30-02, 30-03, 30-04, etc. With our production plan of 750, we must determine how many of these will be the 30-01, how many will be the 30-02, etc., and when they will be built.

When there are many different final configurations a customer can choose, it is often a futile exercise to attempt to forecast each end item. In this case, the master scheduling is done at the next level in the bill of material, which would be the various options that make up the product. This type of scheduling is often done in a make-to-order or assemble-to-order business when there are a number of different configurations for the final end product. In our example, the key options such as housings and motors for the pump would be shown in the master schedule. This would be done with a two-level master schedule, in which the basic production rate for an end item is scheduled and specific options also are scheduled in the MPS.

The important thing is that the master schedule must be accurate; it shouldn't be a wish list. If it isn't a valid simulation of reality, the material and capacity plans which are driven by it will not be valid.

Capacity requirements planning Capacity requirements planning (CRP) helps answer the question "What does it take to make it?" In this case, the "what" is the capacity of the work centers and the people in the shop. Since we have the items, due dates, and quantities of material we will need from material requirements planning, we can use that information to plan capacity. By looking at the routing for an item, we determine how many standard hours are normally needed to make it. Then we take the planned quantities and multiply them by the standard hours to come up with the loads for each job. The loads for all jobs are then summarized to give the expected load on each work center for each week.

The system will highlight any potential problem areas so people can resolve them before they become crises. A possible solution might be to use an alternate work center, a split lot, or an extra shift. Just as material requirements planning gives people visibility into future material problems, so capacity requirements planning gives people visibility into future capacity problems.

Before MRP and CRP, we usually found material and capacity problems when it was too late. Parts shortages and capacity overloads were indications of those problems. Now we can use MRP and CRP to anticipate the problems and help people solve them while there is still time.

Executing material plans Once a plan exists, there has to be a way to communicate it and monitor the planned vs. actual completion of it. That is true of both manufacturing and purchase orders. For the shop, the solution is often a daily dispatch list that shows the work currently in the work center by due date. The supervisor can then schedule the work center in the most efficient way while still meeting the required due dates. It's important to note that it's up to people to determine which jobs will be run in which order, since they are the ones responsible for completing the work on time. The daily dispatch list also shows work from upstream work centers so the supervisors can have visibility into what's coming.

In order to have an accurate daily dispatch list, it is necessary to track the movement of shop orders through the factory, operation by operation and work center by work center. Bar coding can be used to do this reporting. Relatively few companies use that approach today, although it promises to speed up the process, increase accuracy, and automate data entry. In addition, it could be used to speed up and automate the stockroom transactions as well as receiving and inspection. But, like bar-coding for shop floor control, relatively few companies have adopted it.

As for the vendors, the monitoring of planned vs. actual deliveries is done through vendor schedules and the purchased part delivery report. The purchasing people's job is to not only ensure on-time deliveries but also spend money well. If they see that a vendor is consistently delivering late and/or over cost, then they are responsible for working with the vendor. By giving the vendor a schedule showing what will be needed and when, the purchasing people communicate their requirements more clearly to the vendors and the vendors generally improve their on-time delivery dramatically. By eliminating most of the expediting, the purchasing people have more time to do effective purchasing.

Executing capacity plans Input/output control is used to compare the actual hours completed against the plan. Any deviations are an indication of possible problems at either this or an upstream work center. Input/output control allows the capacity planners and shop supervisors to manage by exception.

7.2.4 Finite loading vs. infinite loading

CRP is often mistakenly referred to as infinite loading, but in infinite loading one ignores capacity and loads all orders in the time periods in which they are required. In finite loading the computer makes decisions to prevent the capacity loads from exceeding the work center capacity. The issue isn't infinite vs. finite loading, however. CRP shows the problems to people and lets people solve them. You can't put 10 lb in a 5-lb bag—everyone agrees on that. The question is who decides how to fix a problem, i.e., people or the computer. Computers have not demonstrated an ability to use judgment or creativity, and so they're not good at solving problems. That is a job that people can do much better than a computer.

Examples of finite loading systems are Caposs, Informatics Production IV, and OPT. There are two problems with these systems: accountability and transparency. Accountability means that people are accountable. Yet if the computer makes a decision, how can you hold people accountable? Transparency means that people can see how information was derived. Finite loading systems involve billions of calculations; consequently, when people ask, "How did the computer get these numbers?," there is no satisfactory answer.

7.2.5 Manufacturing resource planning (MRP II)

Manufacturing resource planning is a simple set of numbers that everyone in a manufacturing business can use. The elements of MRP II include all the aspects of closed-loop MRP plus financial planning, simulation, and teamwork.

Financial planning With accurate inventory numbers, inventory valuations and projections can be done simply and accurately. If we take the on-hand balances from the MRP system and multiply them by the costs of the items, we have the value of the inventory. The open order file can be dollarized to give work-in-process (WIP) inventory. Since MRP has a projected inventory for each item in addition to the current on-hand balance, it is possible to use that information to make accurate inventory projections.

The numbers in MRP can also be used to make cash flow projections. Many companies feel they can do that today. However, unless a company is operating MRP well, the basic data will not be accurate enough to support meaningful projections. Cash disbursements can be calculated for the material, labor, and variable overhead expenses. Expenses for purchased material are calculated by dollarizing the planned material receipts. The labor-hours from capacity planning are multiplied by the hourly pay rates to determine the payroll projections. Variable overhead items are calculated by taking standard rates for things like the cost of electricity per hour for a machine or the amount of gas per hour on the paint line and extending them by the projected hours from capacity planning.

Cash receipts also can be projected. In a make-to-order company, the master production schedule is the schedule of shipments. These shipments are extended by the selling prices to give projected billings. In a make-to-stock business, the sales forecast is the projection of shipments. The projected shipments are then multiplied by their selling prices to give the projected billings.

Another decision that is often difficult without good planning information is the make-or-buy one. If it currently costs us $0.25 per unit for a particular part we manufacture and a vendor quotes us $0.20 per unit, then it makes sense to save the $0.05 per unit and have the part manufactured outside. However, if the overhead on our other parts goes up more than the $0.05 we save, then going outside wasn't a good decision. Analyzing the overhead effects of make-or-buy decisions is difficult without the information in CRP. With that information, we can better evaluate and distribute overhead and make better decisions.

Simulation The ability to react quickly to changes in the marketplace can be improved through the use of simulation. With manufacturing resource planning, we can do detailed simulations and see exactly what the impact on the business will be. For example, what would a 10 percent increase in business do to our capacity plans, material plans, cash flows, and inventory? Typically we don't have enough time to gather all the necessary information. With MRP II, we can simulate the situation quickly (usually on a duplicate set of files so the day-to-day operating information is unchanged) and have the same people evaluate the plans and be responsible for executing them. We can then make informed decisions and use the detailed plans to implement the changes quickly. Simulation gives people the ability to see the impact of changes on their operations and then decide what is best for the company.

Teamwork Teamwork is improved dramatically with manufacturing resource planning because the whole company operates from a single set of numbers. In the past, there was no common game plan that everyone could work from, and the fingerpointing developed out of the frustration of working hard and not getting results. In the United States, we have the philosophy of one man, one vote; in Japan, the educational system develops a culture in which everyone works together. But with the proper tools, we can work as a team just as the Japanese do. Our people want to work together, and they want to do a good job, but we have to provide them with the means of making that happen.

With manufacturing resource planning, everyone in the company has realistic goals that they're able to achieve. Although teamwork and quality of life were once thought to be minor considerations, the Class A companies have gained tremendous productivity improvements from the "people" aspects of the system. On-time deliveries, no end-of-the-month crunch, less expediting, and increased productivity play important parts in helping manufacturing people feel good about their jobs. They result in an improved quality of work life which is difficult to measure but may be one of the most important benefits of a manufacturing resource planning system.

7.2.6 Distribution resource planning

In a manufacturing company that supplies finished goods to its distribution centers, a continual problem is how to integrate manufacturing with distribution. Most distribution systems are based on the order point method, by which the inventory at the distribution center is depleted to a certain level and then replenished from the supply at the central facility. However, the demands from the distribution centers tend to be surprises to the factory because manufacturing doesn't have visibility into the specific needs of the distribution network.

The way to solve this problem is through distribution resource planning (DRP). DRP provides the interface between the material plans at the factory and the needs of the distribution centers. It does so by taking the overall manufacturing plan and integrating the branch warehouse requirements at the master schedule level. The result is time-phased requirements for each distribution center for each individual item. The master scheduler can then schedule the factory to meet the needs of the distribution network, rather than assume average rates of usage that can be significantly in error.

By using DRP, the distribution inventory is not only properly located but is also in the proper quantities for good customer service. The distribution centers can have the items they need on hand as opposed to either running the factory well but starving the customers or meeting the customers' needs at the expense of running the factory poorly.

Other planning information, such as transportation planning and distribution center capacity planning, can be obtained from DRP. For example, if we know in advance what we are going to ship to each distribution center (DC) and when, we can consolidate the shipments to take advantage of full trucks or railcars. By determining when the DCs will be receiving material from us, we can do a better job of manpower planning at the distribution centers.

7.3 The Future

7.3.1 Computer hardware and software

There have been major advances in computer hardware and software for manufacturing, and there are certainly more to come. In the mid-1970s there were 40 MRP software packages; 10 years later there were over 200. Microcomputers, as well as minicomputers, are growing as a percent of the total market today. In the late 1960s it took a sizable mainframe to run MRP. In the 1970s minicomputers were able to handle the MRP needs of smaller companies. Today, microcomputers allow even smaller manufacturing companies to have MRP.

Anyone looking at computer software needs to be aware that there are a wide range of packages from excellent to nearly unworkable. The potential customer of any package should compare the software with a standard and ask about other successful installations. In addition, a demonstration of the software, with your own data, is a good idea. If the software salesman can't find successful companies using the software, it might be better to look at another product. Companies continue to experience significant delays because of computer-related problems.

7.3.2 Just-in-time

A concept popular today is called just-in-time. Although the idea is fairly straightforward and has been applied over many years, the term is fairly new in the U.S. manufacturing marketplace. As it applies to inventory, just-in-time means the required part, assembly, or component should arrive "just in time" and be "just enough." Picture a factory in which the components required for final assembly arrive just in time and all of the purchased materials show up just in time to support the manufacturing operations. Imagine also lead times that are dramatically reduced and nearly perfect in quality. If that could be achieved, the impact on inventory turns and fast response to the marketplace would be spectacular. Productivity improvements with increased visibility into what's happening on the factory floor, quality of work life, and improvement in the quality of the product could give us a competitive edge over the competition.

There is a prerequisite to just-in-time: valid schedules. If you don't know what you need and when you need it, how can you bring in just what you need just in time? The way to get accurate and reasonably stable schedules is through MRP II operating at the Class A or Class B level. The only exception is in a highly repetitive manufacturing company that makes the same products every day with a level schedule. For that reason, many Japanese examples of effective just-in-time are highly repetitive, high-volume, level-schedule types of companies. In the United States, however, we are seeing a number of Class A MRP II companies also achieving tremendous success with just-in-time.

7.4 CAD/CAM

The interface between manufacturing systems and CAD/CAM is important to gain the full benefits of CAD/CAM. Without valid schedules, we might have the best-designed product in the world but have difficulty building it. Integrating CAD/CAM with MRP II effectively means that a new product is designed with CAD/CAM and then completely built by using MRP II. It eliminates the frustration of creating plans that are not executed as conceived because of long-lead parts that aren't placed on order in time, for example.

Engineering changes can be effectively implemented, and part obsolescence can be minimized. By knowing in advance what the new and obsolete parts will be, we can plan to phase them in and out smoothly and thereby reduce expediting and scrap. Engineering becomes part of the team with its productivity tools now integrated with those of the rest of the company.

7.5 Manufacturing Professionalism

Systems to help people manage manufacturing operations are available today. Without valid schedules, we're back in the Dark Ages of order launching and expediting with little visibility into the future. But people are the key components of any effective system.

American manufacturing must show what it's made of. We have the tools that manufacturing people need to become professional and compete in the world market. To do that, we need to use the body of knowledge that is standard in manufacturing. Just as there are generally accepted accounting practices such as general ledger, accounts payable, and accounts receivable, so there is a set of generally accepted manufacturing practices.

Manufacturing is no different from any other field; advances are nonstop. The test of professionalism is how those changes are presented and perceived. In the field of physics, Isaac Newton formulated the law of gravity several hundred years ago. In the early part of this century, Albert Einstein formulated the theory of relativity, which added to the basic laws of gravity by refining them. Einstein built upon Newton's law of gravity. He did not come forward with the theory of relativity and present it as the answer. Rather, he reaffirmed the basic laws of gravity and then explained his refinements and improvements to them.

That's a professional body of knowledge, each successive generation refining, improving, and correcting the work of those that have gone before. The latest buzzwords such as Kanban, just-in-time, quality, manufacturing resource planning, and CAD/CAM are for refinements, improvements, and advances in the field of manufacturing. The challenge is to present them as such, not as the latest gimmicks. Manufacturing professionalism will not come from jumping from one rock to another and abandoning what has gone before. It will come from building on previous knowledge and moving the ball ahead a bit further each time.

BIBLIOGRAPHY

Goddard, Walter E., "Kanban versus MRP II—Which Is Best for You?" *Modern Materials Handling,* November 1982.

Landvater, Darryl V., "Banana Peels," *Infosystems,* November 1984.

Martin, Andre J., with Darryl V. Landvater, *DRP: Distribution Resource Planning,* Oliver Wight Limited Publications, Inc., Essex Junction, Vt., 1983.

Wight, Oliver W., *Manufacturing Resource Planning: MRP II—Unlocking America's Productivity Potential,* Oliver Wight Limited Publications, Inc., Essex Junction, Vt., 1982.

———, *Production and Inventory Management in the Computer Age,* Oliver Wight Limited Publications, Inc., Essex Junction, Vt., 1974.

CHAPTER

8

CIM and Communications

Richard Morley

8.1 Introduction

This chapter deals with the communications aspect of computer-integrated manufacturing (CIM). Communications allows us to pass long- and short-term data between system components. It allows us to move data and extract knowledge from the data in order to optimize situations. Communications is an integral part of CIM and will continue to be an important segment of CIM.

Data is defined as the raw, unreduced information that is available on each process component or each system component. Normally, each component wants access to all data to make decisions. Each system component needs to know about its peers, neighbors, and managers. This means that the components in a factory system or CIM system can take advantage of other processes and allow higher reliability and more optimal processing and/or manufacturing. Without communications, the process can be only a grouping of independent units coupled by time and the initial program at installation. Without communications, the only real-time modification or control of the process can be effected by people walking around taking notes and having a time scale of one week to one month. Existing systems, however, usually have a supervisory component that gathers the data and presents it to the decision-making element, which is, in most cases, a human.

Communications theory says that the maximum information rate exists when the information is most chaotic or the entropy of the information is high. That means the harmonic content of the channel is kept at a minimum. Most of the time, communications systems transfer substantial amounts of redundant or useless data, which is harmonic and does not significantly contribute to system performance.

Knowledge, on the other hand, is reduced data overlaid by heuristic rules, or rules of thumb. Formally, knowledge is the application of heuristic rules to a database to help support decisions. This is called a decision support system when employed in a systematic manner. Knowledge can substantially reduce the communications requirement. Communications is the ability to talk to one another and to pass data and knowledge back and forth. Indeed, communications is the base for knowledge, by which we mean that the reduction to knowledge is what communications is all about. If the data is reduced to knowledge by the communications component, requirements such as bandwidth and signal-to-noise ratios are substantially reduced.

Networks, however, are not communications. Networks *allow* channels of communications to exist, and these can be temporary, permanent, or dynamically assigned. Real-time process modification can be effected via electronic communications and a support network. Data from the peer components in and the variables of the system can be passed along in a vertical management hierarchy or a horizontal peer-to-peer network. It is estimated that about 20 percent of the existing data transmissions will be required when good system design is employed. This is called management by exception. Distributed systems also reduce the need for bandwidth. They are, in themselves, smart and make use of information that is passed along or is available to be addressed as needed. The older master-slave concepts of communications required full access to the entire distributed database to make a decision in a computer of at least minicomputer status.

Networks are essential to move knowledge and to allow communications to take place. Networks are the roadways of information in a system. A minimum requirement of this communications system is to find out how we are on a day-to-day, minute-by-minute, and second-to-second basis. Is the system up and running? This information is deduced from the data in a centralized or distributed intelligence. Timeliness is extremely important in this decision. Most applications require the information to be available in less than ten seconds across the entire system.

The primary mechanism for conveying information is a "sneaker net." The sneaker net communications method employs humans wearing sneakers to carry information around on tapes or disks or in paper form. Today, that is, the primary mode of communications in most systems is biological. The trouble with the sneaker net is that it does not get or provide real-time information. The information can be days to weeks late, and maintaining the process in that mode is difficult at best and is impossible at worst. Even humans need a back fence to establish protocols and communications channels to connect the appropriate sender and receiver. The telephone is the most pervasive nonbiological network that has the appropriate protocol and characteristics for establishing electronic networks.

Electronic networks such as the ones used in existing and future CIM systems make data sharing easy, peripheral sharing easy, and information sharing OK. They do not, however, allow for easy sharing of resources. Most people confuse the hardware of communications with the software of communications. The hardware of communications is easy; the hard part is to design a network operating system that makes sense in the applications area. Information and data must be transferred among a number of attached users in an easy manner and controlled by some kind of network operating system. The network must support communications between the required couplings, which need not be one-on-one but can be *n*-ary. Each node needs data to make decisions. This decision making can come from the master-slave relationship, or the node itself may be smart enough in a distributed system to take advantage of its own intelligence and neighboring data. Knowledge can be reduced from data at the source or in the sink.

Wire costs in the hardware are high, and labor is expensive. A local area network has 2 to 10 times more traffic on it than a wide area network. Each node gateway or system bus, if well designed, reduces the communications requirement by the old 80/20 rule: A good design, alone, can diminish redundant communications from 80 to 20 percent. Key items to pay attention to are to pass data easily, in large quantities, on a good availability and reliability basis and to make better use of factory resources. These resources include the system, the machining and processing components, and the information processing components. Resource is the key word here. Resource use in the past, currently, and in the future, is the basic foundation of the need for communications in a CIM application.

Real-time control is one of the specific areas. In the simplest sense, we need on/off status. Where is the system now? What is its status? How can one coordinate peer, subordinate, and supervisory components in the system? To be more practical, programmable controllers are the pervasive mechanisms of control in the modern processing plant, especially in discrete manufacturing areas. Loading programmable controllers with their program and changing the program to create a flexible manufacturing system is a strong use of and need for a network. Right now the sneaker net is the pervasive method in use for the loading of the programmable controller, and it is clearly inadequate at providing the real-time flexibility which will be needed in the near future.

A second key element is maintenance dispatch, and dispatch of appropriate personnel when the machine is not operating is extremely important. The management needs to know the status of the process and the location and type of fault. Then the correct personnel must be sent, whether they be pipe fitters or electronics experts. In order to repair the equipment, management data also must be transmitted down in a meaningful manner. In future factories the maintenance people will have to repair machines that they have never seen before and, in all likelihood, will never see again The same maintenance personnel will have to be "generically" trained, and the specifics of the given machine must be a part of the communications network and be able to be part of any on-line diagnostics.

In data processing, IBM, "Big Blue," needs data to do any processing. Let's not forget that the dollar performance of the system process is being monitored by the financial office or its equivalent. Mainframes, or superminis, need the data in order to report the dollar performance statistics to human management.

8.2 History

Communications for information in the CIM environment is, as Truman said, "similar to the electric industry." He also said, "The only thing new is the history that we haven't learned." In the 1800s, there were "vice presidents of electric power." Each power system had individual components, individual setups, and individual voltages, phases, currents, and cycles. Now the distribution and control of electric power, in effect, come with the building. In the future, i.e., a decade or two, communications also will be built into the building, and we will not think much about it. It will be relatively transparent. Some of the things that are being modified are relays. They are being converted by programmable controllers into their software equivalents. Conveyors are being converted into automatic trucks and other vehicles. Mechanical motion is being replaced by programmable motion and servomotors. More recently, we have seen the conversion of data by information to data by communications networks. In the near future, we will be transmitting, in addition to data, information and, finally, knowledge and how to act on it. We also anticipate that, in communications, electronics will replace people and people will be able to think about their jobs and what they are doing instead of how the jobs are done.

Historically, factory manufacturing was relays, manual materials handling, machine tools, and power distribution. Programmable controllers came along and replaced the relays with state sequencers. Then we began to recover data through simple communications systems, which led to full monitoring. The system became extremely complex. There are factories with 20,000 programmable controllers; each department can have at least 1000 of them. We must know what is going on and be able to make knowledgeable decisions and optimize within that framework. We have to understand how data, knowledge, and information differ.

The vendor problem is severe. IBM doesn't talk to Digital Equipment

Corporation. On a historical basis, each vendor wants to capture the market, by means of its own standard, and exclude other vendors. The standards being imposed by General Motors, with its MAP protocol, are a good move. In the past, communications needs were vendor-driven and diffuse. Finally, they are becoming standardized and user-oriented.

8.3 Local Area Networks

Local area networks (LANs) come in all shapes and sizes. Their topology, access method, medium, and market are different throughout, however. In general, a LAN has geographically local service. It is a private data communications system covering a limited geographical area which is typically less than one mile. A LAN lets a factory network communicate with peers vertically and horizontally. It also allows sharing of high-speed and special-purpose input and output devices and databases. The LAN also allows access to remote mainframes. Many LAN systems are available. The most popular are in use in the personnel or work station product areas; they share data and resources. A second use of networks is to solve applications problems such as airline reservations and maintenance communications needs and electronic mail. The particular area of a LAN application with which we will be concerned is the concept of a LAN viewed as a communications facility that handles a multitude of applications across a wide variety of functions. The LAN becomes a pipeline that supports multimedia applications. Planning becomes extremely important as applications become more diverse. A LAN can be defined as having data rates of greater than 1 Mbits/s and having peer-to-peer communications. It also handles the diffuse requirements and dissimilar devices found in various applications.

The philosophy is that we have to deal with three areas: humans, machines, and databases. We want to stop the turbulence. It has been said that two communications people will decide on and standardize three methods of communication. We must impose standards, not impose change. Communications between the control area and management, which is now by sneaker net, must be by a LAN. A LAN can integrate the material and data on the shop floor; it can supply data, and we can do much with data. An advantage of a LAN is that it allows electronic communications to maintain, optimize, and use other resource databases and process controls with less than 10-s response. It allows, if it's well designed, flexibility. It can be vendor- and equipment-independent. It does appropriate filtering of messages so that bad messages cannot get through, and it allows downloading into things like programmable controllers, numerical coprocessors, and microprocessors with new programs intended to provide flexible manufacturing. LANs allow machines and humans to monitor the entire process by "zooming" to look down into any particular segment in as great detail as necessary. The LAN can monitor quality and make corrections, or stop the process to allow humans to make corrections, before the quality or the yield of the system becomes bad. Tooling for production and for monitoring can also be an outgrowth of use of this system. The major requirements are the questions, "Am I making product?" and "If not, why not?". The product can be established in the tooling, quality, materials component, and monitoring sections and/or everywhere along the production process.

8.4 Network Components

Network components in the CIM system go all the way from cell controls, or islands of automation, to the overall factory and into each department. The primary system and network components must integrate material and information flow together. In other words, the material, the process, and the information must be synergistic and synchronized with each other's requirements.

The physical base for the LAN requirement has many mechanisms. The broadband version puts an analog system onto frequency-modulation channels and delivers multiple channels of voice, data, and video simultaneously. It can run up to 50 km at 100 Mbits/s and is cheaper for a large number of attached devices (Figs. 8.1 and 8.2). Baseband, which is not modulated but uses the wire in an on/off mode, is significantly cheaper but does not run as far or as fast. Typically, baseband can operate only up to 2.5 km and at a 10-Mbit/s rate. It has the advantage of no active, failure-prone components.

Multiplexing, or the transformation of the signal into various dimensions, is a technique which is used throughout. For example, you can multiplex in space, i.e., two or three wires can be used to separate the information. You can also multiplex in time, i.e., select the system, operate and communicate with it, and then select another one. You can also multiplex in

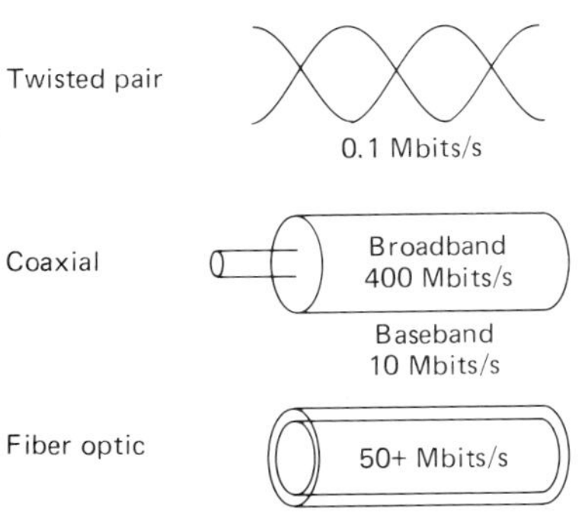

Figure 8.1 Types of cable media for LAN data.

frequency. High frequency might be used for one coupling between the receiver and transmitter and low frequency for the coupling between a second receiver and transmitter.

Many topologies can be used; they are shown in Fig. 8.3. The bus system uses a railroad with stations along it. The ring acts as a conveyor belt on a continuous mechanism that transmits data. The star has a controlling, central supervisor and sends messages out along single wires to various receivers and transmitters. Point-to-point is a wire system that connects server and receiver, more or less permanently, in an initial system design or architecture.

Two access methods are generally being considered for CIM. The first is Ethernet, which works on a CB kind of system. Formally it is called carrier-sense multiple access with collision detect. If no one is using the

		Coax		
	Twisted Pair	Baseband	Broadband	Fiber Optic
Bit rate	100 kbits/s	10 Mbits/s	400 Mbits/s	50+ Mbits/s [up to 5 Gbits/s in tests]
Type of data	Voice/data	Voice/data	Voice/data /video	Voice/data /video
Usual topologies	Point to point Star	Star Bus	Ring Bus	Point to point Ring
Advantages	Lowest cost Good EMI/RFI immunity	Low cost Easy to install	Relative noise immunity Uses CATV components	Immune to EM interference Very secure Low weight and size
Disadvantages	Poor security High loss	Low noise immunity Usually requires conduit	Higher cost Skill required to install RF modems required	Very high cost Mostly limited to p/p High skill required for installation and maintenance

Figure 8.2 Alternative media comparison.

system, the transmitter sends information and the receiver receives it. If the sender detects that someone is using the line, the sender waits until the line is free and then sends another message. Collision is unavoidable and collision detection must be part of both the sender and receiver systems. The second major access method is token passing. In either a ring or a bus topology, this allows each server to take control of the system, make its communications requirements, complete them, and then pass the token along to its neighbor.

The medium used in both topologies and access methods can be a twisted pair such as that used in the local phone lines. The twisted pair typically gives 10 kHz and is prone to interference unless special precautions are taken. Coax, a shielded single wire, is used in most of the existing networks in the factory and can handle up to 300 Mbits. Fiber optics isn't used much now, but we anticipate that it will be in the future. It can go to 4 GHz, but at the moment it's tough to splice and relatively costly.

Protocol layers were decided on several years ago. These standards have been evolved and are physical statements rather than actual implemen-

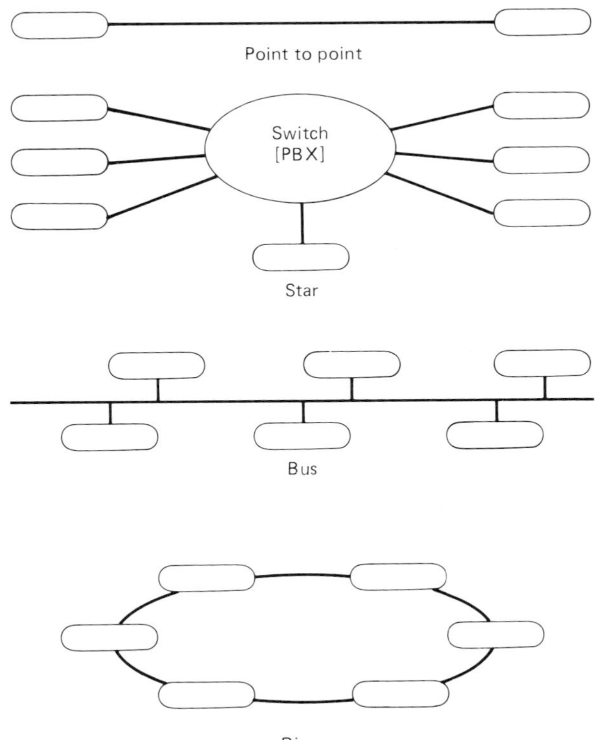

Figure 8.3 Network topology.

tation standards. They were formulated by the International Standards Organization (ISO), and have been controversial ever since their release. The layers are called the physical layer, the data link, the network, the transport, the session, the presentation, and the application as shown in Fig. 8.4.

Each layer invokes the next layer up. Each does a transformation until, finally, the two machines, working all the way down to layer 1 and back up into the other machine, connect. Layer 1 is the actual physical medium, pin connections, cables, and the rules for driving and interfacing to the network. Layer 2 implements conventional protocols. Above layer 7, applications programs can be written as if the computer had fully compatible hardware operating systems and direct communications with the environment.

Typically, two machines are incompatible, i.e., layer 7 in the first machine cannot communicate directly with layer 7 in the other machine. This means that layer 7 must evoke its layer 6; and layer 6 may resolve problems such as the meaning of the floating point format in different machines. Each layer pretends that it communicates directly with its counterpart in the other machine; in fact, it communicates with the next layer all the way down to the actual physical medium called layer 1.

The difficulty is that the OSI model is conceptual and not specific. Vendors design and implement networks that are conceptually similar but incompatible. Some progress has been made on firm standards for layers 1 to 4 by the ISO. The other layers, 5 to 7, are currently in an undeveloped

Physical layer:

Provides for the mechanical, electrical, and procedural means to establish a physical connection.

Data link layer:

Establishes logical data links and transfer data frames.

Network layer:

Transfers data between systems not on the same LAN. Routes packets through intermediate systems.

Transport layer:

Provides for transparent data flow between transport users.

Session layer:

Synchronizes dialog and data exchange.

Presentation layer:

Allows application processes to interpret the meaning of transferred information. Converts syntax. Not specified by MAP.

Application layer:

Allows application entities to coordinate their activities.

Figure 8.4 Protocol layers.

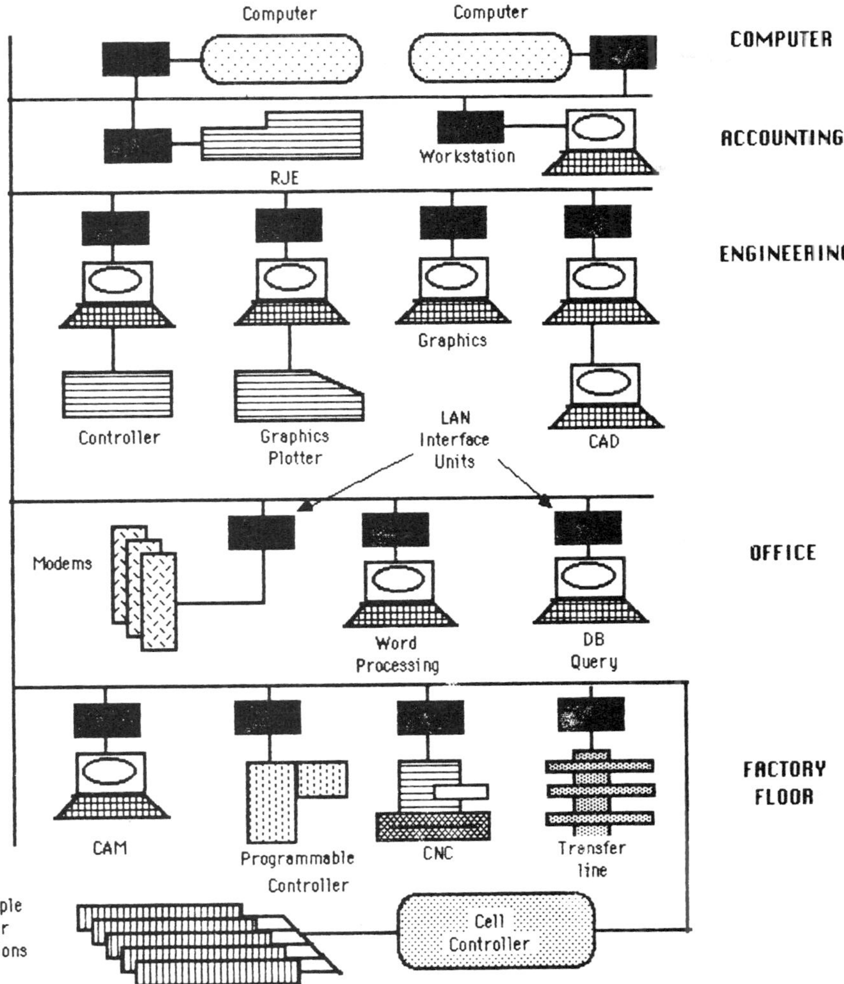

Figure 8.5 Factory-plant communications system.

state. Between LANs, concepts such as bridges and gateways are used. Bridges connect LANs that are similar, and gateways act as translators to allow dissimilar LANs to connect.

8.5 Systems

Systems consist of various levels as illustrated in Fig. 8.5. Consultants and communications experts identify different kinds of levels. The first and highest system level is the corporate level, which needs less timely information than the engineering level. The engineering level deals with design, analysis, manufacturing and the industrial process itself. Each level, from the highest (corporate) to the lowest, must decrease the latency of a response.

At the factory level, the response becomes even more critical. The department level has to have communications in and out of the factory level. They must include sales, order entries, and projections. Finally, we get to the work center level and the operation level. Eventually, there will be monitors or video display tubes on each station with full peer-to-peer communications ability. Programmable controllers will talk to sensors, numerical control machines, and robots, and vice versa, and all materials handling will be done at the operation level. Some of the servers that have to be used as resources on the network are minis and micros on and off the floor. These servers provide access to databases and allow resources such as printers to be used. Material flow and instruction and documentation files are part of the database. Workstations must be made available on the factory floor to provide human communications. They will report the status of the shop and also be a reporting function. Periodic reporting, by humans, on the workstation is necessary, as is displaying the status of the system. Generally, the workstations will be reporting information and data, as well as knowledge, to department and factory management. The user interface is extremely critical and must tend to be transparent with respect to communications and wide area networks.

8.6 Served Components

Served components are numerous. Generally, a served component is a programmable controller, a computer, or a robot. All of motion control, i.e., direct numerical control (DNC) and computer numerical control (CNC), have to be connected. There are also video display tubes with their associated functions. These can be, but are not limited to computer-aided design and manufacture (CAD/CAM), workstations, and terminals for superminis, minis, and mainframes. These are the components which must be served on the network. Specialty microprocessors have to be designed, and they have to be able to communicate on the network. The cell controller and the associated devices and sensors must also talk on the network. As one can easily see, many diverse needs and components must be served by the network.

Networking and networks are matters of some concern. The network we know most about is the public telephone network, and it will continue to be the primary network in the United States. Generally, it is analog, and the use of analog will be predominant in the near future. For data, it is high in cost and low in speed, especially for burst rates. The network connects a server or a transmitter and a receiver together on a temporary basis by making a direct connection. The dynamics of this direct connection are through a private automatic branch exchange (PABX). PABXs are becoming more digital and more LAN-like in their processing. Distributor components, i.e., components having intrinsic intelligence, are a strong, future trend. Central controls make up most of the existing installations. LANs will, most likely, be hybrid and have appropriate bridges and gateways to connect differing vendors. They will operate within distributed database networks with hierarchical schemes and use coaxial cable. In other words, we anticipate that the LAN of the present and near future will operate with coaxial cable, work with distributed networking, and be composed of different types of systems coupled together with bridges and gateways.

Ethernet is a very strong office communications system. It is a carrier sense multiple access with collision detection (CSMA/CD) system. It listens before talking, much as a CB radio does. It has irregular response times and is very good in certain applications. Intel, Digital Equipment, Gould, and others are making integrated circuit chips in order to facilitate such networks. In general, the Ethernet is used in a system that has an infinite number of nodes, each requiring very low communications needs.

Token passing is different; the network allows burst transmissions and provides predictable response times. It does not allow as many users on the network. The network issues a short message, "idle," which is the token. A device grabs the token and then sends a message. The token is then released. Control is centralized by special-purpose nodes which tend to be error-prone. Token passing supports regular and burst transmission and does not practically limit packet size unless a special configuration is implemented.

8.7 Implementation

Planning for the installation of a CIM network and communications needs is difficult; in the short term, it centers around the vendor offerings shown in Fig. 8.6. The capabilities of existing components are limited to communications between devices that differ intrinsically. In the long term, automation must become a flexible manufacturing system (FMS). The networks must be multifunction and multiapplication. The return on investment should be 3 years for an integrated system and must make efficient use of floor space and people. In the short term, we have multivendor problems. The vendor will promise many things but must be asked questions: "What can you do?" "Is the system proprietary?" "Can other vendors get on it?" "Will it service robots, personal computers, mainframes and micros, all from different vendors having different operating systems?" Presently, the answer to all those questions is no!

Planning must be over at least a 5-year period, and the primary costs are labor and wire installation. If you have a green-field or start-up factory situation, broadband communications across coax is good enough. It is sufficiently immune. Planning on optical communications is difficult because of the lack of standards, and the twisted pair method doesn't have great enough capacity. In the retrofit market, baseband is reasonable. It is important to do what is right and not be awed by the technology. A little technology, correctly applied, is better than a perfect system that doesn't work. Do not initially plan on putting sensors directly on the network. The sensor problem is most severe. Sensor data will pass through a multiprocessor, programmable controller, or the equivalent before being put on the network.

Cell controllers will be major system components in the future. On the down side, a cell controller has high communications requirements. The controller is, however, a way to cluster or cell the types of processing to be done. Various types of equipment are used to accomplish a bounded set of applications. A cell controller manages a diversity of processing components in a machining center to carry out a single process function or

Gould:

Modbus
Supervisory master/slave networks, interfaces with Gould minis, micros, and programmable controllers.

Modway
Broadband, master/slave peer-to-peer token passing, 1.54 Mbits, 15,000 ft; gateways to MAP.

Texas Instruments:

TIWAY-I, TIWAY-II, peer-to-peer, 1 Mbit; interfaces with TI minis and micros; gateways to MAP.

Allen-Bradley:

Data highway. Any of 64 programmable controllers in the network can serve as master. 56 kbits, 10,000 ft max.
Software supports DEC's PDP-11 series.
Gateways to MAP.

General Electric:

GE-net conforms to IEEE LAN [802] and specifies physical, link, network, and transport OSI layer. Broadband, CATV coax, 1 to 5 Mbits RS-232, RS-422 twisted pair, 8-bit parallel, serial asynch, gateways.

ISSC:

Universal Communications Processor [COP]

Digital Equipment Corp.:

VAX-750/780. DECdataway-E-net, baseband bus, coax, CSMA/CD.

IBM

Hewlett-Packard

Figure 8.6 A vendor overview.

application. In other words, a machine center uses robots, pallet trucks, machine tools, program motion, and a crane to make one part for an application.

Outside the cell controller, communications requirements are few. A cell controller will have inspection, CNC, and robots, and a single-cell node will treat the assemblage of processing units as a single process. The cell controller must talk to the broadband network as well as to the local cluster. The cell controller is, in effect, the system, or architectural bridge, between nodes of processing and the overall system.

8.8 Management

Management must understand the issues and not be afraid; it must learn the fundamentals of communications as outlined in the brief glossary of Table 8.1. That doesn't mean that it must go out and install the communications networks, but it must learn the standards and benefits of communication. It must also be stable and not allow turbulence to muddy the waters. It must understand that Ethernet is good for light traffic with lots of components on the system, whereas token passing should be used for large, high-speed networks. Baseband is used for short distances and is relatively inexpensive. Broadband is used for long distances, with data, voice, and video all on the same line. Consider medium and method and ask yourself, "What do you want to do, and how fast do you want it to go?". Cost, speed, and features should be matched against the application. Networking communications are not cheap; most of the time, no tangible payback is there. The intangibles are high. A transfer line may go from 50 percent uptime to 55 percent, and that 5 percent converts to 10 percent net on the bottom line. Management must also understand that testing and reliability are important issues. How do you test? How effective is the test? How easy is it to use? These are the questions that management must address.

8.9 Vendors

As shown in Fig. 8.7, there are many computer vendors who offer LANs and communications for the CIM market. They see it as the glue to hold the technology and market position that they have already established. They do not see it as business unto themselves. They anticipate that the market will go from $50 million to $250 million per year in the near future. Most vendors think of selling more of the existing product by offering communications networks. As a caveat, most of the vendor experience is in electronics and mechanics and not in communications. Electronics are the database, and mechanical firms understand the processes, but not the

TABLE 8.1 Communications Acronyms

AMH&S	automated material handling and storage
ANSI	American National Standards Institute
ASCII	American Standard Code for Information Interchange
CBEMA	Computer–Business Equipment Manufacturers Association
CCITT	International Telegraph and Telephone Consultative Committee
CIM	computer-integrated manufacturing
CSMA/CD	carrier sense multiple access with collision detection
DBMS	database management system
DCE	data communications equipment
EBCDIC	Extended Binary Coded Decimal Internal Code
ECMA	European Computer Manufacturers Association
EIA	Electronics Industries Association
FDM	frequency division multiplexing
IEC	International Electrotechnical Commission
IEEE	Institute of Electrical and Electronics Engineers
ISA	Instrument Society of America
ISO	International Standards Organization
ITI	Industrial Technology Institute
LAN	local area network
MAN	metropolitan area network
MAP	manufacturing automation protocol
NBS	National Bureau of Standards

communications. Generally, you can not buy a "CIM" communications package directly from a single vendor or, for that matter, from anyone. The internal purchaser must have enough systems experience to direct the installation. We also must remember that bits do not equal knowledge.

8.10 Standards

Why have standards? They reduce the multivendor problem and turbulence for the user. No true standards exist other than conceptually (Fig. 8.8). Vendors' standards have to address more than the several levels; they include hazardous environment, corrosion, interference, access time, and flexibility. Security and fault isolation are also important. Some of the problems are that the programs of the communications networks are in the roofs of the buildings and very hard to service. Also, the culture of communications service has not been infused into the manufacturing process.

There is a solution. General Motors is a vigorous promoter of manufac-

NEMA	National Electrical Manufacturers Association
OSI	open systems interconnection
PABX	private automatic branch exchange
PC	programmable controller or personal computer
RS-232C	An EIA specification for a modern interface—formally specifies only a 25-pin connector at the modern (DCE). Common usage sometimes implies data rate and asynchronous communication
RS-408	an EIA specification for CNC/NC interface
RS-422	a new EIA specification for DTE and DCE point-to-point communication with differential drivers at a high data rate
RS-423	intended to connect DTE and DCE on a point-to-point basis. Replaces RS-232C and uses RS-449 connectors
RS-449	A specification for DCE connectors—used in RS-422 and RS-423 configurations
SDM	space division multiplexing
TDM	time division multiplexing
SFC	shop floor control
X.25	a standard for communications
802	specifies physical and link layers
802.2	logical link and medium access control—used in 802.3,4,5 standards
802.3	specifies carrier sense multiple access with collision detection—Ethernet
802.4	specifies token bus—MAP and Proway
802.5	specifies token ring
802.6	specifies MAN

turing automation protocol (MAP), a networking standard. MAP is not fully specified yet but is coming on strong. It does have some soft spots. Many of the areas are ad hoc standards, and interface modules cost about $1000. GM needed a set of protocols for all the vendors that they were using, such as Gould, Allen-Bradley, 3M, Western Digital, General Electric, IBM, Digital Equipment, Ford, Boeing, Westinghouse, Intel, Motorola, Kodak, DuPont, McDonnell-Douglas, and Concord Data. All those vendors were very turbulent, and each wanted its own communications system to be used. General Motors got off to a good start and tried to establish a future standard which was broadband and token-based. It made a demonstration of the IEEE 802.4 token bus along with Gould, Allen-Bradley, Digital Equipment, Hewlett-Packard, IBM, and Motorola (Fig. 8.9). Four of the seven layers of communication are being implemented into MAP (Fig. 8.10). Chips are coming by 1986; they will be supplied by Motorola, Gould, and Intel. A secondary standard that does not have the support that MAP has is PROWAY, an enhanced 802.4. It provides for more controlled response time and two or more protocols. The new PROWAY standard will conform to the amended version of 802.4.

8.11 Trends and Futures

It is just a matter of time for all networks to be installed in CIM. No one should consider building or retrofitting a plant without either baseband for retrofits or broadband for new installations. Optical communications are the trend line of the future, and within this decade, they will be significant contributors owing to their superb bandwidth, reliability, and immunity to noise. Virtual networking allows full transparency and requires much less protocol layer work than previously. Full-fledged network operating systems over the broad database that is virtualized across a distributed system can be used throughout. Artificial intelligence, a term that is often misused, is coming on strong and will allow information instead of data to be transmitted. This particular strong trend will allow a substantial reduction in bandwidth requirements, and make automatic protocol conversion, and management by exception possible.

8.12 Conclusion

It is evident that the communications aspect of CIM will continue to be the focal point of networking developments. Each system component needs to know about its peers, neighbors, and managers, both horizontally and vertically, in order to promote a flexible and reliable manufacturing envi-

3M Corporation
Applitek Corporation
Bridge Communications
Concord Data Systems
Excelan
IBM
Interlan
Network Systems Corporation
Proteon Associates
Seeq Inc.
Sytek
Ungermann-Bass
Western Digital Corporation

Figure 8.7 Some communications companies.

ronment. This means that the components in a factory system or CIM system can take advantage of other processes and allow higher reliability and more optimal processing and/or manufacturing. Many communications systems transfer substantial amounts of redundant or useless data, which is harmonic and does not significantly contribute to system performance. Smart distributed systems will permit much of this data to be distilled into vital "knowledge," which is reduced data overlaid by heuristic rules, or rules of thumb. As stated earlier, knowledge is the application of heuristic rules to a database to help support decisions. Reducing data to knowledge will substantially lessen system and load requirements.

A key piece of the communications scheme, which must be addressed, is the network. Networks allow channels of communications to exist, and they can be temporary, permanent, or dynamically assigned. Networks are essential to move knowledge and to allow communications to take place. They are the roadways of information in a system. Today the most pervasive

IEEE 802

Specifies the physical and link layers. Baseband uses the bus scheme with a 50-Ω cable. Devices use RS-232C to interface with network controller. Also runs 1, 5, and 10 Mbytes over 2.5 km. Broadband uses 75-Ω cable with rates of 1, 5, 10, and 20 Mbytes. Broadband uses a single-cable CATV standard for normal use. Operates in half- or full-duplex mode.

IEEE 802.2

Specifies medium access and logical link. Protocol specified is a multipoint peer link. Any compatible physical layer can be used. Supports data rates to 20 Mbytes. Used in all three access methods (802.3, 802.4, and 802.5). Approved in 1984.

IEEE 802.3

Ethernet (CSMA/CD). Uses a carrier sense multiple access with collision detection. The requesting sender is obligated to "listen before using the network." Specified lengths of up to 2500 m can be used. The system has low efficiency for high loads. The specification outlines method, protocol, physical medium (low-noise coax), connections, and cable to I/O device. Works on baseband and broadband. Approved in 1983.

IEEE 802.4

Token bus. Token passing over a bus-type network on broadband cable or on a baseband cable with two allowable types of communications. The specification allows equal access, but devices can transmit only when the token is acquired. The token is not physical, but logical. The system is stable and predictable. The documentation details the token passing, protocols, physical connections, and media (base- or broadband cable). Typical vendors complying are Concord Data, Gould, Allen-Bradley, Western Digital, and 3M. IEEE approved in 1983, and the ISO is reviewing the specification. Fast becoming the de facto standard for the factory floor.

IEEE 802.5

Token ring. Mostly an IBM thrust. Similar to 802.4 but with a physical ring instead of a logical bus. Data can pass in only one direction. One station failure can disable the entire network. Media can be twisted pair or coax. The documentation defines the token management, protocols, physical media, and connections. Approval is expected about one year later than the others.

Figure 8.8 Overview of pertinent communications standards.

form of networking is the sneaker net, i.e. humans walking around and carrying out process modification by hand. Right now, sneaker net is the method in use for the loading of programmable controllers, and it is clearly inadequate at providing the real-time flexibility which will be needed in the near future. Real-time process modification can be a reality with the increased implementation of electronic communications and support networks. Communications between the control area and management, now being done by sneaker net, must be done by a local area network. If it's well designed, it will allow download capabilities into things like programmable controllers, numerical coprocessors, and microprocessors with new programs intended to provide flexible manufacturing.

Resource use in the past, currently, and in the future, is the basic foundation of the need for communications in a CIM application. In the past, communications needs were vendor-driven and diffuse. Finally, they are becoming standardized and user-oriented. We must not impose change; we must impose standards like those formulated by the International Standards Organization (ISO). As one can easily see, there are many diverse needs and components that must be served by the network. The LAN of

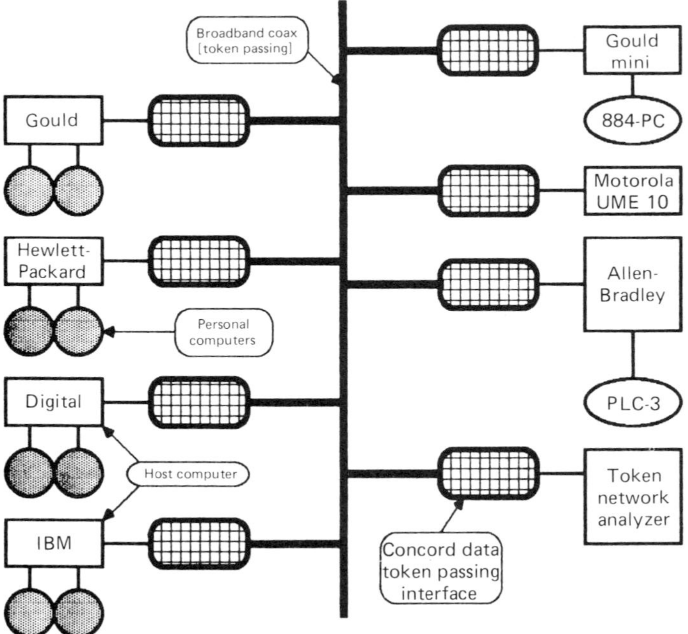

Figure 8.9 General Motors demonstration of MAP protocol at the National Computer Show, July 1984.

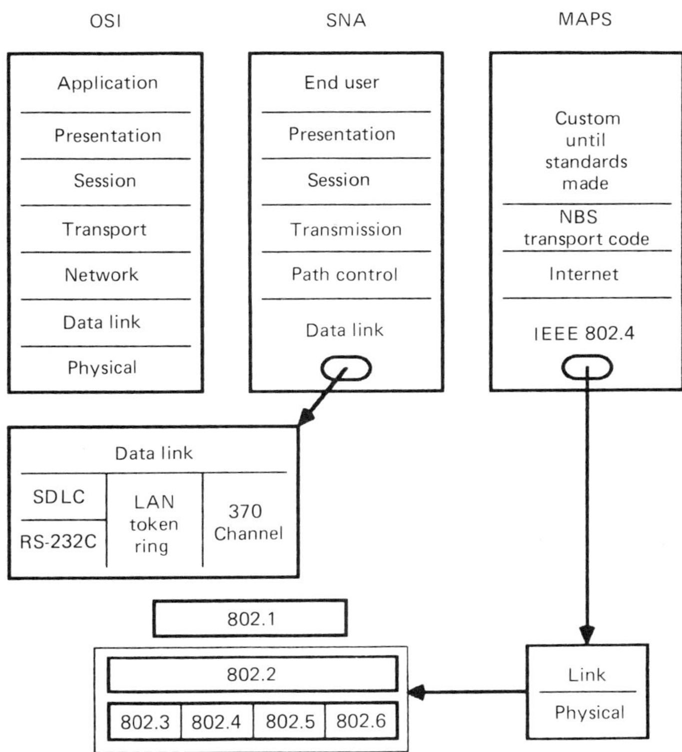

Figure 8.10 Protocol layer groups.

the present and near future will operate with coaxial cable, work with distributed networking, and be composed of different types of systems coupled together with bridges and gateways. Planning for the installation of a CIM network and communications needs is a difficult task. In the short term, it must be vendor-oriented, and we, as management, must learn the standards and benefits of network communications.

CHAPTER

9

The Role of Materials Handling

Leo Roth Klein

9.1 Introduction

9.1.1 Defining computer-integrated manufacturing

There are probably as many definitions of computer-integrated manufacturing (CIM) as there are vendors of equipment and software or consultants who wish to be associated with this latest buzzword. Therefore, a statement should be made as to what definition we shall use in this chapter. There is no consensus on definition. Even the *C.I.M. Dictionary* published by the Computer & Automated Systems Association of the Society of Manufacturing Engineers, although it defines over 500 terms involved with CIM, doesn't define computer-integrated manufacturing itself. For the purpose of this chapter, our definition of CIM will be *the application of computer technology to the tools of production in an interconnected hierarchy*. If we can accept this as a temporary or working definition, then we can proceed.

Certain key concepts are implied in the definition. They are locked into the two phrases "tools of production" and "integrated hierarchy." We can find many cases of tools of production with computers linked to them. Numeric control (NC) is far from a new technology. Equally, we can find integrated hierarchies of computers. They have been used in the world of data communications for decades. Therefore, to simply meet one criterion is insufficient; both criteria must be met. That is integration. To achieve one, and only one, criterion is to create another island of automation. These islands of automation are very much like the floating islands of Swift's *Gulliver's Travels*—they just keep moving around. Islands of automation will not coalesce into an integrated environment unless something holds them together. The technology we will address is especially well suited to the role of linking the islands of automation and forming a manufacturing environment that is manageable and controllable by an integrated hierarchy of computers. It is a valid path to achieving CIM—computer-integrated manufacturing.

9.1.2 Criteria for examination

As we have seen from our definition, there are many aspects of CIM. We will be looking at only one aspect: materials handling. We therefore have to identify what criteria we are going to use to accept technology in this area as being CIM-type equipment. Just because it may have an on-board microprocessor does not make equipment eligible for CIM. That level of technology would cover almost everything including wristwatches. The level of technology we will consider must first of all meet the definition of CIM established above. Second, the equipment must be that which is used to transport material from point A to point B or is required for the transport of that material.

Certain normal materials-handling equipment is not, therefore, CIM type. Standard shelving and racks are not CIM. Conventional forklift trucks, man-aboards, are not CIM equipment regardless of their capability or sophistication. Neither do gravity conveyors, raceways, or towlines qualify. We will evaluate the acceptability of equipment based upon its own level of computer control and points of human intervention.

9.2 The Technology

We are going to look at the various technologies available to us in CIM for materials handling. We will see what they can do—their capabilities—as well as their limitations—what they can't do. We must understand that there is no one answer that will be right for every company. The answer will vary with the environment being addressed. In order for us to develop the answer for our company, we must know the technology. With that understanding we can move forward into the applications arena. We will not make you salesman-proof; that task takes more than this short chapter. We will try to give you the basic understanding that will enable you to ask the right questions. We will explore six technologies: conveyor systems, monorails, guided vehicles, automated storage and retrieval systems, and bar-coding.

9.2.1 Conveyor systems

The first integrating technology we are going to examine is the conveyor, and we are going to refer to the powered systems exclusively. We can distinguish between the two basic types by the conveyor bed; the two types are roller and belt. We are probably most familiar with powered-belt systems. They are similar to what one sees in the baggage claim area at airports. The key features of such conveyors is that the bed almost always moves at a constant rate. The rate may be variable over time, but at any moment the entire belt is moving at a single constant rate.

Belt conveyors are more like fixed than flexible automation. Although one can expand the system by adding a belt or contract a system by taking a belt off, it is extremely difficult to change a floor configuration. Any major movement is a large-scale capital investment, which is not made easily or often. Belt conveyors can move items along a horizontal plane, up an incline, and down a grade. All three movements can, and often are, included along a single path. They have the advantage of maintaining a known rate and an on-board product will keep its position relative to the bed through the journey on the belt. Also, there is little movement of the product while it is in motion. Belt conveyors interface easily with vision systems and as recipients from robots. Because they rarely come to a stop, they are not ideal delivery vehicles to robots.

Roller conveyor systems have many of the characteristics that are found in belt conveyor systems. They cannot be used for slight grades without loss of control over product relative position. Powered-roller systems tend to have the separate rollers spaced closely together. That simulates, at least partially, a belt conveyor. With angled rollers, turns of specified angles (including 90°) can be performed. A special characteristic of roller beds is the ability to stop the product at specified points along the path for a length of time and then set it back in motion. This is done very often either to prevent an unworkable queue at a workstation or to present a specific queue to a work center. Unlike the belt-drive conveyors, roller-drive conveyors can have T or + paths.

To make these systems more flexible in their applications, standard options include gates to vary the path of the product. Although the gate can be sensor-based and react to weight being present, it is far more common to have the gate triggered by a specific product passing an identification station. The latter method employs microprocessors and is totally programmable. At this point we have true CIM-type equipment.

Computer-controlled conveyors operate in the three basic environments: manufacturing, warehousing, and distribution. In manufacturing they are generally used for conveyances of over 6 ft; they tend to be used for long distances or for parts that have been containerized. Conveyors on the manufacturing floor usually have roller-type beds. There is no standard for utilizing gates or other diversionary equipment, but some type of material diversion will be deployed. In the manufacturing environment conveyors offer a quick and reliable method of moving kits and piece parts between workstations. Since the equipment is static and the entire bed area is available, conveyors can be effectively utilized in the repetitive intermittent and flexible manufacturing environments. The only timing issues are identification and location of sufficiently unoccupied space on the bed.

9.2.2 Monorail systems

Probably the most famous monorail systems in the world are those in use at the Walt Disney theme parks. Monorails are, however, workhorses in the manufacturing environment. Depending on the weight load and the building structure, monorail systems can, at times, be relocated within a facility. Monorails can be installed with the load suspended from or mounted on the system; the latter configuration is referred to as an inverted monorail. Powered monorails can be of two general classes. In one class, the power can move the entire monorail track; then the system moves the material at a steady or constant rate as a belt conveyor does. Alternatively, the conveyance device attached to the monorail contains a motor and each load is moved independently.

One of their special advantages is that monorails can move odd-shaped or bulky items, and items with poor balance configurations also are especially suited to movement by monorail. A fairly common application of monorail systems is in hostile environments such as paint rooms. Current paint technology utilizes substances that are toxic to human beings. To have a person move material into and out of paint areas can impose a significant expense in OSHA compliance. Monorails avoid that entirely. When the monorail is connected to a robot paint system, the necessity for human presence is totally avoided. When piece part identity is combined with robot painters and monorail presentation, the savings in terms of paint material can be as much as 30 percent. This level of integration does not require new or untried technology, but it does provide a return on investment that should impress even the most conservative of financial officers.

In the fabrication and assembly environments we find monorails especially useful for maintaining work-in-process queues. The technique employs spurs similar to those found in railroads. In fact, exotic spur systems that look and function similarly to railroad marshaling yards can be created. They can be coordinated with production schedules to present the right parts to work centers without developing large queues at the work centers themselves. Since monorail systems can be tied to part identity systems, the spur track can be used to discriminate piece parts and to store similar items together and/or release items in a predetermined sequence. When the conveyance mechanism itself is powered, the entire monorail system can be paced to the shop and changed as the pace changes. As with conveyors, monorails are very common in the distribution environment.

Regardless of how parts are stored, monorails are an effective vehicle for bringing goods from the warehouse to the point of packing or shipment. That is especially true when they are installed in multitiered storage areas with part selection and conveyance loading under human control. At the packing or shipment area, the marshaling yard concept mentioned above can be used to group parts to an order or orders to a truck dock or to balance the work load in the packaging areas.

In the manufacturing warehouse, monorails provide an acceptable method of stores removal. In the situation of two commonly owned adjacent plants, monorail systems are often used to bridge the gap. It is in the manufacturing environment that we also find inverted monorails, which are well suited to the transport of very heavy or bulky parts. Applications of that type have been used to carry such diverse products as automotive bodies and paper rolls for newspapers.

The monorail systems available today are logical outgrowths of the towline systems that have been employed for over two centuries. They are, however, a flexible technology; and because of the application of computer technology to the proven conveyance system, they have a long-range place in CIM.

9.2.3 Automated guided-vehicle systems

Of the newer technologies, the one that offers the most exciting possibilities for making materials handling a vital portion of CIM is the automated guided-vehicle system (AGVS). AGVS is an outgrowth of the standard forklift truck and towline technologies. The original concept of the forklift was to provide a movable platform to transport heavy items; motorized systems put a man aboard and provided an in-plant truck device. The AGVS, however, takes the man off the truck. Early versions were like the towline systems. They were rather dumb vehicles that followed a wire embedded in the floor and could do nothing more.

Today, the AGVSs are far more sophisticated. Today we can lay paths on, not in, the floor with switches and crossings that enable the vehicle to reach any part of the floor. Vehicles on the system move independently of each other but at fairly constant rates of speed. AGVSs are not independent systems as conveyors and monorails are. They are usually integrated with other devices such as robots and automated storage and retrieval systems units and also with machine tools. To understand the vehicles themselves, one has but to conceive of wheeled motorized platforms that move along predetermined paths. Given that definition, one of the first considerations is safety. What is to keep these vehicles from crashing into each other in a commercial version of bumper cars? What is to prevent them from running into a person or over something on the floor? What is to prevent them from tipping and dropping all or part of their loads in the pathway?

Those are not trivial questions. The bumper car possibility has been eliminated by placing sensors along the floor paths at regular points. When a vehicle passes one of those points, a signal is sent to the systems computer. If it is calculated that two of the vehicles can meet, one vehicle is shut down for a few moments and then proceeds along its way when interference is no longer possible. The vehicles have rather large, but light, bumpers which are placed close to the floor. The impact on those bumpers of something as light as an empty carton is sufficient to stop the vehicle within less than 6 in. That is less than the distance from the bumper to the vehicle proper. Also, sensors on board can determine when a load is out of balance during movement and halt the vehicle.

AGVS units can have a variety of platform configurations to hold the goods they must move. For pallet-type loads, two cross-members are usually sufficient. For container loads, a flat bed can be equipped with side rails that are either static or may be automatically raised and lowered. The materials support may be configured to hold cylindrical or odd-shaped loads. The device is, therefore, highly flexible. The latest option is to offer a cargo bed that can be raised and lowered depending upon either its location on the floor or preprogramming.

As for pathing within the facility, newer systems can have the vehicle leave the guide wire and travel totally under computer control. Although the distances possible at the time of writing were quite short, 50 to 75 ft, we can expect that to change. The AGVSs of tomorrow may well not need a guide path at all. The control of path traffic can be vested in the central computer, which can display the current position and direction of each vehicle. In the computer's simulation mode, various traffic patterns can be tried and an optimization for vehicle availability can be selected.

The adjustable bed offers interesting possibilities for true CIM activity. The AGVS vehicle can be used as part of the holding device for the machine tool; then there is no loading or unloading time for the workpiece. That can dramatically increase the throughput of the shop while reducing the overall manufacturing lead time. Another interesting successful application of AGVS is to use it as the carrier of the machine tool. This is the flip side of the preceding application; rather than presenting the part to the tool, it is presenting the tool to the part. If we then have robotic parts loading and unloading, we have taken a step further toward a lower per manufactured unit labor-value-added environment. At any level of sophistication, its utility in hazardous work environments is obvious, its desirability is inevitable, and its ROI is dramatic.

9.2.4 Robots

Robots are not usually thought of as a materials-handling unit. Yet, wherever you see a robot, it is probably going to be handling piece parts. Robots represent the ultimate in integrating computerized equipment. It can almost be said that their flexibility and utility are limited only by their owners' imaginations. Even so, robots do come in a variety of flavors.

The two basic types of robots are mechanical and servo. The mechanical robot is most limited in its capabilities. It can move its arm only to points designated with mechanical stops. Rotary motion is limited by cams. Any computer control is primitive. The work envelope of a mechanical robot will be either cylindrical or rectangular. Reprogramming will be a mechanical function performed on the physical robot. The primary uses of mechanical robots are for what is called pick and place. In CIM the mechanical robot is an ideal vehicle to transfer parts or containers from one conveyor to another, from an AGVS to a conveyor, or from a workstation to a conveyance device.

Servo robots represent an entirely different level of technology. If the mechanical robot is a complete idiot, then the early servo robots were only morons. Therefore, the robots of 5 years ago were only stupid, and the most recent robots are only "getting smart." The ones to come, continuing the analogy, will be first intelligent and then ultimately brilliant. All robots are characterized by a high degree of repeatability. Servo robots achieve this repeatability while under direct computer control. Unlike mechanical robots, servo robots can move their arms to any conceivable points that can be described by X, Y, Z coordinates. They can also be made to stop at any number of those points in a totally random fashion and then perform a series of wrist movements as simple as turning a piece part upside down or as complicated as loading a fixture from a pallet. If the robot, at least one of the newer models, can be told what part is being presented, it can perform a different series of movements based upon what part is being specifically presented at that time.

Here the limit is the computer controller on board, not the robotic mechanism. This is real flexibility. The servo robot is therefore really only an extension of the computer. There is neither a real nor a functional constraint on the number of parts that can be handled. The system can be associated with a flexible machining center and perform not just parts handling but also parts segregation. At the absolute other end of the manufacturing cycle, robots can be used to load diverse parts into containers. With the current availability of quick-connect fittings for the manipulators, the spectrum of parts to be handled during a short time period is broadened. That brings the robot into a whole new area with an improved ROI.

As we said earlier, the repeatability of the robot is well known; the robot will move a given part to a given point with an accuracy greater than one

one-hundredth of an inch virtually forever. What that means is that its accuracy is at least equivalent to, and in repetitive cases greater than, that of a human. For very sensitive and delicate parts the robot can be the preferred short-haul parts mover. This repeatability is very important when a hostile environment must be dealt with. One of the early applications areas for robotics was the foundry. Foundries are places where toxic fumes abound, the primary raw material is molten metal, and the product is too hot to touch. Robots are ideal unloaders of parts from the casters. They are, furthermore, ideal for the ladling of molten metal. They perform the task simply and steadily and require no OSHA protection. More recently, we have seen the rise of vision systems. Robotic systems can now identify and discriminate piece parts and act according to the parts presented. Vision systems are still new, and no real standard has developed. Before the end of the 1980s a standard should appear.

The most exciting development in robots as part handlers is to utilize the robot as a fixture holder presenting the part to the machine tool. This uses the seventh axis of freedom, traveling, whereby the robot presents the part to a variety of machine tools. It is a further movement toward an integrated computer-controlled workplace.

9.2.5 Automated storage and retrieval systems

For centuries warehouses could also be called "where-houses." The warehouse was a building into which many things entered never again to see the light of day. Executives have been known to wonder if there were not creatures living in the warehouse who existed by eating the company's products—containers and all—without leaving droppings. Probably the ultimate extreme example was a warehouse manager who said to the author, "There is at least $5 million worth of parts in here that we know we have but must be invisible because we can't find it!"

Computer-integrated manufacturing demands that we keep a very tight control on inventory—from raw material right through to finished goods. If it goes into stores, we had better be able to find it. Given that requirement, the manual warehouse, with all its known problems, is unacceptable. The answer is in the computer-controlled warehouse: the automated storage and retrieval system (AS/RS). The AS/RSs come in two modes: unit or pallet and mini- or bin-box load. They are the ultimate in computerized materials handling because they are totally paperless. Now the warehouse, aside from maintenance, operates totally under computer control. It can receive material and put it away. It can bring forth material either on demand or in response to a preplanned schedule of need dates. It can keep track of what has been stored, where it is stored, and what quantities are stored.

The AS/RS has certain characteristics. As a general statement, it is usually not viable to retrofit a building for a unit-load AS/RS. The usual approach is to build the storage racks for the systems first and then enclose them with the building's skin. One of the objectives is to get the maximum use of the building cube. In fact, the building wall curtains may even be attached to the racks, which may also support the roof. AS/RS units can store pallets of standard size and even those that are custom-designed. The control computer keeps track of what is stored and where.

Mini-load AS/RS units can be erected within existing structures. They are ideal for the storage of kits or small loose parts for the manufacturing floor. When components are stored in a mini-load, it is generally because they are relatively large and either are not delivered in pallet loads or are delivered when the pallet load has been broken down. Mini-loads can also be used to hold work-in-process on the floor.

AS/RS units, mini- and unit load, can, and often are, integrated with conveyors or AGVSs. Depending upon the configuration of the entire systems, both pallet and bin-box AS/RSs can be maintained completely unmanned. That is, after all, what CIM is all about! Human interactions to the AS/RS come at two points. The first is identification to the systems, or what part or parts are housed on a specific pallet or in a specific bin-box. The systems will keep track of the pallet or bin-box location until the parts are exhausted. The second point is at the quantity level. Very often a pallet will not be completely used nor will the entire contents of a bin-box. Someone must indicate to the AS/RS what quantities have been removed or added. This count modification must be accurate if the system is to work.

The efficiency of this warehouse, like that of any other, is its ability to receive and dispense parts quickly. In a conventional warehouse we are constantly moving items from bulk to active stores. In addition, we put items not needed for a while in the back. The AS/RS does that also. The process, called profiling, is the central program in the computer control. It provides the needed number of empty spaces up front for incoming stock and also moves to the front the parts which will be needed the next day or shift. Moving around, the actual profiling operation will be done on the third shift or whenever activity is at a low ebb. For on-the-floor, in-process material (the active queue), another option is vertical storage and retrieval (VSR); it consists of small-scale units that are a cross between the AS/RS and the carousel.

9.2.6 Bar-coding

Our final technology is bar-coding. Although the equipment involved is not literally materials-handling equipment, it does integrate all the other technologies with those of production. Along with vision and voice systems, bar-coding serves to identify material. Without identity we are in the position of having modern technology driven by antique methods, somewhat like trying to drive a new Thunderbird with a Model T engine. The equipment associated with bar-coding falls into two types: readers and printers The methodology of bar codes, or as it is called, symbology, is quite diverse. Basically the symbology in a specific bar code is the meaning associated with the placement of wide or narrow dark bars that create narrow or wide white bars.

It has taken us 30 years to have a standard for character generation in data processing—ASCII; we do not yet have a standard for bar codes. In fact, there are over three dozen different bar codes in use today. Most of these are specialized, but many attempt to create a standard. Every success and failure can be shown in the development of the most common bar code, the Universal Product Code (UPC). It started as a simple five-digit number to identify the contents of case shipments. It is now a ten-digit number which identifies not only product but also packaging. For example, Coca Cola—not Diet Coke or caffeine-free Coke, just plain Coca Cola—in 8-oz containers would need different codes for metal cans, plastic bottles, and glass bottles (both deposit and no deposit) sold singly, in six-pack, eight-pack, or twelve-pack, in addition to the case and pallet load identity. That is why we need 10 billion item capacity in the UPC—and that's for supermarkets only! On the reader side, several technologies are available; light or laser beam and magnetic are the two basic ones.

The laser offers the ability to read bar codes moving very rapidly by the station. It is, however, more expensive than ordinary light. Magnetic read is less common. On the printer side the issue is quality of the bar code and the symbology available. A further problem for bar codes is their placement. How and where does one place bar codes on odd-shaped items? How does one read them? Placement is a major issue because the reader cannot scan the entire geometry of the part to find the code; it has to know where to find it. This applies to all parts that a specific reader will have to scan. Lack of placement planning can and will be very expensive for implementing companies.

With the rise of the LOGMARS (Logistics Applications of Automated Marking and Reading Symbols) project of the U.S. Department of Defense (DOD) there has become almost a de facto standard in Code 39 (also called 3 of 9 coding). This code's advantage is its ability to accept alphanumeric data, not numeric data only. The capability is sufficient to significantly increase the number of identities within a specific code length. If in selecting a symbology a company should find that there is an industry-specific code and it does business with the U.S. DOD and also with European industry, it will find itself having to use three codes immediately. It would be far better to work with one of them than try to find another code internally.

Bar-coding brings equipment together. It tells the robot what is coming off a conveyor and therefore where to place it. On the conveyor it tells the system what gates to open or close. It tells the storage unit what goods are arriving so that the profile program knows what it has. It simplifies the paperwork at the dock so that material can be put in motion as soon as possible. Bar codes also tell the materials-handling control system what is where on the system so that appropriate resources can be allocated. Bar codes also relieve the system of the human errors that always creep in. Although they are still a new technology with a lot of definition yet to be done, lack of maturity should not inhibit companies from moving strongly into the area. If CIM is an objective, then bar codes are a necessity.

9.3 Materials-Handling Applications

We have now looked, in very broad terms, at the technology available to us. The hard part comes when we try to put all this together: select the equipment, plan the implementation, justify our actions to finance, and, hopefully, get on with doing the job. Nothing, however, is as simple as it seems. Resistance to any part of the CIM technology will be both entrenched and emotional. CIM means fewer people on the shop floor: a lower labor-value-added content. It is antagonistic to a worker's perceived goals, antagonistic to union goals, antagonistic to a supervisor's perceived goals, and antagonistic to the company's desire to present a "good guy" image to the community. CIM relates to and is not antagonistic to the company's desire to survive. Without survival all other goals and objectives are meaningless.

In this section we are going to cover five big subjects very quickly. Those subjects are selecting the application, selecting the technology, planning the implementation, controlling the implementation, and integrating the systems. Books have been written about each, but we will hit the high points. There is a great deal of work to be done before the first application of CIM is installed. Not the least is management and company education. A rather standard joke is the five steps of system implementation, the second of which is complete disillusionment. The only way to prevent the disillusionment is to prevent the first step—wild-eyed enthusiasm. That is the product of salesmen at work. They all seem to take the approach that they have the best thing since sliced bread, and, even worse, they believe it. To further compound the error they communicate their unrealistic attitude to company managements. It is this unbridled optimism that starts the ball rolling toward unmitigated disaster.

Our first step is to let our management know that there is no free lunch; there is grief with the glory, and both are part of the process. Management must know what the table stakes are. As soon as we start to implement CIM technology we are going to be eliminating job positions. We have to have a plan ready to go that says what our policy will be to those whose job positions are destroyed because of technology. It almost doesn't matter what the policy is as long as it is clearly stated. We have to know how the

ROI will be calculated and whether that calculation is realistic. We have to know what capital appropriation dollars are available or will be considered and some perspective as to where we stand in the company pecking order. If we do not, then our proposals are a waste of everyone's time and effort and will only breed frustration. Once we can live with the ROI calculation and have a realistic view of what we have to do to get the capital dollars, we have to know how many dollars, plus or minus 30 percent, could be available. With those dollars in mind we can select the implementation area. Without them we are only building castles in the air.

9.3.1 Selecting the application

The first time we do anything new it really has to work. This is a fact of life. It seems as if the motto ought to be "If at first you don't succeed, update your résumé." No joke; it's reality. In selecting the first application of CIM materials handling make darn well sure that you can't fail. Do not look for the exotic application. You can spot an exotic in one of two ways. Either it has been a long-term unresolved and irritating problem or it is an application the salesman describes as being "interesting."

The very first step after you get the go signal for a proposal is to take a walk through the facility and list how things are done today, the equipment used, the paths taken, the people employed, and the working conditions. If it is a necessary operation and not overly complicated, we may have a "lay-down situation." Any OSHA compliance savings are guaranteed dollars; they are well known and the dollars are documentable. Unless you are just going to fire the current staff, everyone will be happy to see those jobs eliminated. No one will object because no one really wanted to do the jobs or even liked them in the first place.

For the first implementation you can reassign materials handlers and show avoidance of future expense. Later, as you move forward, you will have to balance head count saving with retraining costs, and the job will not be as easy. Target in on the ruptured duck area and avoid sacred cows. The ultimate rule in identifying application areas for state-of-the-art materials-handling equipment the first time is simplicity. The old saying "Keep it simple, stupid—KISS!" is all too true.

That we are going to presume success requires us to look ahead and identify all implementation areas. You need also to develop a checklist of the kinds of savings that are reasonably achievable with automated equipment. In concert with management and using its real (and, hopefully, stated) policies, you must develop weighting factors for each item on the checklist. The objective is to have in hand a listing of applications areas, some perspective of the current and future conventional expenses that are avoidable, and the priority in which you would like to address those applications areas. Please, don't be so naive as to believe that your priority list is the ultimate priority list. There is a management priority based upon political considerations, public relations, union negotiations, and just someone in authority having a gut feeling. The important thing is to end up with an agreed-upon prioritized list of acceptable applications areas that is written. Before you can move on to the technology selection, you must also get from management two final things: the gross dollars that have been budgeted and the criteria for appropriation approval.

9.3.2 Selecting technology

With the applications areas identified, the next question is what technology will replace current equipment. One does not fit the applications to preselected equipment; rather, one determines what equipment will do the best job for the applications. Most likely alternative solutions will be possible. Naturally, they will have to be identified and prioritized. In order to have viable selection criteria, one must know the equipment with which the technology will have to integrate now and in the future. There must be a strategic plan for the shop or warehouse. The technology to be selected must be in concert with that plan. It may also well be that, with only one application in view, a specific technology is preferred but, when the entire facility is examined, another is found to be better suited.

In looking at the applications area the first questions to be asked are: Is what we are looking at a stand-alone application? Should it be considered a stand-alone? If there is no advantage to integrate with another mechanized system—as, for example, transfer from a packaging area to the shipping dock—then the key consideration is what will do the job best. We may be able to do the job just as well by using conveyors, monorails, or AGVSs. If the application is stand-alone, conveyors may be the simplest and least-expensive answer. If, however, you are filling multiple trailers and the goods will be off-loaded from the conveyance device almost immediately, then maybe conveyors are not so inexpensive. If the main floor is going to be AGVS, then the incremental cost of the shipping dock area is negligible compared to that of an entire conveyor system. Of course, if shipping is to go in first, then you are going to have a few turns around the block with finance. Here the question is going to be how much of the initial computer hardware and software investment will be allocated to the initial implementation. If finance insists on 100 percent to the first implementation, there is little hope of ever finding a pilot area and justifying it economically.

You will have to know the costs involved in implementation. They are more than just equipment. Your other concerns are engineering studies, site preparation, facilities (and utilities) modification, work rules modification (including process plans and procedures), education, training, installation, initial monitoring, software modification, and project management. Regardless of what you submit to management, don't lie to yourself about the real costs. When you have put dollars against the cost items for the alternatives, calculate the ROI according to the ground rules you have been given. It is too late to try to have them changed. When you make a selection of technology, be clear about what technology you are selecting, which vendor and why. Yes, line up your ducks, but don't duck out on recommendations.

9.3.3 Planning implementation

Once you have identified an implementation point and decided on the technology to be used, you have to plan to put it all in place. The old saying, "We don't plan to fail but all too often we fail to plan" is so very true. A well-thought-out productivity improvement will fail if there is no action plan on how things will be done. This plan takes three parts—financial, physical, and people—and identifies the activities to take place. The plan is also a tracking document to assist the implementation team in evaluating its own efforts.

The first planning step is to put in place the company team to be responsible for the implementation. The disciplines represented should include plant engineering, industrial engineering, process engineering, shop operations, human resources, data processing, and, naturally, a project manager. Depending on the size of the eventual implementation, a representative of the facility operating committee may also be appropriate. This committee must be organized with a clear perspective of the short- and long-term tasks at hand. The meetings of the committee should be subject to a regular schedule and placed on the committee members' calendars immediately. The project manager has the difficult task of assuring that "juniors" are not assigned to the committee. Planning the implementation means developing a written plan of what events should occur and when they should occur. The representative of each of the functional areas should indicate which steps he or she will be performing and be responsible for. For each task step so identified there should be a narrative of what will occur, the resources required, what things will have to be completed prior to this task getting started (dependencies), and how long the task will take.

Once this rather laborious process is done, the entire project committee "puts it all together." It is suggested that a PERT chart of events be assembled from the tasks showing all the dependencies. With the estimated time for each task an overall initial time line can be determined. Prior to any submission of this version, the plan should be examined for its reasonableness. Keep in mind that there is a natural tendency to be overly optimistic. Since part of the plan will involve outside contractors and the equipment vendors, the plan should be reviewed with them also. With the time and efforts agreed upon, the events should be dollarized so that finance knows what funds will be required and when. What is being established is a working budget that is related to an approved (hopefully) project and capital appropriation. Once the plan is management-approved and the difficult task of vendor negotiations has been completed, the clock starts ticking. The project has passed the evaluation, recommendation, planning, and approval stages and is now in the execution phase.

9.3.4 Controlling implementation

A project plan is not a volume to be placed on the shelf to catch dust; it is a tracking mechanism. No one is foolish enough to believe that all plans will be executed right on time with military drill team precision. It doesn't happen that way. Murphy works for us all. Probably the simplest method of making project progress visible to all involved is a posted Gantt chart of time and dollars budgeted vs. actual. It might be advisable to break it down by task and groups of tasks. Since project members are responsible for their tasks, the project committee can determine when additional resources are needed and why. Most important, the committee has the responsibility of reporting project progress to management. The project committee cannot just be the carrier of good news but must be prepared to tell management when targets are not being met, why they are not, and what additional costs are involved. It is not always an inside problem. Your vendor may be having problems with one of his vendors. If you cover it up, then you are responsible for missing the dates. If you identify the problem to management, although the time line has not been shortened, at least no one is in the position of having to make excuses. Most managements are reasonable; they know that you don't have total control over everything. They do require honesty and information.

Effective project control means that the schedule is not a tablets-of-stone document, neither is it written in sand. A rather common problem is not meeting dates. Although we must know why dates are not adhered to, we have to put in place a mechanism that encourages adherence. A major milestone event with a target several months out will probably not be achieved. Major milestone events should be targeted loosely. If we place hard but realistic dates on the individual events in between and achieve those dates, the milestone will fall in without further effort. Setting the targets for the individual events, or inch pebbles, takes a lot of work. Tracking them, keeping on top of people, is more work and not likely to be an enjoyable task. There are always day-to-day problems that get in the way, and few people want to be reminded of the project tasks. To control the project, you have to monitor all the individual efforts and see that they are done and done properly.

Control reports should be regularly submitted. The progress or lack of it must be fully reported. Performance to time estimates and dollar budgets must be provided. If you discipline yourself to do your job right, you can discipline others to do theirs. As Benjamin Franklin said, "A job worth doing is worth doing right."

9.3.5 Integrating the systems

American companies are implementing automation technology constantly. That is not because we want to but rather because we have had to. Once, a few years ago, it was pointed out that the United States had, although the largest, the oldest industrial base. We were totally noncompetitive. Financial managements had too keen an eye to Wall Street performance and created an orientation to maximize short-term profit and ignore long-term survival. As we now start to reindustrialize, and we are doing that, we are creating what are called islands of automation. Propagating the shop with areas of high productivity does not make the productivity happen. The shop is paced to the bottleneck. If we are moving to a higher level of automation, there must be a grand design—a strategic and long-range plan. Automated materials handling is an ideal unifier linking these automation islands with automated transfer. The pace of at least one part of the shop will be significantly changed. The automation program must be so planned that no automation island remains an island.

The classic automation strategy is to attack the materials handling at the raw materials end. It is a big project with high potential. Consider, however, going after the back end—finished goods and the shipping dock. When the front end moves rapidly and everything else is at the former rate, the inevitable result is a buildup of inventory. We do our job very well, but we are giving someone else a real headache. If, however, we work from the end of manufacturing cycle back to the beginning, we are pulling inventory along and eliminating queues. This approach is more likely to gain the cooperation of materials management rather than its antagonism. Back- to front-end implementation is compatible with the material requirements planning (MRP) or just-in-time (JIT) programs. Although there will be changes in shop paperwork and standard operating lead time, the change to a routing standard will happen only once. If your implementation is in the middle of the manufacturing cycle, link as many pieces of equipment together as possible.

Regardless of where your automation implementation begins, have a total facility plan. If the company cannot commit to the whole route over the long haul, it is unlikely that any implementation will ever be truly effective.

9.4 Conclusion

Let's wrap this up in what is hopefully a neat bundle. Modern technology for materials-handling equipment does exist. We have seen that we have a respectable stable to draw from. The technologies that exist today have a track record of reliability. We are dealing with equipment that offers a choice of vendors. There is no single dominant force in the industry. It is also reasonable to believe that no single technology has reached maturity, so we can expect advances yet to come in the equipment available and possible new equipment of a type not yet on the market.

We can recognize that we are dealing in major capital appropriation items and major projects. We have to do a superior job of project planning, implementation, selection, and project control. The quality of our efforts relates to not only our personal livelihoods but also to the success of the company. The days of ad hoc planning and blindly charging ahead are gone and not likely to return. We have to be professional in all our endeavors.

We have an opportunity to make major contributions to our companies. The job will not be easy, but it certainly will be interesting. Moving to CIM is a necessary task for both our companies and American industry.

9.4.1 Benefits of CIM

What are the benefits of moving into CIM as applied to materials handling? The nature of the equipment makes one benefit obvious: we get the materials-handling job done faster. If that were the only benefit, it might be sufficient, but it is only one item on a very respectable list. Let's look at the others, the CIM benefits.

The job is done with a higher degree of reliability. Let's not kid ourselves; computer-controlled equipment does not work faster than people at their optimum. But it is more consistent and less prone to error. Therefore, the total reliability is greater. There is an overall increase in capacity. Computers and computer-controlled equipment do not take coffee breaks or have all the other people-oriented disruptions, and most of all they do not object to working a full day—24 hr. With the ability to add a second and/or third shift on demand, we are looking at a flat total capacity increase. Even if we are running three shifts now and not considering weekends, we gain all the break time and shift turnover. Almost as an aside, dramatic reductions in materials-handling damage come with the system. Last, but not in any way the least, we are decreasing the labor-value-added dollars and, probably immediately and inevitably, the total value-added dollars. Squeezing every dollar of cost out of a product while maintaining or enhancing product quality is the major benefit of CIM. CIM should also be called survival.

9.4.2 Potential developments

What is yet to come? Although no one knows the future, some trends are visible. We are moving toward a fewer-people factory. The skill of the workman with a tool is something that is disappearing. The image of a man (or woman) pushing a cart is also an anachronism. We will push no carts in the factory of the future. What is not here now is the discrimination of what ought to be. There is no recognition as to what industry wants and doesn't want. If we don't know where we are going, why bother to get there?

What may yet come is an expert system that will electronically examine the collection of parts for a shipment, determine the container and the method and sequence of container loading, and call for the carton, packing material, and parts to be routed to an appropriate workstation in the appropriate sequence. With the automatic generation of packing and shipment documents which we now have, the role of the dock worker (and it would be in the singular) may well be reduced to the remaining paper sorting, visual inspection, sign-off, and security.

What we will also see is a wider variety of quick-connect manipulators for robots and a reduction in manipulator costs. For too many years the robot industry has been poised for a leap. It is a promise unkept. Quick-connect manipulators may be the advance that pushes it off the edge. When it finally makes the leap one way or another, we will all hear about it.

Regardless of what comes, it will be interesting. It may well be not what we expect, but it may be something we haven't considered or even conceived of yet. Our times will indeed be interesting; they will also be exciting.

CHAPTER

10

The Role of Quality in CIM

Malcolm Macfarlane

10.1 Introduction

We in the United States don't understand quality. The symptoms are everywhere. Managers devote large sums of money and great amounts of effort to quality campaigns but never say exactly what quality is or how to get it. Factories spend increasing percentages of budget on inspection while they preserve old methods of manufacture. Workers are still often measured by production rate or paid on a piece part basis. Control charts are tacked everywhere, but no one can explain exactly what they mean or how to use them. Companies pride themselves on their service organizations. Our trade imbalance with other countries is growing, and we talk of economic sanctions and quotas rather than the fundamental problem of building competitive products. Designers blame marketing. Factories blame design. Field service blames the factory. Management blames the workers. Customers are the only ones to blame management—*and the customers are right*.

We also don't understand what quality has to do with computer-integrated manufacturing (CIM). SPC, MRP, JIT, CAM, CAE, CAD, ATE, CAPP, and other acronyms are discussed in the popular press as if they were independent concepts, like pills to be taken to cure productivity disorders. Automation is justified by worker replacement. Products lose their identity in the field. Major product recalls are issued to repair items *known to be defective when shipped*. The return on investment (ROI) used to justify an automation project never seems to be there, once the project is implemented.

CIM (a cohesive database of manufacturing information) provides an opportunity to implement, promote, and preserve the concept of quality throughout a manufacturing organization. CIM provides a focus for quality—a mechanism by which the costs of quality can be tracked and progress toward their reduction can be measured. But there's a problem in getting there from here. All of the applications implied in the acronyms assume that elements of the manufacturing process are performed without error: orders are complete; designs are exact; and processes are competent. If those assumptions are incorrect—and in most factories it's likely they are—then the entire process, automated or not, decays to a series of firefights and finger pointing.

It's not like that in Japan. Automation of all forms has enjoyed tremendous success, making Japanese products among the world's most competitive. It's well known that, in 1985, the average Japanese automobile had a $2000 cost advantage over a comparable U.S. product, without consideration of its hassle-free useful life. The single most important reason why the Japanese get so many benefits from applying the automation that was developed in the United States is that, in their factories, the assumptions that form the foundation of the automation hold.

Thus, implementation of effective quality programs is right at the top of the productivity pyramid. Without those programs, CIM benefits will remain frustratingly illusive. Building a CIM system without considering the central role of quality in its design is like building an airplane in the basement: it may be functionally correct, but it'll never fly. A clear understanding of quality and its role in the manufacturing process is essential for successful CIM implementation.

10.2 Quality Defined

Quality is neither a slogan nor a fuzzy ideal. It has operational meaning and calculable cost. Misunderstanding of quality in the United States is perpetuated by our quick-fix mentality, by our focus on product rather than process, and by a variety of authors who approach the concept of quality from fundamentally different directions. Quality is defined in several ways:

- "Goodness, excellence, superiority"
- ". . . a relationship between man and his experience"
- Continuing match to customer expectations
- Conformance to standards
- Product uniformity about the target
- Losses a product imparts to society after it is shipped

The first definition, from Webster's *New World Dictionary*, is the one advertising agencies have in mind when they describe "quality products"—

cars, beer, and the like. This definition is not operational or measurable, and it is of absolutely no use in a discussion about quality in manufacturing. Unfortunately, it is this vague idea that most managers have when they discuss the topic, which results in their misunderstandings about the "prohibitive costs of quality."

The second, somewhat metaphysical, definition is from Robert Pirsig's *Zen and the Art of Motorcycle Maintenance.* It comes closer to defining quality in an operational sense, because it involves experience. Experience creates requirements and expectations, and an adequate match to those is a more useful definition.

The third definition, "continuing match to customer expectations" comes from W. Edwards Deming's *Quality, Productivity and Competitive Position.* It implies a process cycle which measures customer expectations, establishes standards, manufactures products, delivers products, maintains products, tests results against expectations, establishes new standards, and repeats itself on a continuing basis. It implies a quality system involving the whole company.

"Conformance to Standards" is the definition at the heart of Philip Crosby's *Quality is Free.* This definition is operational, and it can be absolutely measured in a discrete process such as completing a customer order, design drawing, or purchase order correctly. Each process has discrete results which are either correct or incorrect, and its quality can be measured by the percentage of correct events occurring within a period. The definition is limited, however, in its allowance for variation about a design specification. If standards include an allowable variance to the specification, thus artificially constructing limits which define correct (within limits) and incorrect (outside limits), the factory is motivated to permit such philosophies as allowable quality levels (AQLs), which result in a sloppy adherence to the design.

The fifth definition, from Sullivan's article "Reducing Variability: A New Approach to Quality," is that employed by most Japanese manufacturing firms. Product uniformity about a target is directed at processes which have variable results. The definition produces a philosophy of constant improvement which does not encourage fuzzy thinking about "allowable" defects. It is directly measurable, and it results in a different approach to quality than "conformance to standards." The discrete processes are found in the limit of the definition where there is no variation; thus the definition is of little use in discussing them.

The last definition is Dr. Genichi Taguchi's. His quadratic loss function is the first operational joining of economics of quality and variability of product, which allows designers to actually calculate the optimum designs for their products based upon cost calculation and experimentation with the design. This definition provides the most perspective into the issue of quality. It is the one a CIM system should ultimately fulfill, because the system potentially contains all the information required in its manufacturing and field data to calculate the product life cycle costs to the manufacturer and society.

From the six definitions emerge quality definitions addressing two distinctly different types of processes: discrete and variable.

- Discrete processes produce discrete results. They have no variability—the results are either right or wrong: the correct change from a $10 bill, the correct number of parts ordered, the correct component in the bill of materials, or the correct instructions transmitted to the factory floor.
- Variable processes produce variable results. Their successes are measured in the variance of the results from the target values established by design.

10.3 The Cost of Quality

The costs of quality in U.S. manufacturing are what are keeping the United States from regaining its status as a world class manufacturer. The cost of quality for most U.S. firms is more than 20 percent of gross sales. Quality cost CQ to the manufacturer can be expressed by the following formula:

$$CQ = \Sigma EC_i + \Sigma MRS_j + \Sigma II_k + \Sigma FR_{ii} + \Sigma OO_{jj} + \Sigma MS_{kk}$$

where EC = costs of engineering changes for the period including design, overhead, retooling, documentation, and training

MRS = costs of rework and scrap including material, labor, supervision, documentation

II = interest on inventory required to support the error rate, work in process held up by errors, and defective inventory not yet detected

FR = field repair facilities, parts, and personnel

OO = organizational overhead required to support manufacturing and field functions required to compensate for defective product—repair centers, hotlines, and refurbishment facilities

MS = lost market share on defective product lines (less measurable, not counted in the 20 percent of gross sales numbers, but real)

The achievable cost of quality in a well-run factory is about 2 to 3 percent of gross sales. Every improvement of product quality drops right through to the bottom line, thus giving rise to Philip Crosby's claim that "quality if free."

The problem is that few managers believe Crosby's claim because they don't understand it. Because of confusion caused by the Webster definition, conventional wisdom in the United States holds that "quality" connotes "expensive"—expensive materials, expensive tests, expensive repairs. This view is likely due to the cultural beliefs fostered by television advertising, which for decades has associated "quality" with "expensive," without regard for whether the so-called quality product would actually perform as expected. How often have you seen a "quality" automobile in the repair shop, for example? *Consumer Reports* measures this on all cars. Contrast the incidence of repair record with the so-called quality automobiles made in the United States with the incidence of repair of the "low-quality" imports from Japan.

In fact, the belief that quality implies expensive is exactly counter to what is found to be true in Japan, where quality is found to be the least expensive way to achieve the standard. In his definition, Taguchi points out that quality is a measure of cost to both manufacturer and society. Value added by the manufacturer beyond what is required by the functional requirements of the product is regarded as cost, whereas the inconvenience caused customers by products failing to meet their requirements is also a cost. The curve in Fig. 10.1 illustrates this function.

Taguchi shows that the loss function L at any point y is expressed by

$$L(y) \approx k(y - T)^2$$

where k is a constant and T is the target value. $L(y)$ can be calculated for a specific value of y through the cost of quality calculation for the manufacturer (Fig. 10.1) and an estimate of the cost to society for the specific variance. Thus k can be calculated as

$$k = \frac{L(Y)}{(Y - T)^2} = \frac{L}{\Delta^2}$$

With a value found for the constant k, the loss due to variation can be found for any y. Taguchi shows further that, for multiple products, $(y - T)^2$ is the mean square of error and the loss function becomes

$$L = k(\text{standard deviation})^2$$

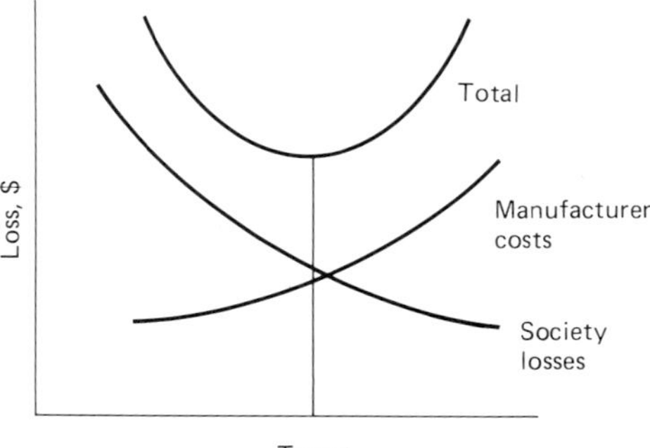

Figure 10.1 Quality as a measure of cost to the manufacturer and society.

Thus variation from target is the antithesis of quality, and it must be reduced whenever k is large.

This union of process variance and cost provides product designers with a powerful tool. The economies of a design can be calculated, and a design which combines the lowest cost with the highest quality can be chosen. Data required to estimate k accurately should reside on a complete CIM system. See the Taguchi reference for a more detailed discussion of this topic.

10.4 Quality in the United States

There are two major thrusts for quality in the United States, and they are on converging courses. The first is the *cultural thrust*, which concentrates on building a business culture to produce high-quality products and services. The second is the *methodological thrust*, which focuses on the explicit methods of exploring and reducing variance in manufacturing processes.

The two thrusts are interdependent. The cultural approach creates an environment receptive to improvement of quality—"doing it right the first time." Unless real means of improvement are forthcoming, however, the factory can return to past means of coping, with a cynical attitude toward "quality" programs. On the other hand, the methodological approach provides a variety of effective means of actually improving processes. But without an environment ready to adopt them, those means are a waste of time.

Quality methods are not a quick fix, as they are regarded in many companies, and fostering a receptive culture can be just another brand of quick fix. Both approaches require understanding and responsible management and workers to implement quality effectively.

10.4.1 Cultural approach

Crosby's approach in *Quality Is Free* and the more recent *Quality without Tears* involves 14 implementation steps to higher quality:

1. Demonstrate management commitment.
2. Form a quality improvement team.
3. Establish real quality measures.
4. Calculate the real cost of quality.
5. Raise quality awareness throughout the company.
6. Implement corrective action as opportunities surface.
7. Establish a committee for a "zero defects" program.

8. Train supervisors in all aspects of the program.
9. Have a "zero defects" day, a corporation-wide event.
10. Establish formal goals for each group.
11. Remove causes of errors in processes.
12. Recognize outstanding acts or contributions.
13. Form quality councils to meet and review regularly.
14. Do it all over again, regularly.

These 14 steps are a prescription for establishing a culture sensitive to quality in a company: quality of any type, whether it is a service, such as taking an order correctly, or a manufacturing process, such as building a circuit board.

Deming also has 14 steps for management to implement in the pursuit of quality:

1. Create consistency of purpose toward improvement.
2. Adopt the new philosophy.
3. Cease dependence on mass inspection.
4. End the practice of awarding business by price tag.
5. Find problems by continually working on the system.
6. Institute modern methods of training on the job.
7. Institute modern methods of supervision; shift from production to quality.
8. Drive out fear so everyone can work effectively.
9. Break down barriers between departments.
10. Eliminate numerical goals, posters, and slogans.
11. Eliminate work standards prescribing numerical quotas.
12. Remove barriers preventing pride for the hourly worker.
13. Institute a vigorous program of education and retraining.
14. Create a structure in which management pushes every day on the first 13 points (keep doing it).

The two lists have a great deal of common ground. Certainly management commitment and continued renewal of purpose are common to both. Deming's list is more useful in detailing specifics which will contribute to developing a culture truly sensitive to quality and thus productivity of manufacturing. The most important common ground is that good or poor quality is more than 80 percent attributable to management rather than employees.

10.4.2 Methodological approach

Japanese authors other than Deming (who can be considered Japanese) spend almost no time discussing the cultural prerequisites of a successful quality program, because the Japanese already have such a culture in place. One of the hardest things to overcome in the United States is worker and engineer skepticism that a new program will work at all. In Japan, the traditionally naive and eager attitude of engineers and workers allows the easy implementation of extensive programs. The Japanese authors assume that management and workers understand the vital nature of quality in their work and concentrate on getting effective tools into their hands.

Dr. Karou Ishikawa's *Guide to Quality Control* is a classic book of quality methodology. The book is written specifically for *factory workers* and covers topics like data collection, histograms, cause-and-effect diagrams (the famous Ishikawa "fish"), check sheets, Pareto diagrams, control charts, scatter diagrams, probability paper, and sampling. The techniques of this book are for the factory worker to improve the process he or she is responsible for. This is possible only in companies in which all employees understand and promote quality improvement. Taguchi's *Off-Line Quality Control* puts economic-based design analysis techniques in the hands of engineers responsible for product design, as previously discussed. Both of his books concentrate on "how to" rather than "why bother."

The methodological school augments the cultural school with operational methods for improving quality of both discrete and variable processes. Techniques of problem identification, experimentation, and monitoring are becoming well known in the United States but have not yet achieved widespread use.

10.5 CIM's Role in Quality

CIM implementation provides an opportunity to build quality identification, experimentation, implementation and monitoring capabilities into the information system. CIM facilitates the automation of the most effective quality techniques and has the capability to put controls where they belong: in the hands of people building the product. It is not a panacea for manufacturing or a shelter from ultimate management responsibility. CIM does not remove responsibility from management to commit to, implement, and continue a company-wide quality program. Any company seriously interested in implementing a CIM system should have two quality programs continually in progress: a discrete process quality control program such as that espoused by Philip Crosby in his *Quality is Free* and a variable process quality program to implement methods which reduce variation in manufacturing processes.

One of the major goals of the discrete quality program is the identification of the processes most susceptible to discrete error, such as order entry, invoicing, purchasing, and inventory, and the construction of a series of internal verification processes which validate the information being transmitted by those processes. Once the information sources, transmission channels, verification procedures, and sinks are identified or constructed, the system of information exchange can be simplified and made more efficient with CIM.

The discrete quality program does not cease with implementation of some of its techniques in CIM. The discrete quality portions of the CIM system are spin-offs from the discrete quality program, and they will be improved as that program continues. The ad hoc quality committee established as part of the process should always have as part of its agenda the consideration of the information system and how it can be improved to obtain zero defects.

The most common attempt at quality programs in a CIM system is the monitoring of variable processes. Fueled by the microprocessor learning curve, technology has produced a wide variety of intelligent sensors to monitor virtually all elements of a manufacturing process. Unfortunately, many factories confuse automated inspection with statistical process control and use the techniques for passing or rejecting product without studying or improving the variance of the associated processes. Building this philosophy into a CIM system and its operators makes it harder to implement a true statistical process control system in the future.

As more and more major manufacturers require quality information from their suppliers, the companies that have genuine automated statistical process control will pull ahead of those that haven't it. There is no question that the trend among major corporations is toward fewer and better suppliers. Suppliers who can prove their superiority on a consistent basis without investment by their customers for verification will rapidly emerge the winners.

Each major manufacturing activity can benefit in unique ways by combining the two quality programs with the capabilities of a well-designed CIM system. The remainder of this chapter addresses the role of CIM and quality in design, process planning, production planning, materials, factory floor, factory support, and field support.

10.5.1 Design

The obvious place for CIM to have deep roots is in the design function of a factory. The computer-aided design, engineering, and manufacturing (CADD/CAE/CAM) systems sold today automate the design function and transmit the finished design to the manufacturing engineering and production departments. Some of the obvious quality advantages of CIM in design are the reduction of multiple occurrences of the same information and the reduction of the number of interpretation steps between the factory floor and design. Both factors are the source of discrete quality problems in the traditional plant. With CIM, the bill of materials can exist in one place and so can the specifications, drawings, and other documents which describe the product. Transmission of the information to the other groups which need it can become a simple matter with CIM.

The goal of the discrete quality program in the design area should be to identify the users of the design information, such as manufacturing engineering, the factory floor, purchasing, and scheduling, and to ensure a clear flow of information to those functions from design. There is a large body of knowledge in this area; it is a major topic in most CIM sessions. A discrete quality program can help the design people focus on the critical sources of error and guide the implementation of the CIM system by addressing the problem through Pareto analysis.

With CIM, the ability of the design department to profit from its past mistakes has never been greater. The traditional flow of information has been outward from design to the rest of the company. Through advances in CIM technology, there is a great potential for information flow the other way. Design departments that take advantage of the factory and field as extensions of their research will profit enormously. Reasons for this statement derive from Taguchi's philosophy of design.

Variability has been shown in the preceding discussion to be the economic antithesis of quality. As explained by Taguchi, three major variables cause variation among products of the same design. These variables are:

1. Outer noise: variations in environmental operating conditions such as temperature, pollution, humidity, and vibration
2. Inner noise: the deterioration of components of the products with age such as bearing wear and resistor losses
3. Variational noise between products: the variation in response between products of the same design, given the same inputs

The most effective means of reducing those effects is through effective product design. The three stages of design are:

1. *System design.* The functional design of the product, which requires expert knowledge of the subject and is almost purely an interaction between the designer and the organization requiring the product.
2. *Parameter design.* The determination of the optimum level of parameters of the system. Some parameters will have a large effect on the final performance of the system, and some will have little or none. This step results in a combination of parameters which dampen the influences of all noise sources while producing the required result.
3. *Allowance design.* The determination of the tolerances of the param-

eters developed in stage 2. Parameters found to have a large influence on the performance of the system are assigned more narrow allowances than those with small influence.

For a detailed discussion of this philosophy and its enabling techniques, read Taguchi's *Introduction to Off-Line Quality Control*.

A CIM system maintaining manufacturing and field histories of products can supply information needed by designers to investigate sources of noise as the product is manufactured and used. This information allows the designers to improve the quality of the design and produce more uniform product under an increasingly wider variety of circumstances.

A goal of the quality program in design must be to identify the sources of information useful to design engineers in understanding their products and ensuring the information is built into the CIM system. The only way this goal can be met is for the designers to participate in sessions of the statistical process control program with the manufacturing engineers. Over time, their roles will merge, so that design, manufacturing, and the field become part of one long continuum of product flow. Only this final organizational structure will permit smooth growth of product quality improvement without artificial barriers of organization or attitude.

10.6 Process Planning

The process planning function can enjoy the same benefits of combined CIM and discrete process quality programs as the design function. That is, the quality program can identify the major sources of error in communication and execution of the process plan, and methods involving aspects of CIM can be put in place for ensuring that the chain of information from the process plan to the components of the manufacturing process is unbroken. Other CIM functions can monitor the progress of the process steps and ensure the implementation of the instructions.

By applying techniques learned in the statistical quality program, process planners can study results from the factory floor, monitor process improvements, and identify the processes which are both controllable and competent. A CIM system which makes the process parameters available to the planners is necessary for the efficient recognition of significant changes in the processes and the subsequent alteration of plans.

A computer-aided process planning (CAPP) system implemented with CIM can monitor and verify the control and capability of the processes on a continuing basis and alter process rules dynamically on the basis of information from the floor on the current state of the factory. A generic, knowledge-based CAPP system could be the source of revised plans which could be automatically implemented if the CIM system extended cleanly from the plan through to the process. A variant CAPP system would have to be made aware of alterations in the processes involved in old plans and revisions carried out to reimplement the historical plans.

One of the major goals of the statistical quality program is to identify the information required for process planning and make it available. CIM becomes the enabling mechanism for this, as it is for the design.

10.7 Production Planning

In Oliver Wight's *Executive's Guide to Successful MRP II*, 20 major steps of a detailed implementation plan are listed. Of those 20 steps, 5 (one quarter!) are required because of potential discrete errors in inventory, bills of materials, lead times, software, and process plans. Many minor steps in the other major items also are concerned with *validation of things which should not have to be validated*.

Errors in production planning are primarily of the discrete nature, and they can be greatly reduced by a combination of a discrete quality program and CIM. A number of companies have implemented manufacturing resources planning (MRP II) with varying degrees of success. The MRP II system is only as good as its inputs, and errors in customer orders, purchasing, inventory, bills of materials, routing, and the like can destroy the effectiveness of the system and the confidence and careers of the people who implemented it. The manufacturing resource planning process should be a primary target for scrutiny by the discrete quality program. Errors of input, translation, communication, redundancy, measurement, and the like must all be identified and corrected. It was recently noted that the Japanese are "enthusiastically importing MRP II" and having very little problem in its implementation. That's because the Japanese have very little problem with defects in their information processes.

The MRP II portion of the CIM system must provide easy access to information for its validation and interpretation by management. A strong work-in-progress (WIP) tracking component of the system can provide a constant check of reality against expectation and give earlier warning when the expected or modeled production does not match what is actually being built. Bar-code readers with simple system interfaces, along with a variety of other techniques ranging from flow meters to photoelectric cells, provide the technology to gather sufficient information to accurately measure factory production against plan. The quality program must have as a goal the identification of the places in which such measurement will be most effective.

A "good factory model" such as Xerox uses in monitoring its production is extremely valuable in highlighting problem areas before they become critical.

10.8 Material

The primary concerns about material are quality, timeliness, and price. Price often becomes the driving force because the vendor choices are made by purchasing personnel who are measured by their financial effectiveness in obtaining the requisite material. Unfortunately, with this orientation, quality and timeliness suffer, often with much greater financial impact than the amount saved by buying from the cheaper supplier.

The CIM system must track the quality history of the purchased material from each vendor to determine vendor control and capability. If the supplier is observed to be out of control, incompetent, and not improving either characteristic, then it's time to switch vendors. A complete CIM system should be able to link purchased material to finished product to measure suppliers' material through the product life cycle.

One of the most important projects for any manufacturer to embark on is a program to improve supplier quality. Ford, GM, Xerox, IBM, and other major companies are helping their suppliers achieve higher quality levels and insisting they achieve those levels to win contracts in the future. An increasing number of companies now require their suppliers to supply evidence of their products' quality and constant improvement in the process used in product manufacture. The evidence is generally in the form of control charts and other statistical information. Of particular interest is the Cp_k index, which measures the process variation around the target. It is the ratio of the width of the variation defined by the specification and the width of the actual process. For example, a specification for a variation of 3 σ and a process variation of 3 σ would have a Cp_k of 1.00. Constant improvement of Cp_k results in constant improvement of product quality. Tracking Cp_k of the critical manufacturing processes is another candidate for the ideal CIM system.

If CIM is to realize its full productivity potential, the material entering the manufacturing sphere it controls must be defect-free. In a recent study of 159 Japanese subsidiaries with factories in the United States, Dr. Martin Starr of Columbia University's Department of Operations found that those factories "cannot implement just-in-time manufacturing in the United States." The reason given is the variability of quality and quantity of the materials received from their U.S. suppliers.

10.9 Factory Floor

The most important effort directed at the quality of the factory floor is statistical process control (SPC). It allows workers to gain control of and improve the quality of the products they build. It builds a sense of teamwork among a wide variety of employees. It increases interest and involvement.

Finally, it is a way to make a process better and keep it that way. The major goals of a factory-wide SPC program are to:

- Identify the processes which must be controlled to build the product
- Isolate the parameters of the process found critical in determining process variance
- Establish values of those parameters consistent with variability requirements
- Develop standard means of monitoring the processes through their critical parameters

The initial implementation of this program may be in manual measurements and control charts; the final implementation is integrated into the CIM system.

Implementing SPC on the factory floor should be one of the first steps in implementing a CIM system extending to the shop floor. CIM quality implementation starts on the floor; it replaces the current inspection equipment with online gages, vision systems, particle sensors, flow meters, thermometers, functional testers, ultrasound transducers, and a wide variety of other microprocessor-based technology capable of hundreds of readings a second. Choice of technology and its placement in the process is dependent on the criticality of the operation and the importance of the variable being measured. A noncritical variable does not need monitoring. The CIM system eventually monitors the critical parameters of each process as determined in the SPC implementation and thereby ensures the processes remain competent and in control.

Companies specializing in implementing SPC programs are becoming more common and available as recognition of SPC's pivotal importance to productivity increases. Because SPC is so critical to the success of a quality system in CIM, the major stages of its implementation by several such companies are described below. SPC is implemented in five major stages:

1. Orientation
2. Education
3. Investigation (pilot project)
4. Implementation (pilot project)
5. Expansion

The orientation stage involves briefing factory management on the SPC program. What benefits can management expect from the implementation? What is expected of it and its people during that process? The goal of this stage is a management commitment of active participation by itself and its people. If a program of discrete quality is already being conducted at this factory, commitment is easy, because SPC relieves management frustration with applying the discrete techniques to the variable problems on the factory floor.

In the next stage, engineers responsible for the processes on the factory floor are educated in the statistics of problem identification, isolation, and removal. Techniques taught are practical rather than theoretical; they are geared toward the problems of the production engineer. Pareto analysis, control charts, cumsum techniques, and experimental design are among the subjects covered. Labs every day are based on real data gathered in the plant. Following the instruction, engineers are able to examine and test any of their manufacturing processes.

During and after the education stage, engineers are encouraged to identify the most troublesome processes in the plant: those which have a high failure rate, low yield, unpredictable results, or other problems which add significant cost to the manufacturing process without adding value. Techniques learned in the class are directly applied.

One or two of the most visible problem areas are selected as pilot projects for SPC implementation. Within these one or two troublesome processes, engineers apply techniques of experimental design (often called "troubleshooting" to remove the stigma of the laboratory from the process) in development of hypotheses, the design of the experiment to be performed, and the actual implementation. Once the critical variables are identified and adjusted, technology is implemented to monitor them. A series of procedures are developed to do the monitoring to ensure the process remains stable at new settings. A manual is developed for the use of the shop workers, and the workers are trained in the monitoring of the process through the use of control charts or the CIM system if it is in place.

Once the pilot projects are recognized as successes, it becomes easy to introduce the same implementation steps to the other areas in the plant most in need of control. This implementation, like the discrete quality program, is a continuing effort for the life of the company if the company is to become and remain truly productive. We have seen that reduction of variability of process has a direct economic impact through the quality function. A continuous cycle of test, improvement, and test institutionalizes process improvement in the factory and thereby guarantees continued productivity growth.

After critical parameters are identified and instrumented, a CIM system can be extended to monitor the processes and alert the appropriate personnel when the parameters are drifting. The design of the CIM system, therefore, should make allowances for the addition of data groups and real-time analysis after the system is installed. The availability of this information to the design department is priceless because it gives the personnel hard results on the manufacturability and stability of their designs.

10.10 Factory Support

Maintenance and facilities data may contain parameters critical to the manufacturing processes. The airborne particulate count is critical to the yields of semiconductor manufacturers. The filtering of coolant is critical to tool life. Ambient light is critical to vision systems and human assembly work. A regular maintenance schedule may be important or unnecessary, depending on the equipment being maintained.

Both during the discrete quality program and the implementation of SPC, factory support functions critical to the process must be identified, instrumented, and monitored. The job of monitoring these variables can be allocated to the CIM system, again provided the system is designed to accommodate the variables.

10.11 Field Support

The field is an extension of the test laboratory. Products can be instrumented (like automobiles) to provide valuable information on their operation by actual customers. Repairs and returns must be documented and reasons recorded, which can be traced to quality failures in the manufacturing process, design, or material.

The advent of the bar code and similar marking techniques allows a manufacturer to uniquely identify and track the life cycle of individual products. The ideal CIM system should accommodate life cycle data as it develops over time and in that way build a base of information which can be revisited in later design cycles to determine better design, manufacturing methods, or materials.

10.12 Conclusions

Quality is integral to CIM. The quality of discrete processes, such as the correct completion of a form, and the quality of variable processes, such as the roasting of a peanut, can be monitored and controlled through a CIM system. Construction of such a system, however, has the prerequisite of the implementation of both types of continuous quality programs described in this chapter. If these programs are not instituted, the associated CIM system will not achieve its full potential.

Unless the United States improves the quality of its goods and services, it will become a second-rate country internationally. The distance we have to go appears great when compared with Japan, but it really isn't so far. The amount of effort we have to exert to cut our cost of quality in half is very small compared with what the Japanese have to do to cut theirs in half again. Through the use of innovative tools and techniques within a new environment of quality awareness and effort, U.S. manufacturers can close the gap at an accelerating rate. The most effective tools we have at our disposal are found in the union of CIM and quality.

BIBLIOGRAPHY

Crosby, Philip B., *Quality Is Free*, McGraw-Hill, New York, 1979.

———, *Quality without Tears*, McGraw-Hill, New York, 1984.

Deming, W. Edwards, *Quality, Productivity and Competitive Position*, Massachusetts Institute of Technology, Cambridge, Mass., 1982.

Ishikawa, Karou, *Guide to Quality Control*, Asian Productivity Organization, Tokyo, 1976.

Persig, Robert M., *Zen and the Art of Motorcycle Maintenance*, Bantam Books, New York, 1974.

Sullivan, L. P., "Reducing Variability: A New Approach to Quality," *Quality Progress*, July 1984.

Taguchi, Genichi, *Introduction to Off-Line Quality Control*, Central Japan Quality Control Association, Tokyo, 1966.

Wight, Oliver W., *The Executive's Guide to Successful MRP II*, Prentice-Hall, Englewood Cliffs, N.J., 1982.

PART 3

Planning for CIM

CHAPTER 1

Getting Started

Don Ewaldz

George J. Hess

1.1 Introduction

It's difficult to overestimate the potential value that computer-integrated manufacturing (CIM) (or is computer-integrated management a more appropriate term?) holds for businesses today. CIM offers firms an opportunity to streamline their operations, slash operating costs, drastically reduce production lead times, and compete with subsidized offshore producers.

Unfortunately, there's not a large body of knowledge to guide firms in implementing CIM. One reason is that so few firms are actually doing it. Many firms that could benefit from CIM don't seem to recognize the dramatic gains to be had. Others seem to be floundering about with disconnected elements of CIM but never coming to grips with the real issues. They're *not* getting started toward CIM!

A second reason for the lack of information on implementation may be misunderstanding the variations possible; there seems to be some belief that CIM is a "generic." All one need do is buy the box and plug it in! Of course, that's not at all true. There are as many versions of "CIM" as there are firms; each has unique characteristics, is subject to different influences, is composed of a different set of computer-based elements, and supports a different management style. All those characteristics, and more, go into determining what form of CIM is best for *that* firm.

However, there are some general recommendations which will help point firms toward developing and implementing CIM—getting started. In the following pages, we will discuss those recommendations: what senior management's role should be, how a firm should go about planning for CIM, and why incremental implementation is advisable. Finally, we will offer some guidelines for the firm in first selecting the various elements which make up CIM and then putting them in place.

1.2 The Prerequisites

There are a series of prerequisites which a firm must recognize and meet before a CIM implementation can be expected to be successful. That doesn't mean a firm cannot buy a very large computer and a host of software without meeting the prerequisites. As a matter of fact, if the prerequisites are not met, the firm is more likely to buy several computers and a great deal of software but still not achieve anything like computer integration! Meeting the prerequisites goes a long way toward preparing the firm to accept the streamlining CIM makes possible.

The first, and perhaps most critical, of the prerequisites is to have the commitment and ongoing support of top management. History is strewn with the wreckage of grand plans which went awry because managers lost interest in them or quailed when something went wrong. The firm is going to change—perhaps change a great deal—because of CIM. It would be unrealistic to think that everyone involved is going to be happy with the change. Each tradition, regardless of how costly or how inefficient it might be, is going to find a champion. There must be a real commitment on the part of senior management to withstand criticism, which is going to mount. There is also probably going to be some lag between incurring the costs for CIM and reaping the benefits. Some things are probably not going to work!

Senior management must *believe* in CIM. That doesn't mean blind faith or blank checks. It does mean ongoing support, commitment, and enthusiasm. The firms which have effectively implemented CIM have done so because the senior managers have had the courage to make a commitment to CIM and have stuck to that commitment. It's also important that the system designers keep senior management involved throughout the design and implementation of the system; it takes a lot more than faith and confidence to continue support in a vacuum!

The second most critical prerequisite for a successful CIM installation is the participation of all the user managers and the systems group in a central coordinating committee. The CIM system design must be the joint product of those who will use it and those who will operate it. (It's unrealistic to expect an integrated product from a fragmented creator.) The committee should be formed as soon as CIM arises as an issue. The concept of the committee is to create a "user-driven" system design with operations management as the dominant force. The members of the committee should act together to develop the strategic plan for the CIM system. The systems group should act as the coordinating function for translating the plan into the most advanced hardware and software. In other words, the members of the committee representing operations management should decide *what* the system should accomplish and the systems group should decide *how* to go about getting it done.

The leader of the coordinating committee should be a senior manager with a background in both systems and operations so he or she can understand the concerns and interests of both groups. The other members of the committee must be senior managers representing all of the user groups; the committee won't be effective if the responsibility is delegated far down the line in any one of the groups. It's even less likely to work if it's delegated far down in two or three groups, excluding the rest. *The responsibility for CIM must reside at the same level as the authority needed to make the changes necessary to implement the new system.* The committee members must have enough power across the various functions in the firm to make things happen!

The committee must not be allowed to assume dictatorial powers, however. Its function is to represent and serve the operating functions of the business. On the other hand, it should not be a "servant organization" to be subjugated to interests of a single strong operations executive. It should act as a coordinating organization that can lead with the "authority of knowledge."

Another prerequisite for successful CIM implementation is to build a systems group capable of making needed modifications, evaluating and implementing software, and otherwise supporting the system. It is sometimes possible to rely on software and hardware suppliers in the beginning. However, the suppliers generally aren't able to "get inside" the firm like regular employees, nor do they have a personal interest in the firm. They are not often objective, either. Selling their own products almost always supersedes answering customer needs.

The user firm itself should have the capability to do extensive tailoring of the purchased software, which will be necessary in most cases.

The systems group should also have the ability to design and implement special applications programs which meet the needs of the user managers. Some firms have gone so far as to custom-design a major share of their software. Given the proliferation and quality of software, that approach seems unnecessary and expensive. However, it *is* necessary that the firm be capable of supporting the software as changes and upgrades become necessary. A strong, capable systems group also helps integrate the CIM system into the firm, since the system is clearly a unique product of the firm rather than the work of outsiders.

The last prerequisite is to provide training in computer science to the firm's employees to build basic computer skills and remove, to the greatest extent possible, any fears of computers individuals might have. The training should start with a basic familiarization course. *Informed* people are more likely to become enthused with and use the tools CIM provides. The training should continue with a user's course to introduce the people in the firm to the data available and the means of obtaining it.

All employees should have the opportunity to learn how the system operates and how it can serve them in their work. "Mystique" may make the system charming, but it also severely limits the potential level of use. Training programs often provide a fringe benefit: they can become an important source of systems personnel.

1.3 Understanding the Firm as It Exists Today

Each firm is made up of a myriad of strengths, weaknesses, and traditions, some of which make sense and some of which don't. They must be recognized before the CIM system is finalized; the strengths must be highlighted and the weaknesses must be strengthened or corrected. Certainly the firm is going to be much different after CIM; the data that is going to be readily available—in usable form—could once have been gotten only by digging through mountains of detail. Managers are going to be making decisions with far more knowledge than ever before.

However, that only resolves the need for data; it doesn't of itself eliminate any structural shortcomings. CIM is not likely to resolve any managerial deficiencies. Poor operating policies are probably not going to be any less poor because of CIM. They may even become more obviously poor. Management must be ready to make changes when they are needed. In addition, the firm's senior managers should be aware of where resistance to CIM is likely to surface and be prepared to address it. It's not impossible that some personnel changes will have to be made if particularly strong reaction is offered.

The firm should be aware of what it is that has made the business viable and of what makes up the "dominant force" in the business. Sometimes it is the ability to serve a particular market need; e.g., prompt deliveries might be the firm's strength. Most firms have achieved short delivery lead times by maintaining extremely high levels of inventory. The CIM system must be designed to further strengthen the ability to support customer needs but to do so without the demand for high inventory levels that are funded with large amounts of working capital. The CIM system should accomplish the same end by expediting order entry and eliminating as much of the initiating paperwork as possible to shorten the time and effort between the customer's order and the shipping papers!

Sometimes technical innovation is a particular strength; CIM should enhance it. Maintaining the engineering database in computer memory means that the database is available to far more people at the same time. Tying in finite element modeling and a computer graphics package may eliminate a great deal of routine work and significantly enhance the firm's capacity for technical innovation. It may be determined that some other factor should be the dominant force. If that's the case—and it sometimes is—the senior executive group should determine what factor would be more desirable and develop a plan for achieving it. The CIM system must be designed to accommodate the new objectives.

The CIM system must be designed to solve *real* problems if it is to make the contribution of which it is capable. This means, of course, that the coordinating committee and the designers of the system must be aware of what the problems are. If operating cost is a significant problem, the system must provide a means of reducing it by attacking its cause. For example, if long production lead times are creating unacceptable inventory levels, the system should address the reasons for the lead times. The system cannot cure an unknown problem, unless through purest coincidence! Getting the most out of CIM requires that the firm know what it wants to get and what is necessary to get it. CIM is a powerful management tool, but it's not a miracle drug.

1.4 Developing the Plan

Getting started in the most effective manner means first determining what the firm is getting started at. An absolute necessity to implementing a CIM system successfully is to first develop a detailed picture of the end product. That doesn't necessarily mean the detailed picture produced at the beginning must be rigidly sought forever, nor does it mean that every detail must be defined. It does mean the plan should be complete enough to provide a direction to be followed and a reasonable set of guidelines as the plan develops. The plan should also set down aggressive, specific, measurable goals such as cost savings, improved responsiveness, or increased production capacity. Some examples:

- Direct labor costs reduced 30 percent
- Overhead reduced 25 percent
- Production lead time reduced 50 percent
- Engineering costs reduced 20 percent

The plan is particularly important because most CIM systems are installed piecemeal for several reasons:

- Putting in the entire system at one time is going to take a good deal of capital, perhaps more than the firm has available or, even more likely, the senior manager is willing to commit, even given enthusiasm for and belief in CIM. It makes more sense to go at it on a prioritized basis and install the elements which have a particularly good return and offer little risk. It takes some success to build credibility; by the same analogy, it takes just a little failure to destroy the credibility that might have existed. It just makes sense to start with the opportunities that don't put everything on the line, regardless of how desirable they might be.

- CIM will be met with some resistance, just as any other major change would be. The resistance may not be dramatic, but it will be there. Spoon-feeding change lets the people involved get used to it gradually. Again, it makes sense to start with something less threatening, as by first changing over something that wasn't very popular anyway or that wasn't a cultural issue.
- The senior managers in the firm, though they may be ecstatic about the CIM concept, aren't likely to want to risk major capital on it until it's proven—at least beyond the theory stage. There's a much better chance to get smaller amounts than larger ones until there's a good track record to justify the investment, regardless of how much publicity a concept is getting.
- *The firm must be prepared to accept CIM on an elemental basis.* It doesn't make a lot of sense—nor is it very good economics—to put a high-efficiency machine in the midst of a group of low-efficiency machines, especially if the former must rely on the support of the latter. That holds true whatever the anomaly; the elements of CIM build on one another.

The answer is to implement CIM incrementally, to put pieces in place one by one:

- Incremental implementation reduces period capital layout and also lets the implementer pick and choose his targets. The very chancy ones can be left until a strong credibility has been established. That also gives the implementer a chance to develop some experience in CIM, which will further reduce the risk of failure in the later stages.
- Starting with the less threatening elements of CIM lets the implementer "steer" the culture. Building confidence in the workforce can't be done quickly; it takes even longer if the introductory element costs a lot of people jobs. Start slowly and modestly; the more dramatic things can come later.
- The cost reductions gained from the initial installations can fund the later, more costly elements. Success brings success; a senior manager is more likely to sign off on phase 2 if phase 1 has produced a measurable and predicted profit, even if it's not a very large one.

Of course, this approach further emphasizes the need for a CIM strategy, a CIM plan. The plan is the mortar that holds all the parts together; it assures that everything is going to fit as it's added and no "sunk cost" will result from having to throw something away because something else doesn't match it. There are two points which must be made about the plan:

- The plan must recognize the needs of the business and be shaped to strengthen weaknesses and build on strengths. The goal should not be CIM for CIM's sake; the goal must be a stronger, more competitive firm.
- The plan must be under the control of the coordinating committee, which should judge the value of any changes in it. Changes will be made; there must be some order in which they are made.

The best approach to defining the plan depends on the plan itself, the needs of the specific firm, and the plan's incremental horizons. However, there are some general recommendations:

- The factory plan should be depicted as a scale model. The minor expense of constructing the model is more than offset by the advantages of seeing the plan in three dimensions. It's far more difficult to visualize the plan from a two-dimensional layout, not only for the board of directors but also for the engineers and technicians developing the plan. How often do rigging crews find the planned location for a machine tool has a column in the middle of it? Walk through the average factory and count the number of 12-in-diameter holes through walls—holes needed to load material into machines—and strange cupolas here and there in otherwise flat roofs!
- The information model should be defined somewhat as shown in Fig. 1.1, which depicts the interrelations of data sets and functions. The number of data sets and how they should be related obviously depend on the individual firm and the style of its management.
- The level of detail in the plan should be stratified. The nearer the planned time of completion, the more detailed the plan. If the event is to be completed soon, each detailed step of the event should be defined and scheduled. In events far off, only the general description of an event should be defined. Because any details developed too early are certain to change, they may unfavorably influence decisions to be made later when better data is available.

1.4.1 Incremental implementation

Earlier discussion has suggested that a number of criteria must be considered in deciding the relative priority of the many elements and that the criteria are not entirely economic. For example, it's very important to establish credibility early on. Regardless of the amount of enthusiasm and confidence surrounding CIM at the onset, a failed task can dispel a lot of it in a hurry, whereas a success can gather even more supporters than were originally about. What makes up CIM in a particular firm is unique to that firm; however, there are some aspects of CIM that are so often elemental parts of a system that they are almost givens and can be used as examples without distorting the unique properties of CIM in a particular firm.

The first concerns *factory arrangement*: the arrangement of the various production resources in a factory building, or, more often, buildings. It's typical of factories in the United States that they have grown without apparent plan. Figure 1.2 shows the result; equipment with related production uses is scattered throughout numerous buildings, sometimes even miles apart. The results of such geographic anomaly are clear:

- Armies of people—some moving material, some storing it, *lots* of people trying to keep track of it, and more than a few making it over when it gets lost. And all costing a lot of money.
- Fleets of forklift trucks hauling the material around, being maintained, breaking down, and also costing a lot of money.
- Acres of storage space, some high-rise, some computerized, some unmanned, but all expensive.
- Immense amounts of working capital, often more costly than plant, property, and equipment, needed to fund the inventory.

The modern response to the problem of factory arrangement has been to attack it with sophisticated control packages, usually computer-based, al-

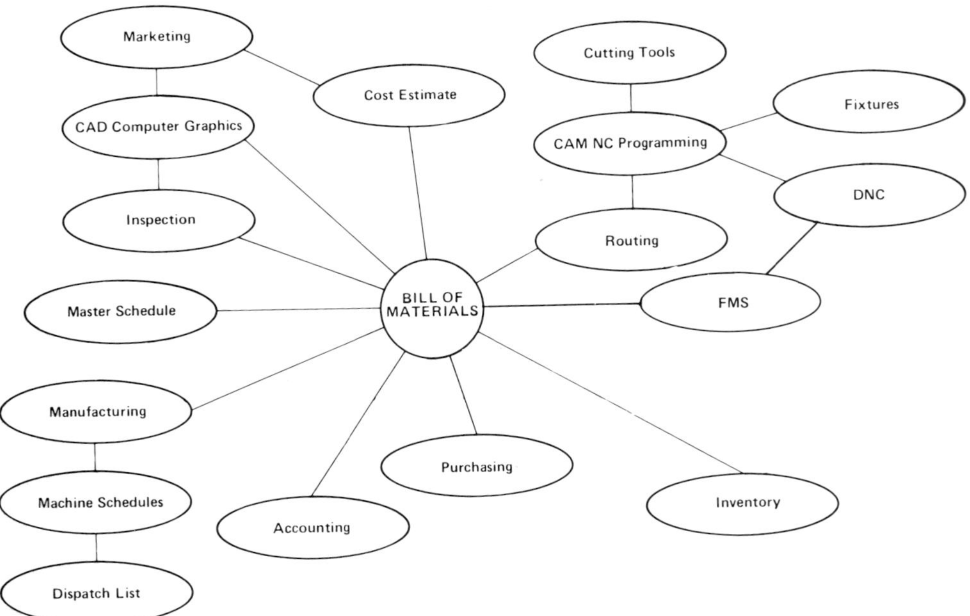

Figure 1.1 Database file structure.

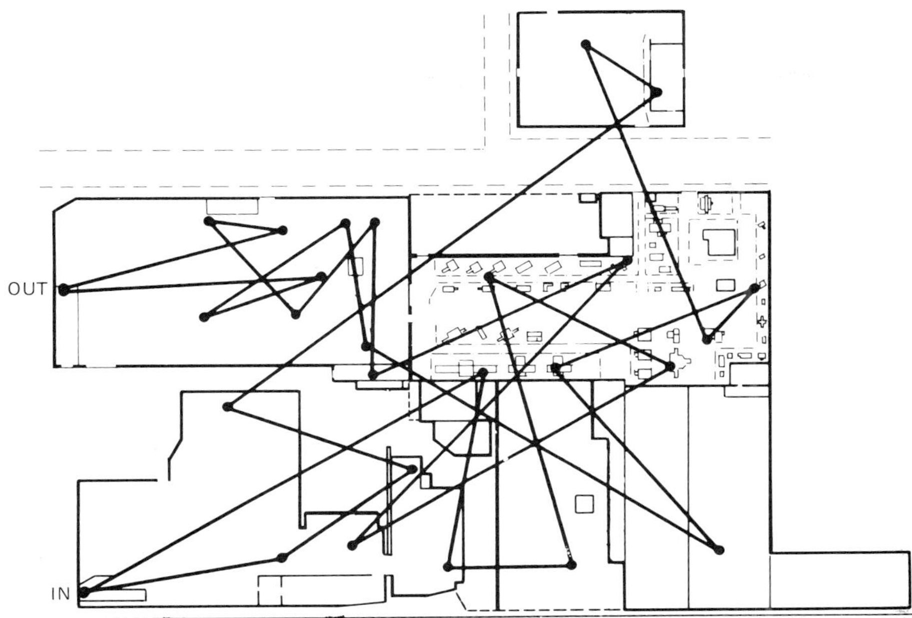

Figure 1.2 The results of decades of evolutionary, haphazard, unplanned growth.

ways complex, and most often failures. The situation is no different with our CIM system; a cumbersome, inefficient factory arrangement isn't going to get a great deal better because of CIM alone. Fixing the factory arrangement *first* meets most of the criteria set down for a good first step:

- It probably isn't going to offend too many people or cross many traditional values. The average factory person is usually aware that moving product several miles between operations is nonsensical and won't be upset if it's changed.
- Rearrangement is generally not very expensive, but it usually pays rich rewards in terms of reduced numbers of materials-handling people, fewer trucks, and the potential for lots less inventory.

Rearrangement is a good candidate to be a first step in CIM implementation. It's almost certain to produce valuable, visible, and measurable results quickly. If it is properly thought out, the rearrangement can position the factory to go quickly to one of the next steps, such as operation of the various groups of machine tools or equipment as cells or the development of flexible manufacturing systems.

The second element of CIM which should be a prime first-step candidate is the establishment of a data base management system (DBMS). The concept of DBMS was talked about for over 20 years before it became a practical reality in the late 1970s. A DBMS is a powerful set of software programs that organize and control complex file structures which make up all the information base for the firm. Special application programs permit users to go into the files and extract data in the form and level of summary the users need. A DBMS, if properly implemented, gets all the information of the firm into a single data source and gives everyone access to everything at exactly the same time and, with good applications programs, in the most usable form.

The importance of a good DBMS to the firm's managers is immense: the DBMS lets them manage with a foundation of accessible, reliable information. The economic benefits also are attractive. If the firm already has information stored on a computer, the odds are overwhelming that, in the absence of a DBMS, the files are a disorganized mishmash of redundant, inflexible, conflicting data sets that are difficult for even the programmer to use and are *unusable* by anyone else. In many non-DBMS firms, as much as 70 percent of the system group's effort is devoted to trying to maintain redundant files. The DBMS concept eliminates redundant files. Given the scarcity and price of systems people, freeing up a major share of the systems staff could be a significant contribution.

Developing the database is a tedious assignment, however. It may take as much as a year, and sometimes several years, even in a firm moderate in size. It is not unusual to encounter some resistance by operating groups to releasing their data. The support of senior management becomes extremely important in such cases, because the DBMS should represent the integrated information base of the entire firm. Omitting some data sets here and there could quickly break down the entire system concept.

The structure of the completed DBMS provides the basis for the addition of new systems; e.g., a particularly viable material requirements planning system can easily be integrated into the DBMS. The DBMS gives it access to the bill of materials structure, the purchasing function, the production planning function, the labor-reporting mechanism, and so on.

An element of CIM which has perhaps been more enthusiastically (but not necessarily more wisely) advanced as a first step in implementing computer-based systems is computer-aided design (CAD). CAD offers some real benefits such as the elimination of repetitive work through use of a graphics database. It also has eliminated the tedium of drawing storage and retrieval by permitting storage of graphics in computer memory. There are some distinct advantages in using CAD as a first step toward CIM. Not least of them is that computer graphics are not conceptually different from paper graphics; traditional graphics methods are still in place. Only the means of producing them has changed; resistance is usually minimal.

Unfortunately, many users see CAD *only* in terms of computer graphics. The most dramatic contribution which can come from CAD is the solid foundation it provides for computer-aided manufacturing! This stems from the opportunity to quickly and accurately produce numerical control (NC) programs and the ability to distribute graphical data instantaneously to all using functions.

The value of CAD to an individual firm depends on the value of each of the separate contributions. If they match the needs of the firm, CAD also should be a high-priority element of incremental CIM. However, they may not! A 1983 study of computer graphics users found that about 75 percent of the installed systems did not meet the anticipations of their users. That does not imply that CAD systems don't work. Given the right application and concept, CAD can be a powerful tool, a critical and valuable element of overall computer integration of the firm. A poor installation, or one which doesn't match the needs of the firm, can be an expensive failure.

1.5 Some Guidelines

CIM systems are unique to individual firms. However, there are some guidelines which the managers of the firm should follow in developing a CIM system:

- The system should be designed to fit the firm. The operating methodology of the firm shouldn't be distorted to fit a particular piece of software. This also implies that the software selection should be based on the needs of the firm, not the technological beauty of a software system. This is nearly always a problem in using a software consultant in designing a computer-based system. It's not at all difficult to understand why a consultant would be motivated to recommend his or her packaged software, regardless of its fit to the needs of the firm.
- The information system should be made an integral part of the operation of the firm. The need for accurate, timely information is too critical to be treated as a secondary issue. The coordinating committee, composed of senior managers and led by a senior manager with skills in both operations and systems management, should play a key role in assuring the integration.
- The top executives of the firm need not understand the details of how CIM operates, but they must be willing to take quick advantage of the opportunities it offers. This includes making personnel reductions where they are possible and changing operating policy to reduce mandated levels of inventory when the streamlined system shows it can answer customer needs more quickly than its predecessor. Too often, firms decry the benefits of CIM, when in fact they've just not acted to capture them!
- The firm must adhere as much as possible to the plan while recognizing that improvements and corrections can be made. Any changes, however, should be considered and authorized by the coordinating committee. The

purpose of the plan is to provide structure to the implementation of the various stages of CIM, not to force the firm toward a predetermined but unworkable "solution."

1.6 Conclusions

The concept of computer-aided manufacturing, or management, is one of the most important opportunities offered businessmen since the emergence of the computer. If it is properly employed, CIM can give the U.S. firm a competitive edge over competitors in the United States and offshore. It can give domestic businesses an edge even though their offshore competitors may be directly or indirectly subsidized, have dramatically lower labor rates, have a favorable exchange rate, or have any of the other advantages competitors are said to enjoy.

However, there's a lot more to CIM than buying a computer and a large library of software. Getting started toward CIM is a lot more than just acquiring the ability to process data electronically. It relates as much—perhaps more—to changes in the firm that take place before the installation of the computer as it does to the form of the data entering computer memory or the operating system to be used.

The basic requirements of getting started toward CIM include preparing the existing organization to be ready for streamlining:

- The ongoing and enthusiastic commitment and support of the firm's senior management.
- Having the strength in systems staff to support the maintenance needs of CIM.

The most critical need, however, is often a comprehensive and well-thought-out plan to design and implement the various elements which lead up to CIM. The plan is needed because CIM is rarely installed as an entity; some elements logically precede others for economic, technical, and social reasons.

- In many firms, rearranging physical resources to better reflect the actual sequence of use makes sense, and, as an important by-product, it often provides some dramatic economic returns. A rational rearrangement often also provides an important credibility builder for the system designers.
- An integrated database structured in an effective DBMS is an early priority in almost all cases. It makes an important contribution by providing managers with a reliable, timely, and appropriate foundation of information on which to base decisions and on which to plan strategy.
- Many firms look to CAD as a first step toward CIM. They are disappointed, not because CAD systems don't work, but because they don't

"change" the firm. They only replace a system already in place with one which, as a stand-alone, is an electronic duplicate of the preceding one. Although it can be an important tool in its own right, a major importance of CAD is its capability to interface with other electronic systems.

The most important aspect of getting started is recognizing that capturing the benefits of CIM means changing the firm as it existed before CIM. Achieving the kinds of operating economies which are possible can be accomplished only by developing a rational facilities plan, eliminating unproductive staff positions, and purging operating policies which limit the effectiveness of using the computer to streamline the firm.

Finally, there are some guidelines which can be applied universally. They advise tailoring the software to the firm rather than the firm to the software. They also advise taking advantage of the opportunities offered through CIM; policies which permit and even encourage large inventories must be reconsidered on the basis of the improved responsiveness CIM provides.

The firm should get started toward CIM by developing first a picture of the firm as it will be after CIM and then planning a strategy to get there by installing prioritized elements. Getting started sounds easy, and it is easy. The firm must take the first step, which is also the most difficult one: make the decision to go forward!

CHAPTER 2

Technology Management and Factory Automation

William Muir

2.1 Introduction

The implementation of state-of-the-art manufacturing processes in a typical U.S. manufacturing company will not, in and of itself, improve productivity or reduce costs. Similarly, although intangible and noneconomic factors are very important in the *capital investment justification process,* the ultimate success of an advanced manufacturing technology must be measured by the impact on manufacturing cost at expected levels of production volume, capacity utilization, product quality, and production mix. As a result, cost-reduction (containment) success will be accomplished only by making certain that financial planning and control practices properly support the analysis, identification, implementation, and management of advanced manufacturing technologies and their revised cost-behavior patterns.

It is entirely possible that the implementation of advanced manufacturing technologies has been hindered by antiquated capital justification (cost-benefits) processes that emphasize only direct labor savings. The reality of advanced manufacturing technologies is that cost-behavior patterns are shifted to a lower percentage of direct labor and a higher percentage of other "value added" costs, principally indirect labor and equipment amortization. This trend becomes even more pronounced when one considers current forecasts of the factory of the future: a computer-integrated manufacturing (CIM) facility. Therefore, if technologies that reduce direct labor are introduced and current cost management processes allocate overhead costs to product costs on the basis of direct labor, overhead costs may seem to be reduced for individual products while, in fact, total "facilitywide" costs may not have been as favorably impacted.

The objective of emphasizing the importance of *cost-benefits analysis* and *cost-benefits tracking* in the factory automation justification process does not suggest replacement of present cost accounting and performance measurement information systems. Rather, it is to encourage support for an expanded production and financial management viewpoint. This expanded viewpoint includes the following:

- *Cost-benefits analysis.* The analytical process which assists in (1) identifying the operational areas (function groups) in which the introduction of an improved manufacturing technology will have a substantial financial benefit, (2) evaluating the life cycle cost profiles of potential cost reduction projects, (3) preparing an analysis of and a plan to manage project risk, and (4) identifying cost-benefits tracking requirements so that the planning and control loop can be closed to monitor whether benefits are realized.
- *Cost-benefits tracking.* The continued determination of the actual level of project investment and the associated recurring costs and savings. This information is compared to what was planned during the cost-benefits analysis phase and, subsequently, to current estimates based on any changed circumstances.

2.2 Cost-Benefits Analysis

Cost-benefits analysis begins with *determining a matrix* of as-is costs vs. performance (efficiency and effectiveness) of the existing *manufacturing functions* on a *facilitywide basis.* Based on this matrix, proposed manufacturing improvement programs are then identified and analyzed according to their productivity improvement potential, economics, and implementation risks, also on a facilitywide basis.

The following guidelines have been prepared as aids in the cost-benefits analysis process. Each guideline contains (1) a statement of the guideline, (2) background on the need for the guideline, and (3) a brief suggested approach for applying the guideline. The guidelines cover the topics of:

1. Defining manufacturing functions (activities) and function groups
2. As-is cost baseline preparation

3. Evaluating the efficiency and effectiveness of the function groups
4. Assessing project prioritization (needs matrix)
5. Evaluating technologies for productivity implementation
6. Developing to-be cost-behavior patterns
7. Assessing intangible factors, risk, and integration
8. Time-phased economics

2.2.1 Guideline 1

Structure the cost-benefits analysis of a factory modernization strategy on a manufacturing function (top-down) and function group basis.

Background Most cost-management systems are accounting-oriented (cost of goods sold and product and inventory valuation). They do not adequately measure operational performance or the *value-added* costs of the manufacturing functions. Also, the current focus of many cost-performance measurement systems is at the product–work order–customer order level. Although that is generally consistent with objectives described in the financial budgets, there is no assurance that the resultant cost centers are structured consistently with manufacturing functions performed as defined in a top-down type of analysis, the level at which improvement programs are applied to reduce costs.

Suggested approach A well-executed manufacturing technology improvement program requires the use of a top-down manufacturing function identification process. Typically, this involves developing a nodal tree type of structure (refer to the U.S. Air Force IDEF process). This requirement is necessary to identify the manufacturing functions currently performed (as is) or proposed (to be) that will be impacted by the improvement program(s). The use of a nodal tree type of functional structure assists in providing consistency between capital expenditure planning, production planning and control, performance monitoring, and financial monitoring. An example of a nodal tree is shown in Fig. 2.1.

In many companies, an as-is facilitywide nodal tree will encompass several hundred functions (nodes). The next guideline in this cost-benefits process suggests using the nodal tree to develop an as-is cost baseline. Doing that with several hundred nodes is highly impractical. It is, therefore, appropriate (necessary) to group the nodes following the preparation of the initial as-is nodal tree. They should be grouped according to what work center a *potential improvement project* may impact.

These groupings are called *function groups.* Each low-level node should be assigned to a function group, even to the point of setting up a "dummy" group for nodes that will not be impacted by any technology improvement project.

2.2.2 Guideline 2

Identify and analyze all *significant* costs incurred by each manufacturing function group. *Develop an as-is cost baseline.*

Background Manufacturing cost can be thought of as the monetary quantification of the purchased and value-added factors employed to produce a product. In other words, the cost of direct labor, direct material, machinery and equipment usage, and information systems usage as recorded in financial statements represents the unique mix of those factors used by a particular company. It is the goal of most companies to continually analyze the mix, identify where changes could reduce total operating cost, and recover, within a defined time frame, the expense of implementing the necessary changes.

For example, direct labor might be significantly reduced by replacing selected manual processes with automation by using robots. However, implementing robotic technology will impact manufacturing operations in areas other than just direct labor and amortization of the cost of the robots. Indirect labor, manufacturing throughput time, product quality, and the ability to respond to changes in product volume and mix will also be

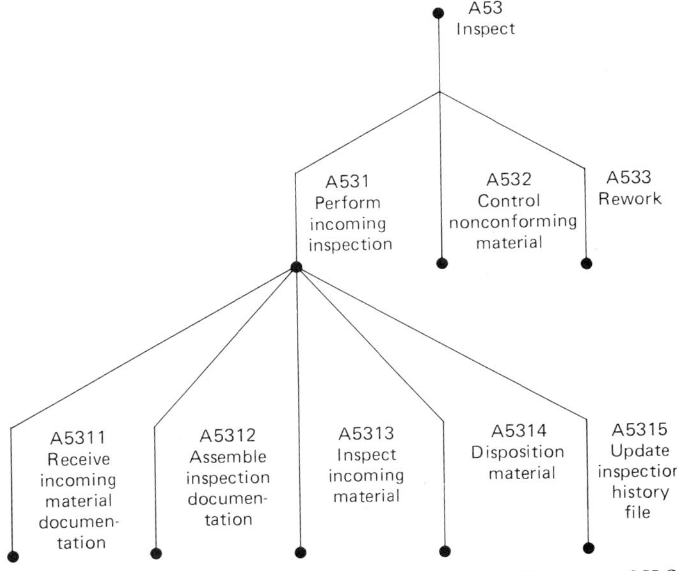

Figure 2.1 Standard ICAM aerospace nodal tree. *(Courtesy of U.S. Air Force)*

affected. In fact, it has been observed, on occasion, that total manufacturing costs have risen despite demonstrated reductions in direct labor by using robots.

This apparent contradiction can be attributed to traditional cost-accounting systems, which typically require collecting direct labor and direct material only and allocating all remaining costs as a percentage of direct labor. The assumption is that indirect costs will change in the same proportion as direct labor costs change. *That assumption is usually not valid when advanced manufacturing technologies are implemented.* Generally, new manufacturing technologies involve the application of computerization and automation to processes which were previously controlled manually. The change usually results in a decreasing percentage (vis-à-vis cost of goods sold) of direct labor, an increasing percentage of indirect expenses, and a significant decline in manufacturing steps in which the operator controls the pace of the process.

Suggested approach First, develop a manufacturing cost model common to both the facility and each manufacturing function group. (Table 2.1 is a sample manufacturing cost model.) This model should identify *all significant summarized manufacturing cost categories*, as shown in the chart of accounts. It should be structured by using cost classifications that can easily be identified with the *critical success factors* of production. At a minimum, manufacturing technology improvement projects should analyze the following (high-level) cost classifications (on an equivalent basis):

- Direct labor
- Material utilization (unplanned scrap)
- Materials handling
- Machinery and equipment amortization (including tooling)
- Operations support
- Engineering support
- Plant and facilities amortization
- Information systems support
- Inventory financing
- General and administrative support

The change in the *cost behavior pattern* of a function group in which an improvement project is implemented describes the impact of the project on the value-added costs identified by the cost model. It is, therefore, important that the assumptions and definitions underlying the cost behavior patterns (both as is and to be) be explicitly stated and documented. It is also im-

TABLE 2.1 Master Company List; Companies 001 through 001 Selected

COMPANY: 001 Example Contractor, Inc. TIME FRAME: January 1, 1985 to December 31, 1985

Cost elements					Performance measures		
Code	Description	Infl	Cost type	Source	Code	Description	Comment
DPL	Direct production labor	3.00	V	Pyrll regstr	PRDLBR	Direct prdctn hours	Product productn labor
INDL	Indirect labor	3.00	SF	Pyrll regstr	PRHDCNT	Direct prdctn headcount	Product productn labor
SUPLS	Supplies & indirect matl	4.00	V	Gen'l ledger	INDHDCNT	Indirect headcount	Other mfg labor
SCRP	Production scrap	4.00	V	Scrap report	EQPHRS	Equipment usage hours	Equivalent units produced
UTIL	Utilities used	5.00	SF	Gen'l ledger	SQFTG	Square footage	Functional total
MTLHND	Matls hndlng & logistics	3.00	SF	Gen'l ledger	INVVAL	Inventory value	By function
MIS	Infmtn systems support	5.00	F	Gen'l ledger	UNITS	Units of production	Equivalent units
INV	Inventory support	4.00	F	Gen'l ledger	CofGS	Cost of goods sold	Current dollars
EQMNT	Equipment amortization	4.00	F	Gen'l ledger			
FACIL	Facilities amortization	2.00	F	Gen'l ledger			
CORP	Corporate allocations	4.00	F	Gen'l ledger			
G&A	Finance & admin support	3.00	F	Gen'l ledger			
ASTUSG	Asset usage interest charge	2.00	F	Gen'l ledger			

SOURCE: Courtesy of Price Waterhouse.

portant to remember that the cost classifications within the model will vary from company to company.

Second, by using the function groups from the previous guideline and the manufacturing cost model, develop an as-is cost baseline for the entire facility. (Table 2.2 is a sample as-is cost baseline by function group.) This baseline, which should be reconcilable with an accepted financial statement source (i.e., the general ledger), represents the segmentation of annualized operating expenses by function group by manufacturing cost model classification. It describes the cost reduction *potential* for manufacturing modernization projects.

2.2.3 Guideline 3

Determine the performance (efficiency and effectiveness) of each manufacturing function group. *Develop an as-is performance baseline.*

Background Besides cost, the other dimension of identifying improvement opportunities is determining how efficiently and effectively the manufacturing function groups are currently performing. (Efficiency is the measure of how well current resources are being used; effectiveness is the measure of productivity improvement possible from employing more advanced resources.) The opportunity for improvement is greatest for the function groups which currently have inefficient and/or ineffective performance and are high in value-added cost.

Suggested approach Each function group identified in the top-down model should be analyzed to determine how efficiently and/or effectively (relative to current company and/or accepted industry critical success factors) it is being executed. The analysis should emphasize the quantification of problems, diagnose their causes, and provide benchmarks against which to measure improvement. Some of the measures (critical success factors) which could be used, other than cost, to measure efficiency and/or effectiveness include:

- Work-in-process inventory
- Throughput time
- Schedule adherence
- Equipment utilization (as scheduled)
- Labor utilization (as scheduled)
- Scrap and/or rework volume
- Capacity utilization

TABLE 2.2 Cost Baseline Summary Report; All Function Groups Selected

Function group	Direct production	Indirect labor	Supplies & indirect	Production scrap	Utilities used	Matls hndlng & lo	Infmtn systms supp	Inventory support	Equipment amortizatio	Facilities amortizatio	Corporate allocations	Fince & admin sup	Total cost
AS/RS	776,250	682,084	147,270	307,750	279,145	82,633	134,999	393,084	482,802	115,428	30,331	60,586	3,492,362
CAD/CAE	80,000	853,067	67,918	36,666	66,396	3,000	15,866	20,000	153,043	34,546	9,128	18,227	1,357,857
FMS(A)	509,687	912,492	78,483	236,895	256,781	17,953	67,862	138,833	299,490	91,729	23,949	47,584	2,681,738
FMS(B)	2,352,563	1,816,466	265,570	1,097,188	1,120,532	65,806	146,449	496,500	922,308	370,645	95,821	191,417	8,941,265
MRPII	85,312	459,681	34,621	37,520	43,952	7,860	131,032	56,855	82,863	18,476	4,879	9,673	972,724
NOGRP	696,188	1,276,210	156,138	283,981	283,194	72,748	178,792	394,728	561,394	119,176	31,492	62,513	4,116,554
	4,500,000	6,000,000	750,000	2,000,000	2,050,000	250,000	675,000	1,500,000	2,501,900	750,000	195,600	390,000	21,562,500

SOURCE: Courtesy of Price Waterhouse.

2.2.4 Guideline 4

Develop a *needs matrix* for *prioritizing improvement opportunities* by function group.

Background Most manufacturing technology improvement projects are directed at the manufacturing functions, or function groups, which have the highest gross cost, irrespective of how efficient or effective they currently are. Instead, improvement funds should be channeled to the function groups having the greatest *cost reduction potential.* They may or may not be those having the greatest gross cost.

Suggested approach The identification of high-cost and low-productivity function groups is accomplished by developing a matrix of the current performance of the manufacturing function groups (as measured by their associated critical success factors) and the gross cost of each function group. Developing a needs matrix of this nature assures that each improvement project is viewed on a facilitywide basis and avoids the islands-of-technology syndrome.

2.2.5 Guideline 5

Evaluate technology alternatives for improving the productivity of the function groups with the highest potential payback. Select the alternatives which offer the highest probability of attaining the potential.

Background Many companies starting out on a program to upgrade the productivity of their manufacturing operations establish the solutions they want to implement before they have completely defined and analyzed their problems. This "solutions looking for problems" scenario often results in the implementation of technologies which only relocate costs but do not result in an overall *facilitywide* cost reduction.

Suggested approach Once the function groups having the greatest cost reduction potential have been identified, all available technology modernization alternatives (within reason) for reducing costs within that function group should be identified. The technologies that, on the basis of their implementation history or technological promise, seem to offer the greatest potential should be selected for further investigation.

2.2.6 Guideline 6

Determine, on a function group and facilitywide basis, a to-be cost-behavior pattern for each improvement alternative.

Background As previously mentioned, many organizations expect a manufacturing improvement program to reduce only direct labor cost. The impact of a new manufacturing technology on support costs, such as indirect labor, energy, and maintenance, is often neglected. As a consequence, many improvement projects "push down" costs in one cost classification (e.g., direct labor) within a function group only to have costs in other cost classifications within that function group "jump up." Thus, *total costs* are not reduced.

Suggested approach By using the manufacturing cost model approach such as that suggested by Guideline 2, develop a manufacturing engineering projection of the to-be cost behavior pattern for each improvement alternative being investigated. By using an analytical tool such as the model described, there is greater assurance that all costs will be analyzed and cost reduction will actually be attained, not just cost relocation. (See Table 2.3.)

2.2.7 Guideline 7

Analyze the impact of intangibles (such as implementation risk, inflation, human factors, and legislative trends) to the extent that they can be quantified. Assure that all to-be analysis is done on a facilitywide basis to avoid the impact of double counting.

Background Varying economic conditions make any capital investment difficult to quantify. In particular, the U.S. economy has recently been subjected to a number of variations, such as:

- Modulating rates of inflation
- Modulating interest rates
- Increased offshore competition (with its better productivity growth rate)
- A long-term trend toward greater leveraging of capital investments

TABLE 2.3 Cost Savings Summary Report; Cost Baseline Compared to Model "First"

Function group	Company		Variance	Percent variance, %
	As-is	To-be		
AS/RS	3,492,362	3,462,746	29,616	.85
CAD/CAE	1,357,857	1,273,301	84,556	6.23
FMS(A)	2,681,738	2,679,028	2,710	.10
FMS(B)	8,941,265	7,948,179	993,086	11.11
MRPII	972,724	1,077,996	105,272 –	10.82 –
NOGRP	4,116,554	4,116,554	0	.00
	21,562,500	20,557,804	1,004,696	4.66

SOURCE: Courtesy of Price Waterhouse.

In addition to these environmental conditions, each manufacturing technology improvement alternative is subject to a number of *risk factors* related to its successful implementation. These factors include:

- Technological threshold
- External competition
- Education and training of current staff
- Other human factors and support requirements

Finally, manufacturing advances have typically been incorporated on an individual basis that has been characterized as islands of technology. Realizing the full benefits of advanced manufacturing technologies will require a technology management strategy that integrates each new improvement project with each other one being considered on a facilitywide basis.

Suggested approach Each alternative improvement project must be individually assessed to determine the risk of not achieving the anticipated results, whether because of intangibles, economic conditions, or fate. The important point is to recognize most of the environmental factors that will come to bear on the project(s) and:

- Identify the most significant *risk factors* that could inhibit achieving the anticipated savings. By performing an elementary sensitivity analysis, a best case–worst case condition can be identified for each project.
- Calculate the worst case projected cost savings for each alternative.
- Assess the technological risk for each alternative.
- Develop a risk management plan that would increase the probability of success by focusing project management attention on the identified risk factors. (Do not necessarily give up the project if the worst case is not acceptable. Improve project management first.)

The cost-benefits analysis process should also quantify (to the extent possible) the synergistic impact of all projects on each other to determine their effect on total manufacturing cost. This can best be accomplished through the use of the function group approach already described.

2.2.8 Guideline 8

Determine the time-phased economics of each of the selected improvement projects.

Background Although intangible and noneconomic factors are important in a capital investment justification process, the ultimate success of an advanced manufacturing technology must be measured by the impact on total manufacturing cost over time.

Suggested approach Cost-benefits analysis should provide management with several levels of economic analysis, including:

- Annualized savings (over the planning horizon)
- Internal rate of return (IRR)
- Payback
- Risk-adjusted savings

These can be calculated by using the incremental (difference) between the as-is and to-be function group cost baselines, appropriately discounted based on the method used.

2.3 Cost-Benefits Tracking

The purpose of cost-benefits tracking is to measure the *actual impact* of advanced manufacturing technologies on total function group–facilitywide manufacturing costs and related performance measurements as compared to the to-be baseline as forecast by the cost-benefits analysis process. Cost-benefits tracking information should be developed from the database used to report financial results.

Current cost-benefits tracking and cost accounting procedures are steeped in traditional accounting practices. The tradition is that manufacturing cost data should be collected by production work order for direct labor and direct materials and all other costs should be collected in aggregate and allocated to product production on a quantifiable basis (typically direct labor). This approach was developed for a manufacturing environment in which direct labor was the dominant portion of value-added manufacturing cost and the direct laborer controlled the pace of the manufacturing process.

Today, however, that model is no longer universally appropriate. The proposed factory of the future will have even lower direct labor costs than the factory of today. This newer environment dictates a revised approach to measuring the actual costs and benefits to be attained from implementing advanced manufacturing technologies.

There are four guidelines to cost-benefits tracking vis-à-vis factory modernization projects. They cover the topics of:

1. Capturing actual cost and performance data
2. Compatibility with accounting practices
3. Verification and auditability
4. Time horizons

2.3.1 Guideline 1

Capture the actual costs and performance measures which have been identified as critical success factors by the cost-benefits analysis.

Background Most current cost accounting systems collect direct labor and materials by work order or by department. Also, although factory data collection hardware has evolved to higher levels of sophistication, the type of information collected (direct labor and material transfers) has remained the same. One of the primary reasons for the information stagnation is the cost of collecting data in a labor-intensive environment. This can be significant, since data collection activities are performed in lieu of production work. In that environment, it is often difficult to justify the cost of extensive data collection.

The factory of the future, however, will be highly automated and dependent on computerization. Thus, it will be possible to access a significant amount of actual performance information which has not been previously available economically.

Suggested approach Develop a benefits-tracking plan at the conclusion of the cost-benefits analysis steps. The plan should identify the:

- Information required to monitor the status of the critical success factors
- Sources of the actual data

As advanced manufacturing technology is incorporated into a manufacturing process, the volume and detail of information available from the process should improve. The information required for the cost-benefits tracking of an advanced manufacturing technology will thus be more readily available from the new process than from present reporting systems (as a general rule).

Once the actual data sources and the data to be captured are identified, an actual cost baseline similar to the as-is and to-be baselines used during the cost-benefits analysis should be prepared. This actual baseline is then compared to the to-be baseline. Significant variances are identified, and steps should be taken to either resolve them or understand the reasons why they have occurred.

2.3.2 Guideline 2

All cost and performance measures should be compatible with internal and external reporting requirements, including generally accepted accounting principles (GAAP) and cost accounting standards (CAS).

Background Through the years, traditional financial reporting practices have resulted in a significant superstructure of data compliance requirements. The purpose of these requirements is to provide both reporting standardization and the information necessary to determine a fair price and adequate profit (for public sector contracts). Unfortunately, a significant cost is associated with this standardization of reporting.

Suggested approach The cost-benefits tracking process should be structured to use the common database already being used to satisfy other compliance-reporting requirements. One method of ensuring this congruence is to develop a matrix of data elements vis-à-vis benefits tracking, operational management, and financial reporting requirements. Any data element not currently being reported on should be evaluated to determine whether it is at variance with proper management objectives.

2.3.3 Guideline 3

The information provided by the cost-benefits tracking process must be verifiable and auditable.

Background The development of an auditable, reliable, and accurate cost management methodology is extremely important to the success of a manufacturing cost reduction program. This process must provide feedback to management for judging the progress of the program. Problems can be highlighted and solutions implemented only if the assumptions underlying the cost projections can be scrutinized. A well-structured audit trail will facilitate a detailed analysis of variances from forecast.

Suggested approach Document all sources of information that were previously identified in the cost-benefits tracking plan, including report number (by date), person, or document number. Each assumption should be explained in detail and appropriate supporting documentation provided.

2.3.4 Guideline 4

Cost-benefits projections should be tracked for a sufficient time to determine whether actual savings have been achieved.

Background Traditionally, advanced manufacturing projects are justified by using corporate capital investment procedures the selection of which is based on IRR and/or payback. An ongoing determination of whether the actual savings have been achieved is normally not an integral part of the process. Instead, the projects become part of normal manufacturing procedures and are reported on through the traditional cost management systems.

Suggested approach Since cost management systems are not usually structured on the same basis as a company's capital investment policies/procedures, an ongoing tracking of savings can best be achieved by modifying the existing cost management system, developing a dual system, or performing a periodic review of the project results. Whether to restructure the cost management system should be based on the perceived use of the information within the management decision-making process and the cost of modifying the system. Even if the decision is made to restructure, the change should be evolutionary and be integrated into the corporate strategic factory modernization plan.

If the decision is against modifying the current cost management system, periodic industrial engineering studies could be used to validate savings. In that case, the cost-benefits tracking plan should include the timing and scope of the audits and the procedures used for verification of the results.

CHAPTER

3

Planning for a Competitive CIM Environment

Leonid Lipchin

3.1 Introduction

Tough foreign competition, falling productivity, and scarce engineering resources are accelerating the need for manufacturers to implement automation through computing technology. Tremendous breakthroughs in interactive computer graphics, software engineering and production applications, and relatively inexpensive mini- and microcomputers have hastened the proliferation of manufacturing automation technology.

In the first half of the 1980s, American companies invested more than $100 billion in manufacturing automation technology. The major components of this technology were computer-aided design (CAD), group technology (GT), manufacturing planning and control systems (MP&CS), automated materials handling, computer-aided manufacturing (CAM), and robotics. Efficient management of data resources via computing technology [e.g., computer database management systems (DBMSs) and sophisticated communication interfaces] provide the necessary synergy among those components [1,2].

When basic automation technologies are integrated into an overall manufacturing plan, the result is a new form of manufacturing automation—computer-integrated manufacturing (CIM). Joseph Harrington coined the term in 1973. Since then it has become a popular subject for discussion and exploration by academia and industry [3,4]. For the purposes of this discussion, CIM may be defined as follows: *CIM is an environment to support a companywide operations management infrastructure designed to improve certain manufacturing performance characteristics in accordance with a company's long-term competitive strategies.* The support is achieved through a synergism of three major CIM capabilities:

- Translation and communication among diverse company disciplines
- Enhancement of creative functions
- Automation of repetitive operations

The CIM configuration and the degree of CIM penetration and synergy are defined by the company's competitive position, business strategies, operations management background, manufacturing performance status, and economic and human resource constraints. Therefore, each CIM system must be custom-made.

3.2 Cost-Effective Implementation

Companies can use the CIM concept to eliminate many of the internal problems that cause uncompetitive manufacturing performance. These problems are usually associated with an earlier manufacturing performance pattern, the organizational structure, the type of product, and other internal parameters.

Manufacturing cost is one of the most important competitive parameters of a manufacturing company. An industry cost learning experience in manufacturing a new product can be characterized by a cost experience curve. Mass producers commonly perceive the slope of the curve to be only a function of a company's manufacturing experience with a product and scale of production. However, the company's process technology level of manufacturing automation is also an important consideration. A company that can enter a mature industry with a steep experience curve quickly and establish itself as a cost-price leader usually possesses unique process technology. Unfortunately, it is not always possible to keep process technology a secret for long. For that reason and because of the market trend to a greater variety of low- to medium-volume products, other strategies should be used to secure sustained manufacturing cost competitiveness.

A focused manufacturing automation concept allows a company to respond efficiently to the competitive requirement of the low- to medium-volume segment. This concept is based on the automation potential of the experience curve for major manufacturing operations such as mechanical and electrical design, component manufacturing and assembly, and testing and packaging. Overall cost competition is very often based on the few critical manufacturing operations that make the highest contribution to manufacturing cost.

Therefore, it is important to analyze the contribution of manufacturing operations structure to cost competition and the potential for experience curve improvement through focused implementation of automation technology or computer-aided design for low-cost production. In building a highly competitive CIM environment, a company should analyze the current cost experience status of its major operations and then compare that status to the cost experience status of similar operations of its competitors (normalized for equal production value-added).

Analysis of automation and integration opportunities of noncompetitive manufacturing operations provides a basis for developing a company's CIM strategy. Figure 3.1 shows sample experience curves and their automation potential for major manufacturing operations of an electronics manufacturing company. X and Y coordinates are cost per unit in constant dollars and accumulated production volume subsequentially plotted on log-log scale.

The practical penetration of manufacturing automation technology into a company moves through six stages of CIM growth [5,6] (Fig. 3.2). Manufacturing automation expenditures over time take the pattern of a growth curve. The curve also represents the organization's learning to use automation to improve manufacturing performance. The growth process evolves along four parallel lines:

- Building a portfolio of manufacturing functions subject to current or potential support by manufacturing automation technology (i.e., the manufacturing automation portfolio)
- Integrating the systems components of manufacturing automation technology companywide
- Building up automation resources in personnel, technology, and information
- Controlling the benefits achieved through the use of manufacturing automation

The correlation between tangible measures of each dimension of the growth process defines the state of a company in terms of manufacturing automation implementation (current or planned). The first three stages of growth represent managing the individual engineering (CAD) and production (CAM) applications and associated resources, and the last three stages represent managing CIM and business resources from a companywide perspective. The transition takes place sometime during stage 3 and may result in a dramatic increase of a program's budget and sudden organizational disturbance [7].

Naturally, it is important to determine the current stage of growth for a company's manufacturing automation and to assess how the automation contributes to overall manufacturing performance. Knowledge of that stage, associated growth data on technology, organization, and business performance, and careful analysis of a company's problems with automation growth provide a foundation for developing an appropriate strategy for further

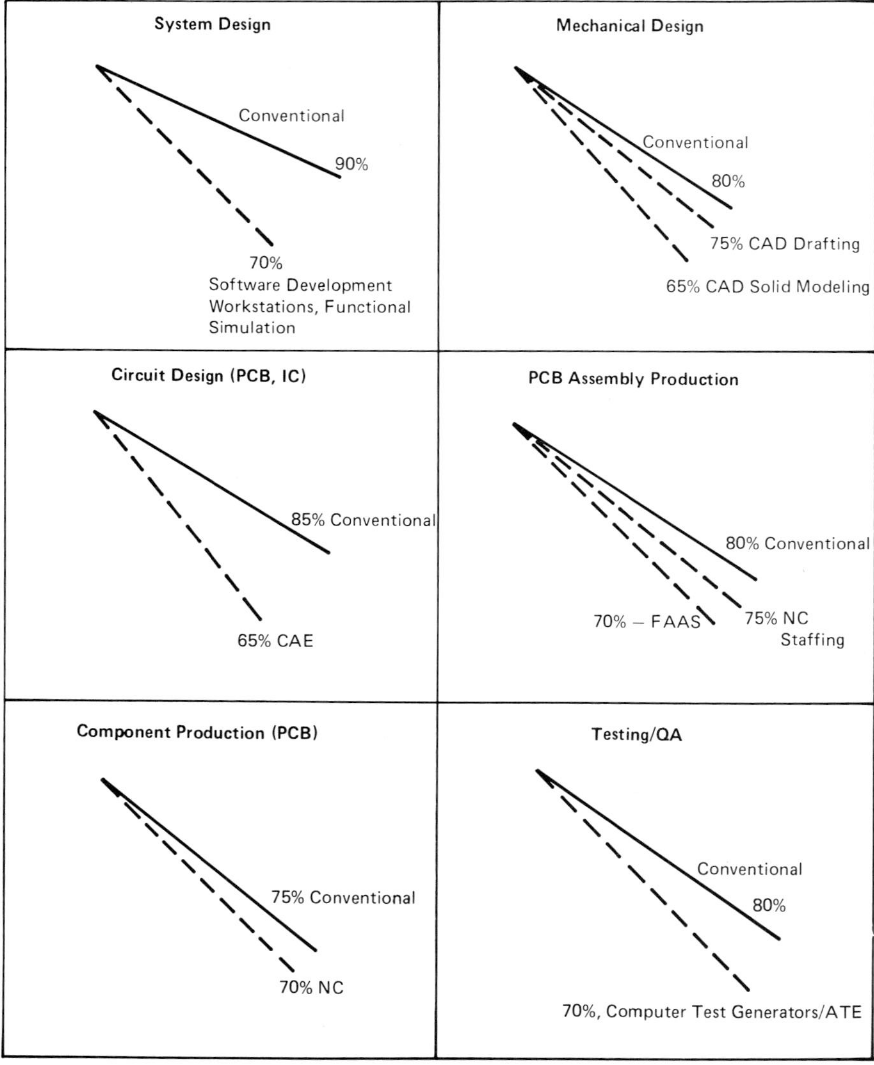

Figure 3.1 Sample experience curves for major manufacturing functions of electronics company performed in conventional and manufacturing automation environments.

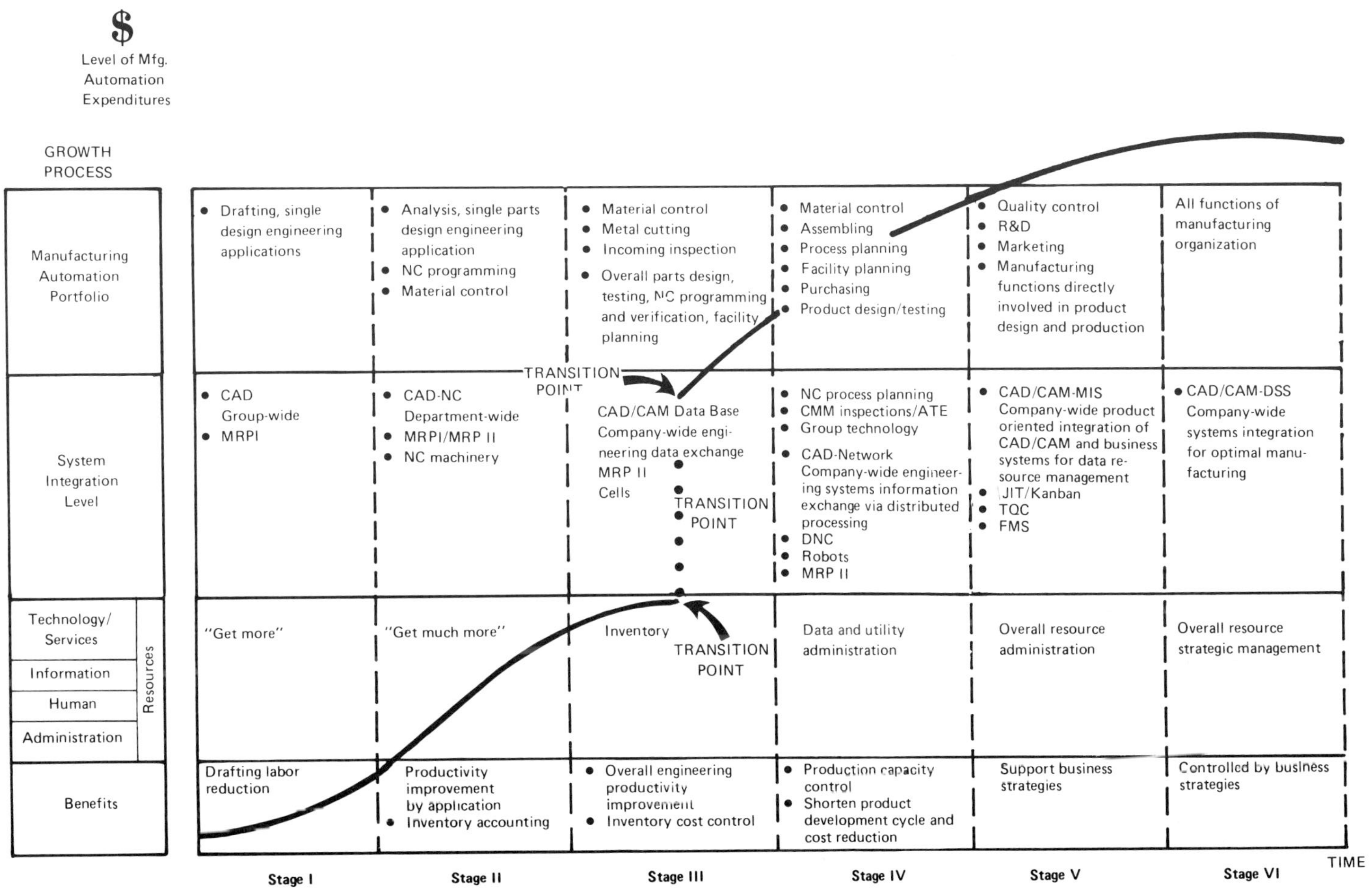

$ Level of Mfg. Automation Expenditures

GROWTH PROCESS	Stage I	Stage II	Stage III	Stage IV	Stage V	Stage VI
Manufacturing Automation Portfolio	• Drafting, single design engineering applications	• Analysis, single parts design engineering application • NC programming • Material control	• Material control • Metal cutting • Incoming inspection • Overall parts design, testing, NC programming and verification, facility planning	• Material control • Assembling • Process planning • Facility planning • Purchasing • Product design/testing	• Quality control • R&D • Marketing • Manufacturing functions directly involved in product design and production	All functions of manufacturing organization
System Integration Level	• CAD Group-wide • MRPI	• CAD-NC Department-wide • MRPI/MRP II • NC machinery	TRANSITION POINT CAD/CAM Data Base Company-wide engineering data exchange MRP II Cells TRANSITION POINT	• NC process planning • CMM inspections/ATE • Group technology • CAD-Network Company-wide engineering systems information exchange via distributed processing • DNC • Robots • MRP II	• CAD/CAM-MIS Company-wide product oriented integration of CAD/CAM and business systems for data resource management • JIT/Kanban • TQC • FMS	• CAD/CAM-DSS Company-wide systems integration for optimal manufacturing
Resources: Technology/Services, Information, Human, Administration	"Get more"	"Get much more"	Inventory TRANSITION POINT	Data and utility administration	Overall resource administration	Overall resource strategic management
Benefits	Drafting labor reduction	Productivity improvement by application • Inventory accounting	• Overall engineering productivity improvement • Inventory cost control	• Production capacity control • Shorten product development cycle and cost reduction	Support business strategies	Controlled by business strategies

TIME

Figure 3.2 Manufacturing automation evolution toward CIM.

implementation of manufacturing automation. This strategy is focused on a long-term contribution to competitive manufacturing performance improvement. If, for the first three stages, the appropriate slogan is "link manufacturing automation strategy with business strategy," for stage 4 that slogan might well become "secure business through CIM" and for the last stage it must be changed to "compete in the marketplace through CIM."

Unfortunately, CIM growth in the United States often does not follow that pattern, primarily because U.S. industry has failed to base its CIM development efforts on competitive analyses of manufacturing strategies and operations. The author's working experience with a number of Japanese manufacturing companies exposed him to their concept of CIM (called factory automation in Japan) and its practical implementation. The Japanese automation efforts, in contrast to those of American companies, emphasized the development of a factory automation program to optimize the synergy between design and production to achieve a company's competitive market goal.

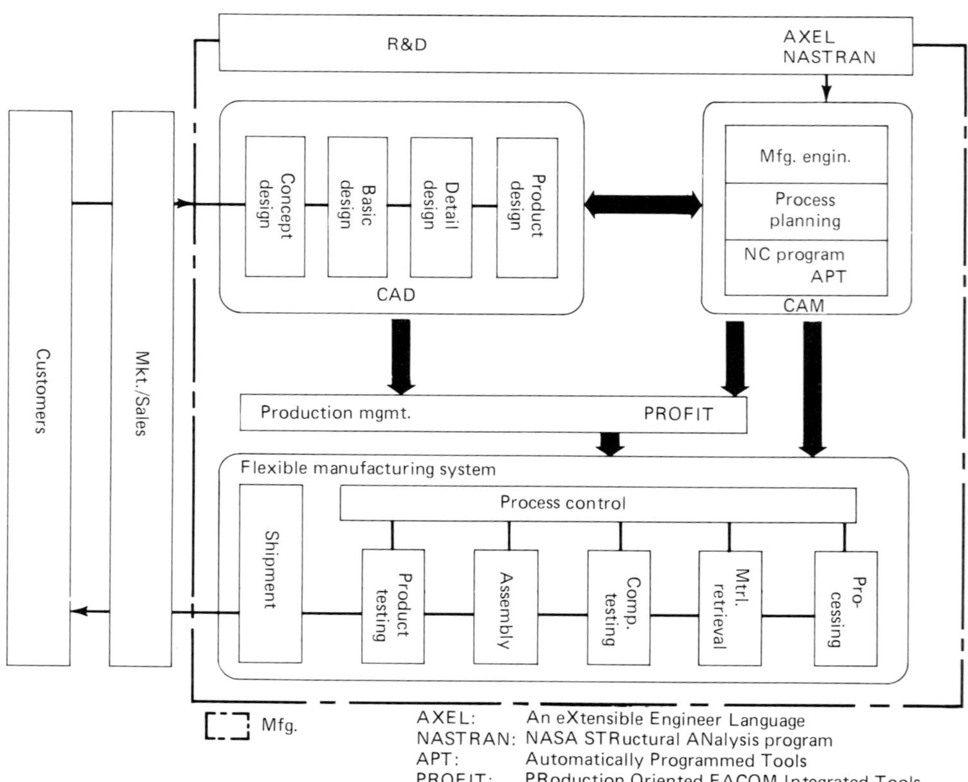

Figure 3.3 Fujitsu's concept of CIM.

Each of the Japanese CIM efforts has three significant features in common. First, all combine an appropriate custom-made production management and control system with a product-oriented engineering and manufacturing database. Second, all share the goal-oriented use of a flexible manufacturing system (FMS). For example, Fujitsu Ltd.'s CIM approach (Fig. 3.3) involves developing an engineering environment that will create a product or family of products specifically designed to be manufactured by the Fujitsu designed and made FMS. Here the objective is to shorten the entire product development and manufacturing cycle. This approach is based on a careful investigation of inefficiencies in synergy between the engineering and production departments involved.

Third, CIM systems, especially those used by electronics companies, are designed to be tested by using CAD/CAM and computer-aided testing (CAT), or so-called CAD/CAM/CAT. This scheme gives designers of printed circuit boards (PCBs) and integrated circuits (ICs) the information they need for preproduction design modification. It also offers test engineers the flexibility of automated production program development. CAD produces all the circuit and PCB information already engineered for testing; a computer can readily use the CAD database to generate automated test equipment (ATE) programs. CAD/CAM/CAT networks linked to management information systems (MIS) make it possible to analyze production test reports which are generated automatically from data collected by the ATE during a preceding 8-hr shift. The networks can then effectively modify the operation as needed within a few hours. Currently in the United States, only IBM, Hughes Aircraft, and a few other companies have implemented full-scale CAD/CAM/CAT systems.

3.3 Key Aspects of the Implementation Process

From the CIM definition, one can see that implementation of a CIM environment succeeds only if its business, technical, and human aspects are well thought out, planned, and budgeted together. Each of these CIM aspects has key factors to be considered. Those of the business aspect are:

- Market competitiveness and penetration and the role of the manufacturing system in determining those factors
- Economics
- Budget and schedule requirements and constraints

The key factors of the technical aspect are:

- Product and process strategies
- Design and production background
- System integration

The key factors of the human aspect are:

- Personnel requirements and constraints
- Operations management background
- CIM leadership and organization

For some reason, these three equally important aspects of the CIM subject have not received equal attention from academia and industry. The CIM business aspect has received the least attention. Few articles on research and industry experience have been written for a senior management audience in spite of the fact that those managers face a tough capital investment dilemma: Is CIM a fashionable engineering toy or is it a necessary manufacturing tool? This dilemma is amplified tremendously by the risk of undertaking a CIM program of 5 to 8 years' duration with a budget ranging from $15 to $100 million (for small and medium-size manufacturing enterprises) for the sake of an unpredictable impact on a company's competitive position and internal economics.

3.4 Management Considerations

Obviously, prior to implementation, it is essential, or even vital, to analyze the possible significant management and financial implications of turning a company from a mature conventional operation mode to an embryonic CIM environment and how this process may (or may not) contribute to the company's competitive business improvement. However, whether the required senior management's feasibility analysis takes place strongly depends on how a CIM program is initiated and whether engineering or business management leads it. A company may find itself at risk if engineering management and business management find they cannot cooperate and an engineering management seizes a strong leadership role. In this case, implementation of a CIM environment may become a CIM technology show for the rest of the industry rather than a tool to achieve the company's original goal—to make money and compete efficiently over the long term.

The 15-year proliferation of an already mature component of manufacturing automation—CAD/CAM—demonstrates that this implementation phenomenon occurs quite often [8,9,10]. Typically, CAD/CAM is introduced to a company by a head engineer (e.g., design, manufacturing, or reliability) who has some knowledge of its benefits and decides to bring it to his department. He persuades senior management to allocate the necessary resources by arguing that the sophistication of his specific engineering function is important for the entire manufacturing process. The argument is persuasive because it cannot be disputed unless senior management performs a careful analysis of the economic relations among all engineering functions and the company's overall manufacturing performance (e.g., lead time, productivity components, and product cost structure). However, because no one is looking at the situation from an overall perspective, the company's performance may well not benefit from a CAD/CAM investment. In addition, if CAD/CAM grows in a haphazard fashion, engineering weaknesses may be obscured.

Who should weigh all of a company's engineering functions to determine their potential for benefiting from CAD/CAM? Ideally, a strategic planner from an advanced systems technology group who is responsible for overall planning. Unfortunately, because many companies are decentralized, the decisions are being made by technical individuals at various levels without the necessary comprehensive analysis or an overall top-down plan. Also, engineering management typically must prove CAD/CAM's implementation benefits within a rather short time. This environment leads to a selection of CAD/CAM alternatives on a first-come, first-serve basis.

Even manufacturing organizations with established CAD/CAM technology planning processes (according to some studies, less than 10 percent of the total CAD/CAM user community) translate companywide, long-term business objectives into engineering and manufacturing strategies and, as the next step, into a CAD/CAM plan. Figure 3.4 shows the problem of a CIM planning process which is initiated, revised, and approved by a com-

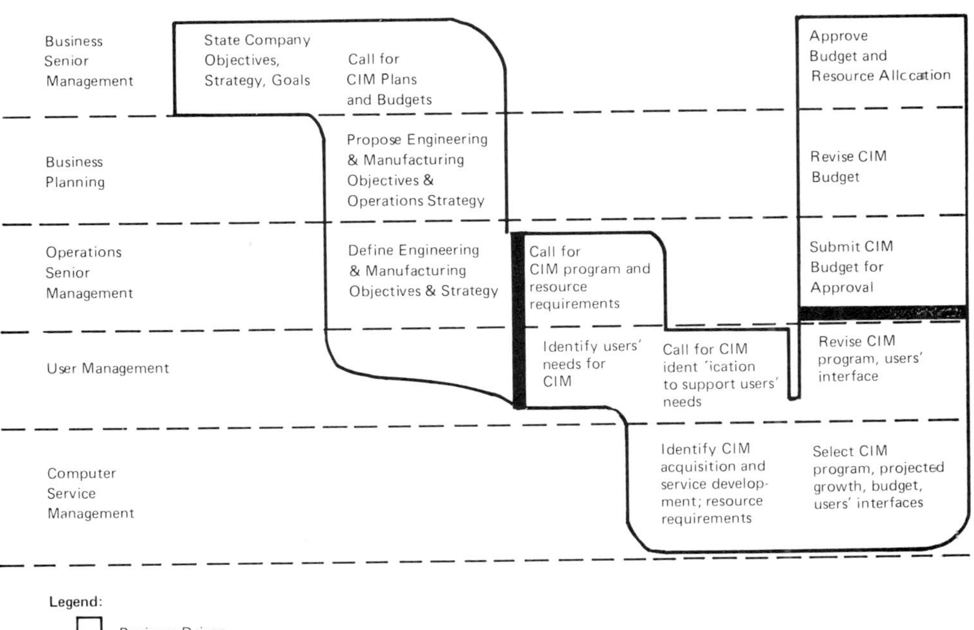

Figure 3.4 Current problems in CIM planning.

pany's senior management concerned with improving competitive market position and manufacturing performance. The process is developed and executed by users concerned only with having better tools for their applications. The gaps between "business driven" and "technology driven" portions of the planning process may result in a plan that, in spite of high return on investment for the individual applications, will lead to inefficient capital investment in CAD/CAM or CIM from the company's overall business standpoint. Thus, implementation will not lead to the desired improvements in manufacturing competitiveness or business performance.

Figure 3.5 shows two alternative patterns of CIM growth based on a business-driven (Fig. 3.5*a*) or technology-driven (Fig. 3.5*b*) planning environment. The alternatives are based on different sets of critical success factors and result in different CAD/CAM programs based on different technologies, function support, resources, and/or organization. The technology-driven plan (Fig. 3.5*b*) does not take into account historical problems of noncompetitive manufacturing performance. Even successful implementation of this plan would not guarantee that desired improvements in manufacturing performance, such as lower cost per unit and short product development cycle, would be achieved.

Sometimes a company begins to implement manufacturing automation

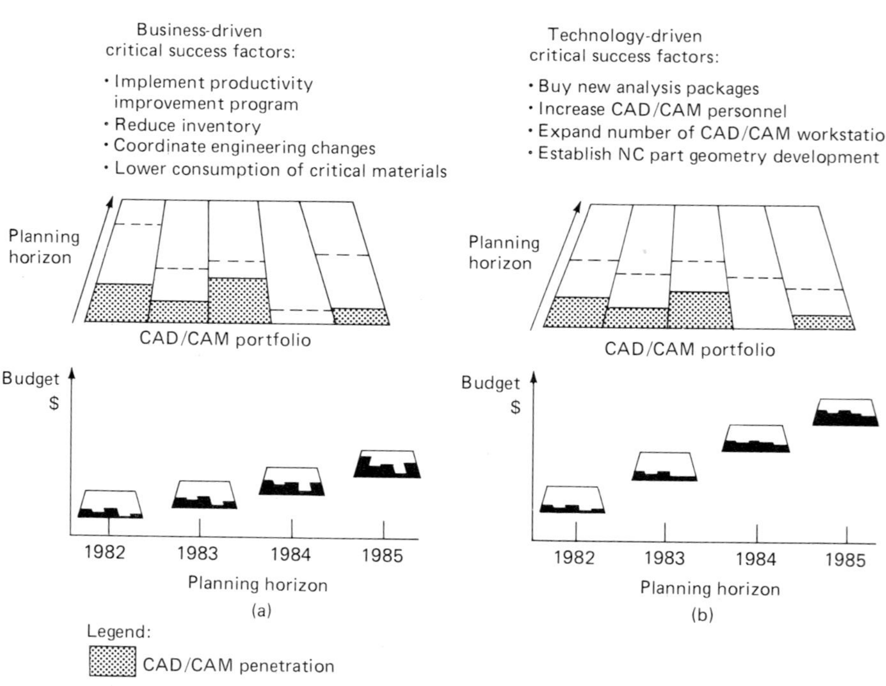

Figure 3.5 Two alternatives of CAD/CAM growth in a manufacturing organization.

technology without a management entity to direct the effort; instead, it relies on voluntary cooperation among the technical users. In that case, differences among the individual manufacturing function budgets, manpower resources, and willingness for change can create problems. Because of the company's organizational and business background, various functional groups have varying amounts of power inside the company, which gives them varying ability to influence the manufacturing automation program. It is important to understand the current and potential balance of power in order to design an appropriate structure of coalitions with respect to each stage of manufacturing automation growth. The realities of company politics can make that a very difficult task. One practical solution is to bring in an outside management consultant to assess the balance of power and make recommendations to the CIM program.

During the stages of manufacturing automation growth, relations between major management groups are characterized by intensive dependence on the use of power to resolve conflict. Therefore, selection of an organization to lead and manage automation implementation programs can be crucial to the total effort. The selection should be made only after careful analysis of the existing balance of power between the data processing (DP) and engineering and manufacturing organizations, centralized and decentralized structure of a company, the company's traditions, and so on. Figure 3.6 shows a typical power group balance and represents a coalition paradigm.

Ultimately, the chief executive officer must accept the role of planning leader and manager if a CIM program is to succeed. That is because an understanding of the real economics of CIM's impact on the entire company is necessary if the CIM tool is to serve as a strategic weapon in a competitive environment. Each dollar invested in CIM must bring a high return in terms of improving a company's competitive manufacturing function. The improvements include shortening the product development cycle, reducing cost per unit, and improving quality.

The experience of many Japanese and U.S. companies that have successfully implemented large-scale manufacturing automation strongly supports that view. The senior executives of those companies realize that they are the ultimate users of the CIM environment. The major reasons for the current low level of participation by senior business managers in CIM programs are:

- The tendency of companies to use technology-driven vs. business-driven planning for manufacturing automation
- Senior management's fear of showing a lack of understanding of CIM technology
- Senior management's underestimation of the potential of a well-designed CIM environment to improve a company's market position

		Power groups balance %	Stage 1	Stage II	Stage III	Stage IV	Stage V	Stage VI
Senior management	Type	Company	—	—	—	5	15	30
		Local	—	—	10	15	5	—
DP management	Type	Company	—	—	40	50	45	30
		Local	20	40	10	—	—	—
User management	Type	Designer	80	60	30	10	10	5
		Manufacturing	—	—	10	10	10	5
		Service	—	—	—	10	10	5
		Marketing	—	—	—	—	—	5
		Customer	—	—	—	—	2	5
		Suppliers	—	—	—	—	3	5
		Manager	—	—	—	—	—	10

Figure 3.6 Manufacturing automation coalition paradigm.

- A mistaken belief that the company can buy an off-the-shelf CIM program in the future
- Lack of formal methodologies and documented planning experience of successful CIM programs

3.5 CIM Planning and Implementation Process

The following is a practical approach to CIM implementation that is based on the author's consulting experience with planning and implementing large-scale manufacturing automation programs worldwide. The approach has been published and introduced to European, U.S., and Japanese industry [11, 12, 13, 14]. A CIM plan that is designed by using the approach will:

- Define the areas of a company with the highest potential for business improvement
- Establish a control mechanism to monitor the success of CIM use
- Determine the proper level and schedule of investment in CIM
- Structure and develop the CIM organization and an ongoing management program

This plan, which is based on assessments of the business operations of the company, the product, and the current status of CIM growth, leads to the selection of a CIM program and an implementation strategy that ensures a competitive manufacturing performance improvement. The plan has four phases.

3.5.1 Phase 1. Feasibility study

During phase 1, a company assesses its manufacturing competitiveness in the related market segment, evaluates how its existing manufacturing automation tools contribute to its competitive manufacturing performance, investigates major operations management and automation factors responsible for poor competitive performance, and identifies operations management, technology, and personnel requirements for a CIM environment with high competitive potential. At the end of this phase, the company should be able to determine how technologically feasible and economically beneficial a CIM concept is for the company's competitive manufacturing strategies.

3.5.2 Phase 2. Launching

Phase 2 actually initiates the practical development of a CIM program. Based on the results of phase 1, the company explores how a CIM concept can be converted into an efficient CIM approach that will improve poor competitive manufacturing performance of the company, exploit positive features of current manufacturing automation technology, and meet the company's budget, time, personnel, and organizational constraints. A custom-made CIM configuration with specified operations management, technology, and personnel features, parameters, and capabilities is created. At the end of this phase, a capable CIM organization is established to implement identified CIM strategies under a designed CIM plan. This phase also starts a gradual buildup of CIM resources and companywide CIM awareness.

3.5.3 Phase 3. Micro-CIM implementation

It is difficult to overestimate the organizational and business impact of premature change from a predictable conventional manufacturing environment to an uncertain, embryonic CIM environment. Therefore, the CIM configuration developed in phase 2 should first be applied to a limited business area. The results of this trial, micro-CIM implementation, will provide the basis for a CIM configuration and implementation approach that can later be expanded over a larger business area.

3.5.4 Phase 4. Transition to full-scale CIM

The final (but indefinite) phase of a CIM program is to convert and expand a micro-CIM environment into a full-scale CIM environment. First, the efficiency of the micro-CIM environment is assessed in an analysis similar to that of phase 1. This assessment may result in significant modifications of the initial CIM plan and may indicate that new forms of management and organization are needed for a full-scale CIM program. Figure 3.7 shows a schedule of a CIM program of a communication equipment manufacturing company.

3.6 Eight Design Steps

During the four phases, a company must perform eight design steps:

1. Assess the role of manufacturing as a competitive factor in the targeted market
2. Analyze CIM feasibility
3. Develop an organizational form and a management control mechanism for the CIM program
4. Design a custom-made CIM configuration

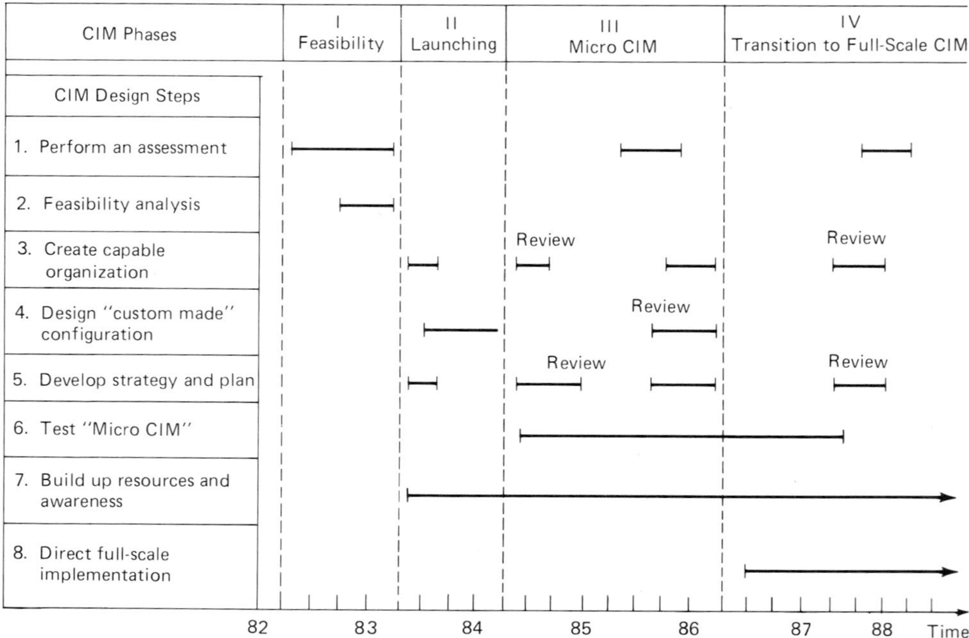

Figure 3.7 CIM program schedule.

5. Develop a CIM plan
6. Test micro-CIM
7. Build up necessary resources and companywide awareness of CIM
8. Conduct full-scale implementation

3.6.1 Step 1. Assess the role of manufacturing as a competitive factor in the targeted market

This step begins with a manufacturing assessment of the company's current and planned business and marketing strategies. This exercise distinguishes the role of the company's manufacturing capability as a competitive factor in a targeted market segment from other factors, such as marketing and distribution capability. This assessment should be performed separately for each market strategy (e.g., defend or expend current product market share and penetrate new market segment). The results should portray the manufacturing profile and competitive status in targeted market segments with reasonable accuracy.

The manufacturing profiles of major current or potential competitors (domestic and foreign) should include competitive strengths and weaknesses of:

1. Advanced manufacturing characteristics:
   - Product originality
   - Process technology used
   - Manufacturing automation penetration and forecast
2. Manufacturing output parameters:
   - Product turnaround time
   - Product quality
   - Flexibility of product volume and variety
3. Manufacturing economics and organization:
   - Experience curve
   - Manufacturing cost structure
   - Budget (R&D, capital expenditures, and so on)
   - Organization and management infrastructure
   - Manufacturing size and degree of consolidation
   - Capacity utilization

The next part of this step involves distinguishing the competitive "might" of the competitors' manufacturing automation from the impact of their product originality and unique process technology capabilities. It is important to estimate not only what the competitions' manufacturing automation and CIM environment looks like but also how effectively it is used to achieve specific manufacturing output parameters and manufacturing economics. All this comparative data will describe *CIM market share,* the contribution CIM makes to a company's market share and competitive position. This can be converted into CIM competitive market share potential, the potential CIM has to contribute to defending or increasing market share. By combining CIM market share potentials for the major competitors, it is possible to characterize a CIM market share, the potential required for successful competition via manufacturing capabilities in a specific market segment.

Boston Consulting Group has created a well-developed two-dimensional strategic planning approach for balancing a company's business portfolio [15] (Fig. 3.8). This approach can be extended to three dimensions to demonstrate the degree of manufacturing automation "might" required (best and worst case) in a real competitive environment to convert a star business into a cash cow (Fig. 3.9). Figure 3.8 portrays the star business of an analyzed company (circle 1) vis-à-vis its major competitors (circles 2 and 3) with cash cow businesses. Figure 3.9 shows how the competitors' CIM maturity and automated manufacturing environment (*B* and *C*) help them to defend their cash cow market shares against any challenge from another

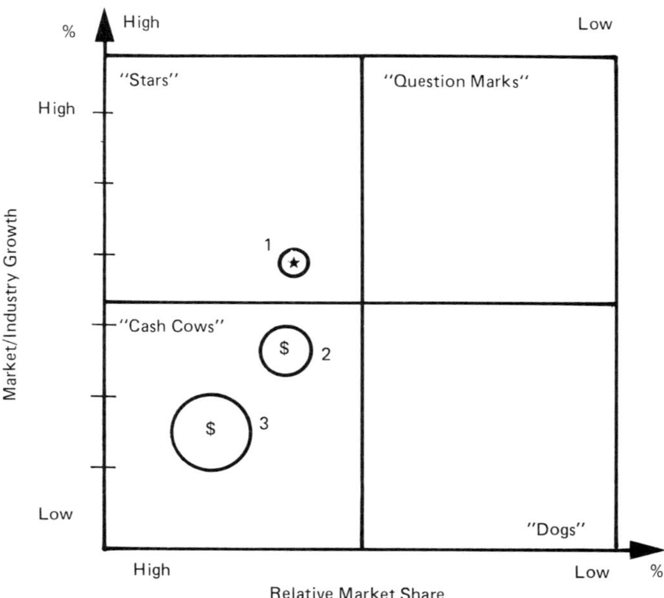

Figure 3.8 Market growth share comparative matrix.

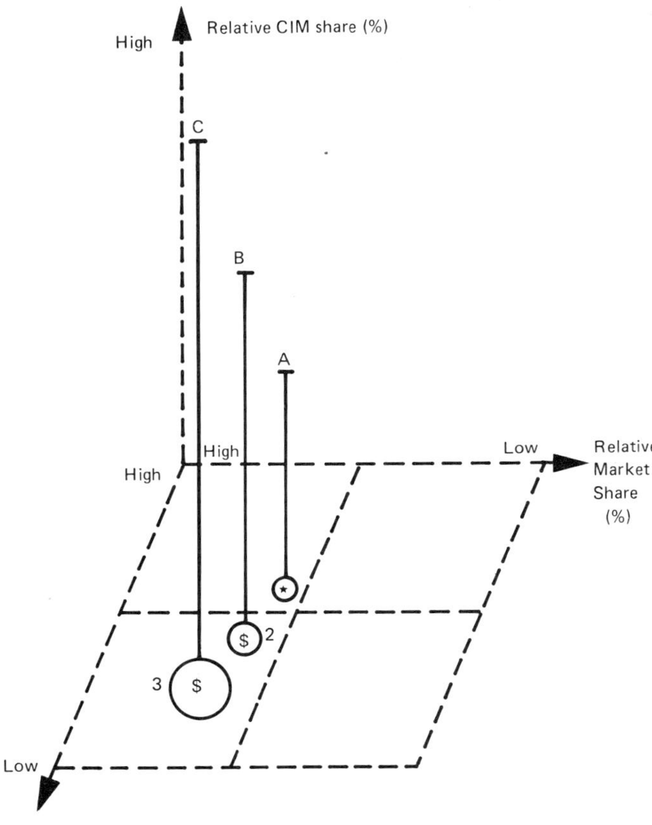

Figure 3.9 CIM growth share comparative matrix.

company with a current CIM market share (*A*). This analysis may discourage a company from undertaking this business "conversion" strategy because the required CIM market share cannot be captured within a certain time and budget. The analysis will also help a company to set realistic expectations for an appropriate return on investment, a step that is vital for developing a focused CIM program.

3.6.2 Step 2. CIM feasibility analysis

The major objectives of step 2 are to:

- Assess the company's current manufacturing automation status vis-à-vis the status required for successful competition in a targeted market with the CIM market shares potential identified in step 1

- Determine noncompetitive manufacturing performance parameters and identify the major bottlenecks in the manufacturing operations flow and lack of key automation technologies responsible for those parameters
- Design the structure of the CIM environment required to eliminate manufacturing performance parameters and estimate the resources and time required to implement that environment

In order to achieve the first objective, a company should conduct the following assessments.

Audit company manufacturing automation expenditures Collect historical, present, and projected financial data. This data will enable the company to use trend analysis to calculate the level of automation expenditures for the affected manufacturing functions. It will also enable the company to perform ratio analysis of automation resources and engineering applications activities (in the form of systems to support manufacturing).

Figures 3.10 and 3.11 demonstrate the background of CAD/CAM expenditures over 4 years of design and production of a new communications system (designed in 1979–1980). In 1982 a consulting company was called in to assess the new CAD/CAM plan for the 1982–1984 period. A detailed analysis of the technical and financial backgrounds of the CAD/CAM program for 1979–1982 and the program's contribution to planned manufac-

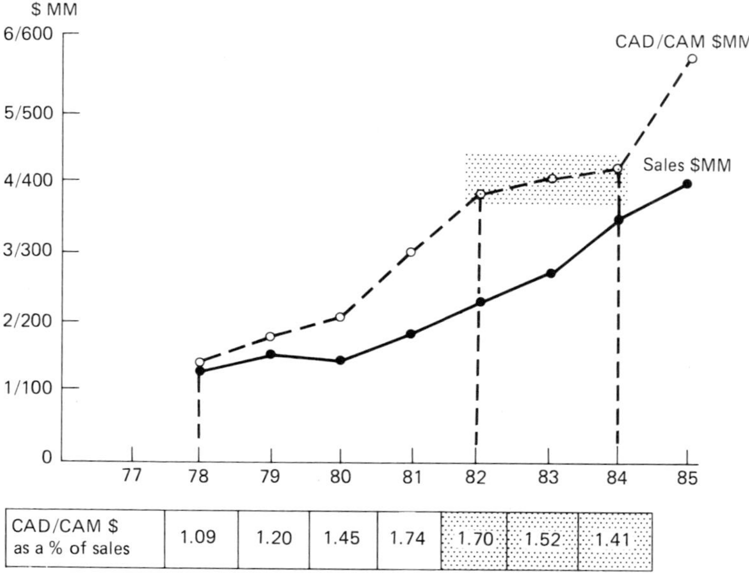

CAD/CAM $ as a % of sales	1.09	1.20	1.45	1.74	1.70	1.52	1.41

Figure 3.10 CAD/CAM expenditures vs. sales.

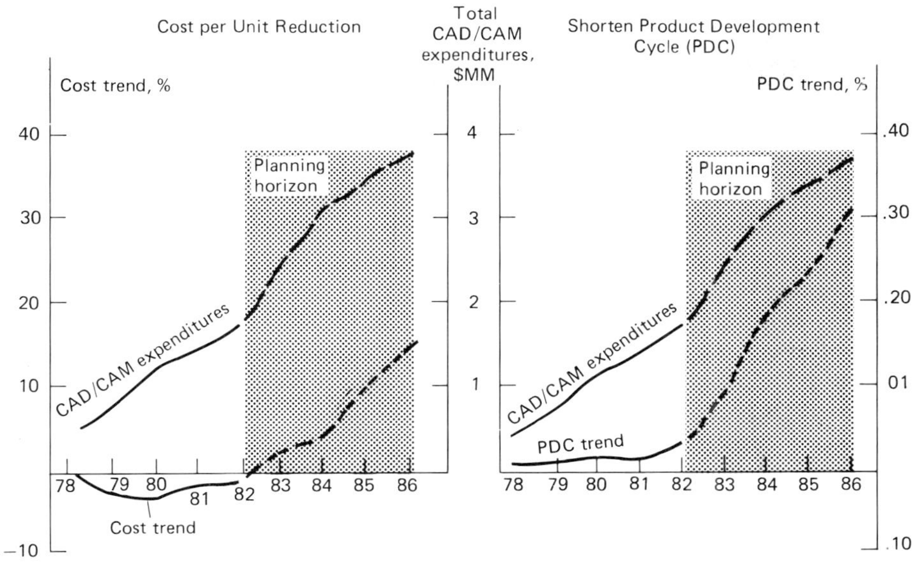

Figure 3.11 Impact of CAD/CAM expenditures on company's business objectives.

turing performance improvement uncovered an anomaly in the program's budgeting. In spite of the program's demonstrated contribution to shortening the integrated circuit (IC) and printed circuit board (PCB) component development cycle (especially in 1981–1982) and reducing the cost per assembled unit in 1980–1981 (Fig. 3.11), the program budget increase (that had been growing 20 percent annually relative to company's sales) was planned to relatively level off (Fig. 3.10, shaded area). The major reason for this decision was completion in 1980 of the installation of 32 CAE turnkey workstations (primarily for IC and PCB design). The next project requested by mechanical design and manufacturing engineering operations—additional installation of 24 mainframe-based terminals and interface with to-be-networked CAD turnkey workstations—was practically sacrificed in favor of some additional CAD turnkey workstation buildup.

Meanwhile, the 1982–1984 period was critically important for product cost competitiveness because of the beginning of the product's large-scale production (after small-scale production of 1980–1982). The company planned an extremely ambitious cost reduction program (Fig. 3.11) intended to result in significant market expansion and sales growth. CAD/CAM's role in achieving these business and manufacturing objectives was not specified. Figures 3.12 and 3.13 show the recommended alignment for resource and technology management of the CAD/CAM program to support cost reduction in 1982–1984. The alignment also resulted in a steeper increase in the CAD/CAM budget for the 1982–1984 period than was provided for in the company's original plan. However, the increase was consistent with increases in 1980–1982 as a percentage of the company's sales.

Determine manufacturing automation portfolio Analyze the major product manufacturing functions typical for the company and independent of the administrative structure. The company's strategic planner needs tangible information that will help him or her to identify the manufacturing functions with the highest potential for overall manufacturing performance improvement. It is possible to inventory major companywide engineering and production functions and identify those that are amenable to current or potential manufacturing automation assistance. The results of this business (vs. systems) inventory can then be mapped in the form of a manufacturing automation portfolio. This portfolio can be presented graphically in the form of Anthony's triangle [16, 17]. The triangle comprises three layered sections depicting the subjects of the manufacturing functions involving strategic planning, management control, and operational support. These sections are arranged hierarchically with strategic planning at the top. Each section is segmented into manufacturing functions with the potential for performance improvement via manufacturing automation technology. For example, the segment of the operational support section representing industrial engineering consists of routing, time standards, material standards, group technology, and labor standards (Fig. 3.14).

Assess existing manufacturing automation technology Evaluate the degree to which existing automation systems meet the requirements of users of the functions which are related to manufacturing automation portfolio. This

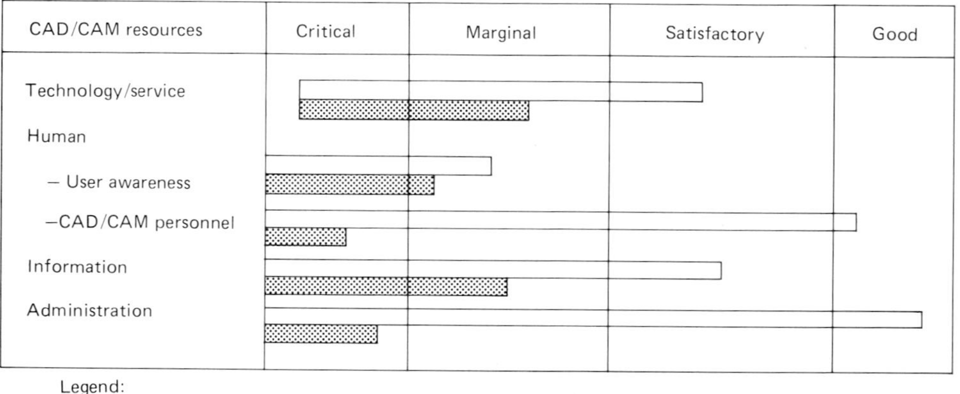

Figure 3.12 CAD/CAM resource management chart for 1982–1986.

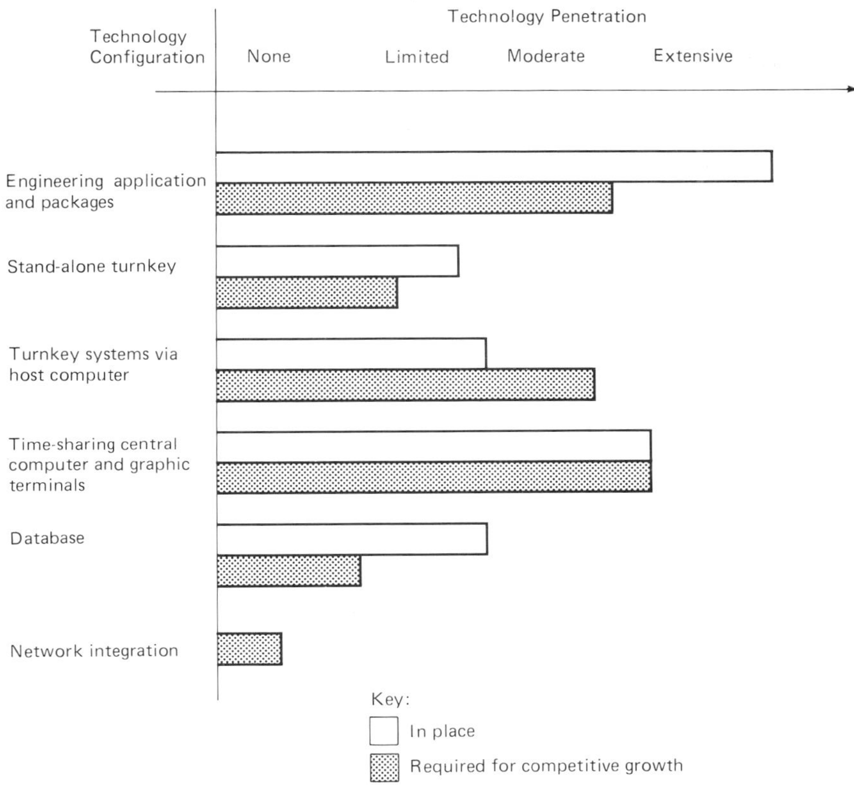

Figure 3.13 CAD/CAM technology management chart for 1982–1986.

evaluation is performed with respect to the system's age and technical functional quality.

Assess the efficiency with which the information and data generated by each major manufacturing automation system companywide are managed. This assessment should emphasize database management and communications issues.

Determine the degree of manufacturing automation technology complexity at the present level of system integration, as shown in Figure 3.2. Assess the effectiveness of the automation systems in supporting engineering and production functions. Assess the contribution of each manufacturing function to selected parameters of a company's manufacturing performance (e.g., lead time and cost). This comprehensive economic and functional assessment of manufacturing automation applications leads to a full understanding of what manufacturing automation is doing for the company's business and how it is doing it [18]. Figure 3.15 is an example

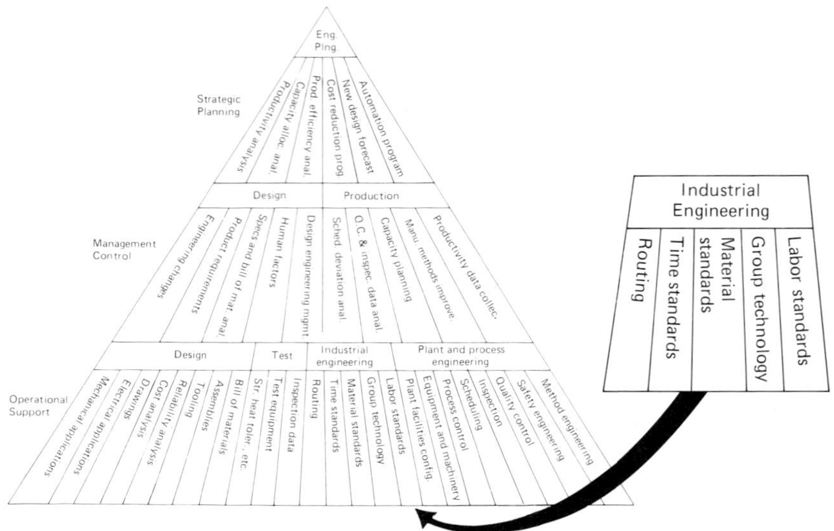

Figure 3.14 Manufacturing automation portfolio identifies opportunities for computer assistance.

of an analysis of CAD/CAM expenditures and their impact (by major engineering functions) on overall company productivity.

Define an organizational form and specify a management control mechanism for automation technology implementation Determine an optimal balance between centralization and decentralization and integration and differentiation of manufacturing automation resources. Evaluate the state of planning, administrative management, application management, steering committees, production services, and the data resource management process in order to design an organizational structure for the effective management of allocated manufacturing automation resources. Evaluate the current and potential balance of power among senior management, DP management, and manufacturing automation users in order to create a workable coalition paradigm (Fig. 3.6).

Assess CIM organizational awareness First assess senior management awareness of and attitudes toward the company's future CIM environment. The results of this assessment should portray (by major manufacturing operations and overall) the senior management's perception of the company's current and projected competitive manufacturing automation status—as is and to be. Major items in the as-is part of the assessment may include:

- Main areas of concern about the company's overall business performance

- Critical areas of manufacturing operations (e.g., R&D, DP, engineering, production, and procurements)
- Names and priorities of the critical manufacturing performance issues (e.g., lead time, quality, and cost)
- Perception of the efficiency of the company's existing manufacturing automation tools and their impact on competitive manufacturing performance (overall and by major manufacturing operations)
- Perception of competitors' manufacturing automation statuses and their competitive impact
- Perception of major causes of poor competitive manufacturing performance (e.g., level or management of manufacturing automation resources, operations and management infrastructure and operations synergy, products, and market position)

Major items in the to-be part of the assessment may include:

- Envisioned status (5- to 7-year horizon) from marketing and operations management standpoints
- Envisioned manufacturing automation level required to achieve desired competitive operations management environment (e.g., technology and integration)

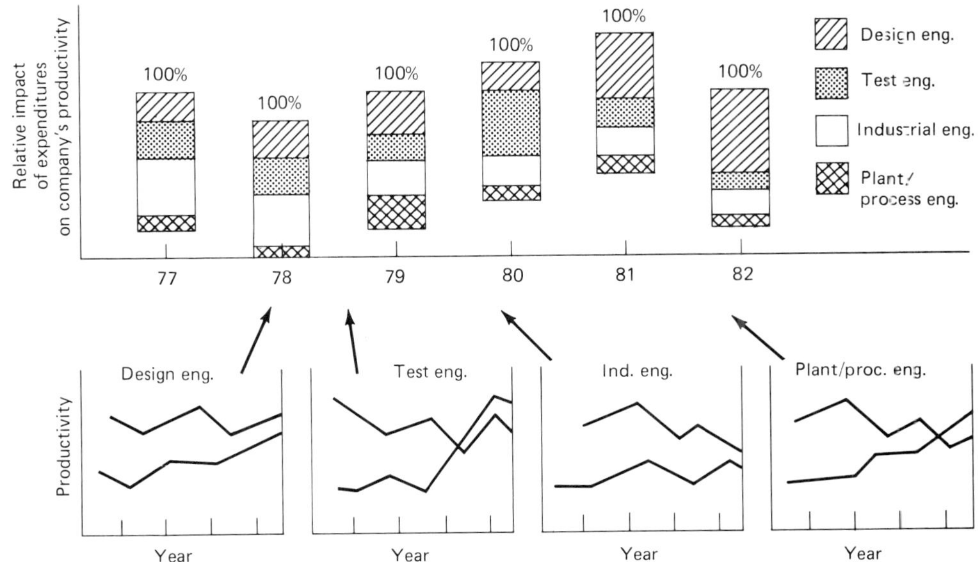

Figure 3.15 Impact of CAD/CAM expenditures by major engineering functions on overall company productivity.

- Known CIM technologies: other companies' implementation efforts that ought to be adopted or implemented by the company and associated benefits
- Estimates, schedule, and justification for a reasonable magnitude of CIM investment over the next 5 to 7 years
- Description of an efficient CIM organization (e.g., leadership and management)

It is extremely important that the key executives who will become CIM management users reach a consensus on what the future CIM environment will be like and how it will facilitate their operations management environment, who will provide CIM management and leadership, and how to measure and justify the CIM program's contribution to the company's competitiveness. Ultimately, overall senior management expectations of a CIM program must be documented in order to provide coherent, long-term guidance for the design and implementation of a companywide CIM environment.

The second part of the assessment of CIM organizational awareness is performed for future CIM technical users.

- Assess, both inside and outside the company, the personnel available to support CIM technology
- Assess the degree of automation technology awareness, needs, and capabilities of all CIM technical users and the budgets and resources available to support a CIM implementation schedule
- Define technical management skill requirements
- Assess performance management and control programs for ongoing automation tasks

Figure 3.16 represents an example of a CAD/CAM technical user awareness assessment.

A comprehensive analysis of consolidated results of the five assessments can identify the status of the four CIM growth components (Fig. 3.2) and their current balance. This information, combined with appropriate industry benchmarks and competitors' statuses, determines the company's stage of CIM development vis-à-vis the CIM position of major competitors. This comparison identifies major differences between the organization, design, implementation approaches, and budgeting of the company's CIM programs and those of its major competitors. It also identifies alternative CIM program alignments which the company could use to achieve competitive manufacturing performance.

To complete the objectives of step 2, the company must investigate key parameters of noncompetitive manufacturing performance (e.g., cost, lead

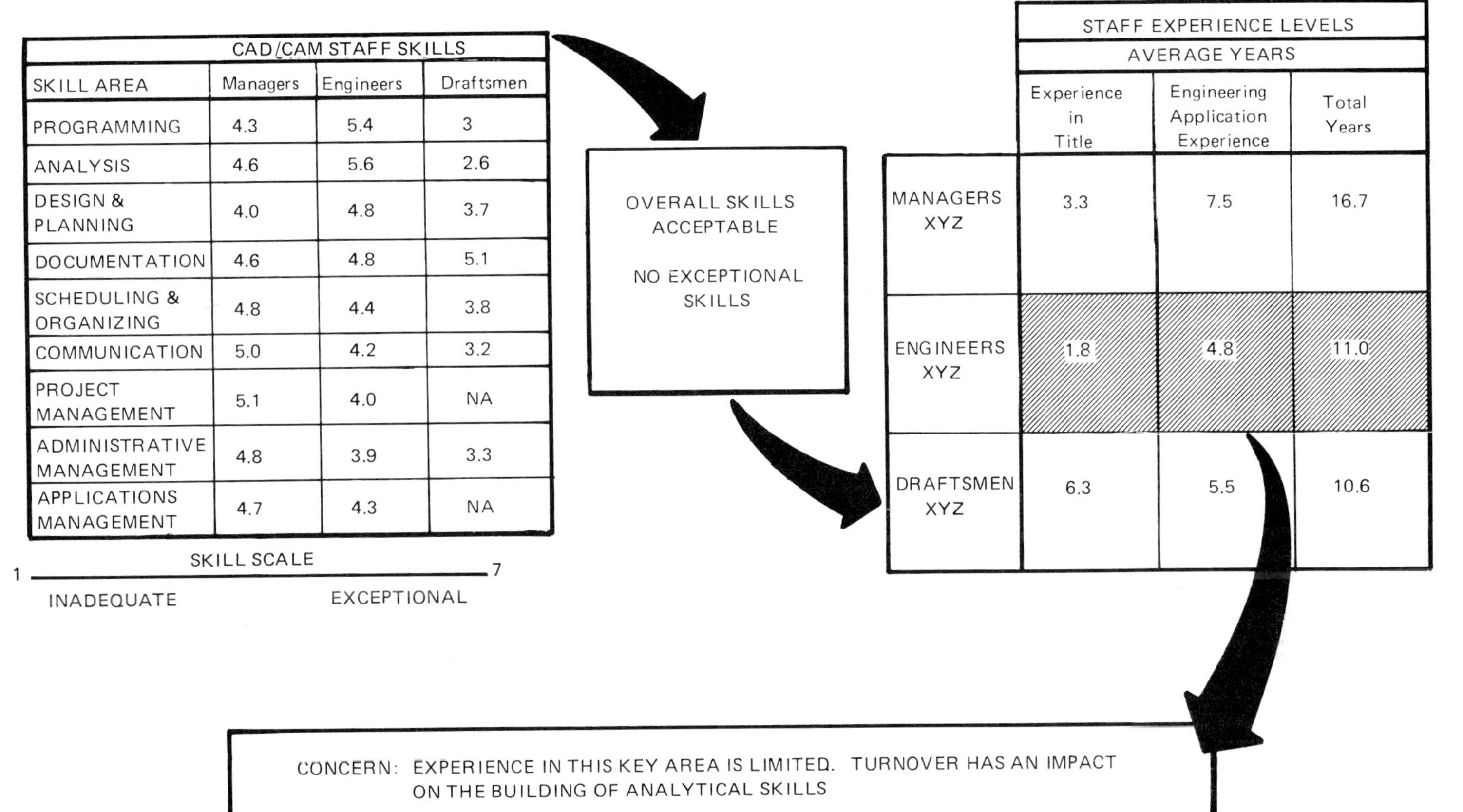

Figure 3.16 An example of CAD/CAM personnel skills and experience assessment.

time, and productivity), identify the major bottlenecks of the current manufacturing operations flow, and determine how implementing new manufacturing automation technologies and/or better integrating existing technologies can facilitate desired improvements.

An analysis of product manufacturing records permits creation of manufacturing performance patterns (e.g., a pattern of manufacturing labor costs and engineering change costs) for the company's major manufacturing operations [19]. This analysis will identify cost or lead time bottlenecks in the operations functions. A value-added analysis of the product cost records will produce an understanding of what manufacturing functions are doing for the company's competitiveness (e.g., cost and lead time). This analysis can focus on one of a few manufacturing performance characteristics, such as cost, productivity, or lead time. Figures 3.17 and 3.18 show the results of cost and lead time value-added analysis performed for an automotive component manufacturer.

Integration of these focused analyses can identify the strengths and weaknesses of specific manufacturing performance characteristics of the manufacturing organization. For a manufacturing automation portfolio, the assessment should be conducted from two perspectives:

- To assess the effectiveness of manufacturing automation systems in supporting the manufacturing functions involved
- To assess the contribution that each manufacturing function makes to the required improvement of overall manufacturing performance

The results of this two-part assessment can be used to balance the company's automation portfolio by distributing resources to the functions that will make the greatest contribution to the company's competitive improvement.

Figure 3.19 shows the assessment procedures that focus on overall productivity improvement via CAD/CAM technology. There is a difference between the engineering management perspective and the business management perspective. In the first perspective an attempt is made to implement CAD/CAM technology for the engineering functions which are most receptive to or experienced with automation, as well as those which will be supported effectively by the CAD/CAM system. This "domestic" point of view does not take into account overall company parameters such as productivity. An assessment of an engineering function's contribution to a company's productivity quantitatively justifies the real value of that function for a company's business. With a balanced automation portfolio, it takes into account company productivity improvement after the investment of CAD/CAM resources into the identified high-potential engineering functions.

Finally, for each manufacturing function in the automation portfolio, the

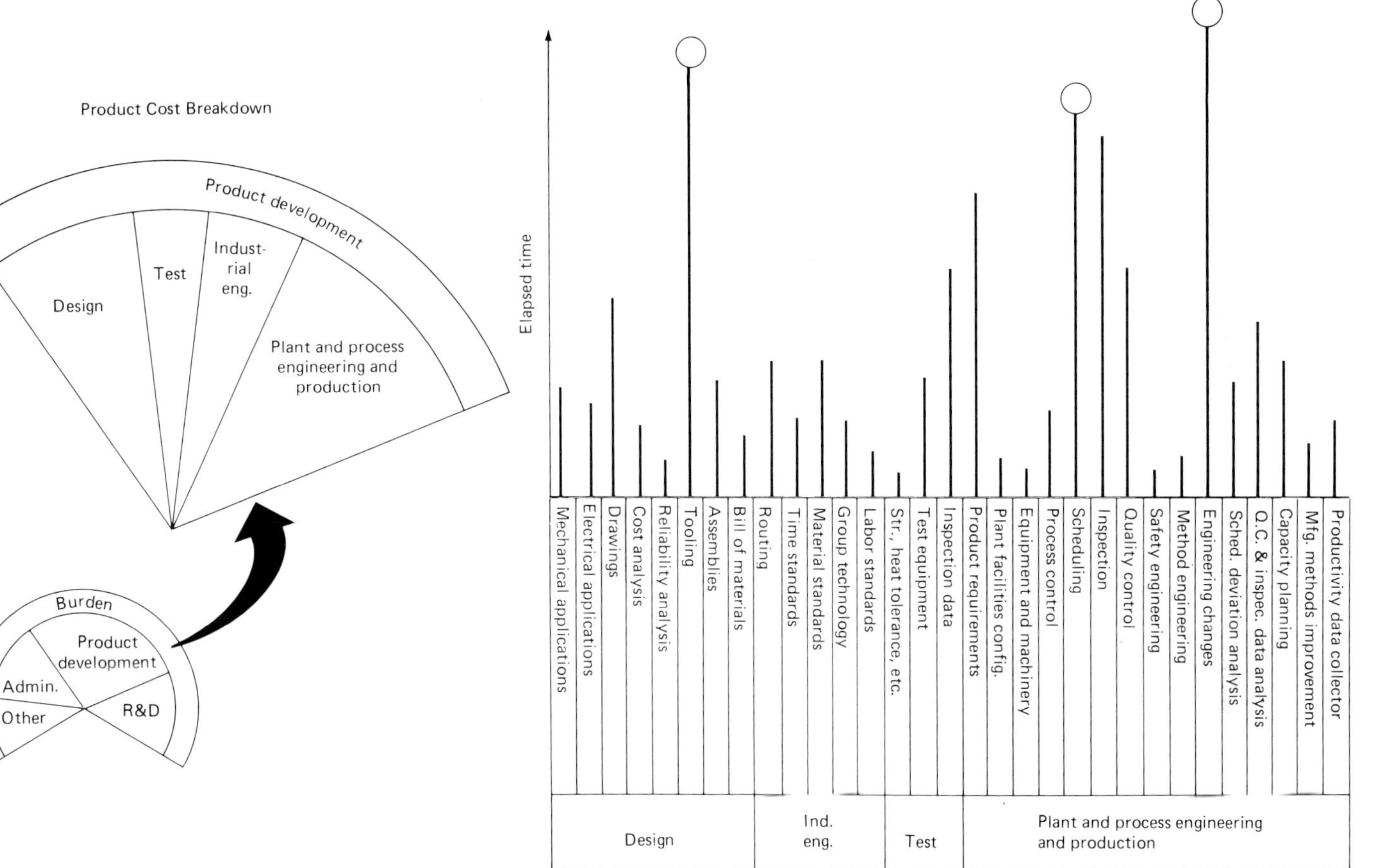

Figure 3.17 Lead time value-added analysis.

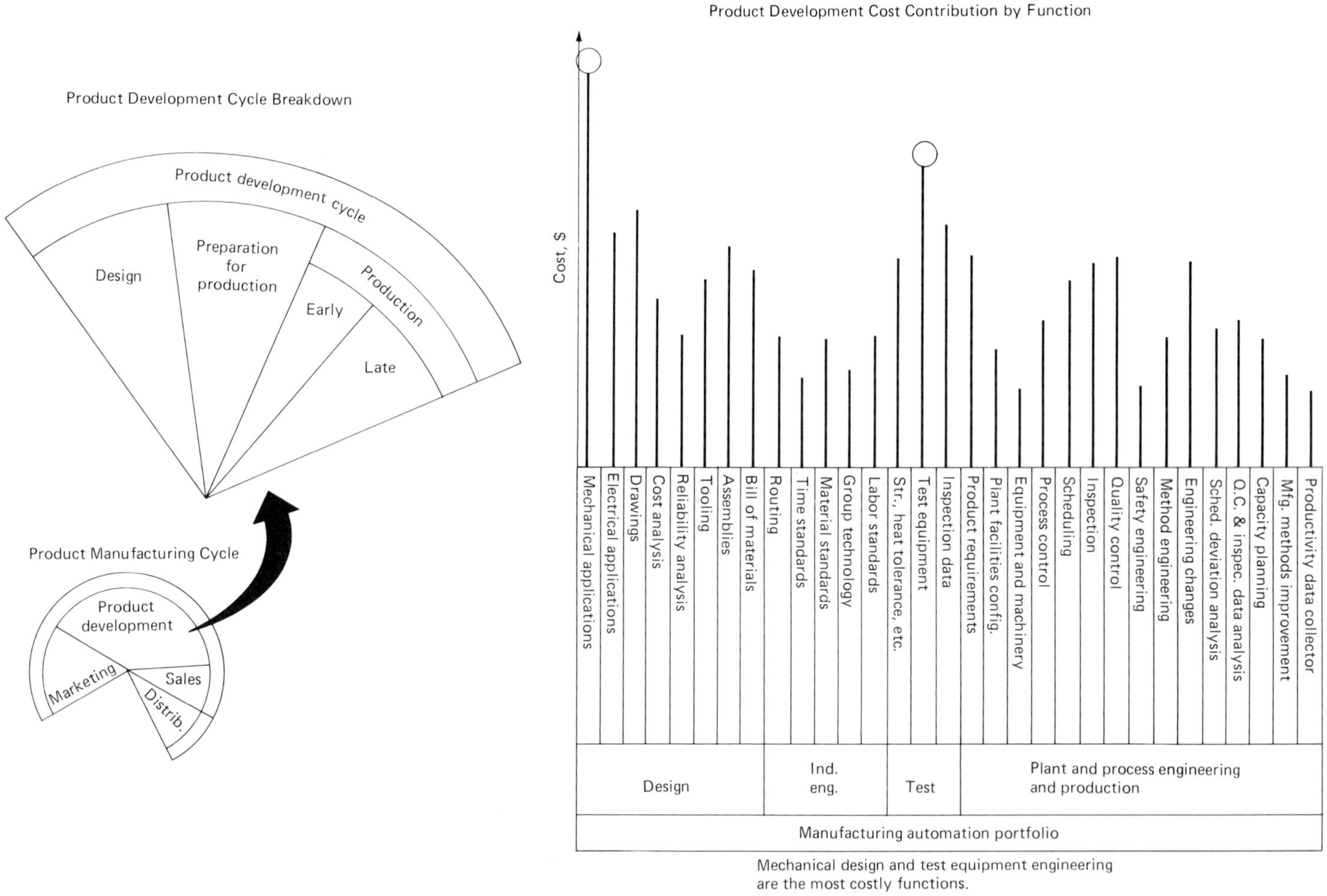

Product Development Cycle Breakdown

Product development cycle
Design
Preparation for production
Production
Early
Late

Product Manufacturing Cycle

Product development
Marketing
Sales
Distrib.

Product Development Cost Contribution by Function

Cost, $

Mechanical applications
Electrical applications
Drawings
Cost analysis
Reliability analysis
Tooling
Assemblies
Bill of materials
Routing
Time standards
Material standards
Group technology
Labor standards
Str., heat tolerance, etc.
Test equipment
Inspection data
Product requirements
Plant facilities config.
Equipment and machinery
Process control
Scheduling
Inspection
Quality control
Safety engineering
Method engineering
Engineering changes
Sched. deviation analysis
Q.C. & inspec. data analysis
Capacity planning
Mfg. methods improvement
Productivity data collector

Design
Ind. eng.
Test
Plant and process engineering and production

Manufacturing automation portfolio

Mechanical design and test equipment engineering are the most costly functions.

Figure 3.18 Cost value-added analysis.

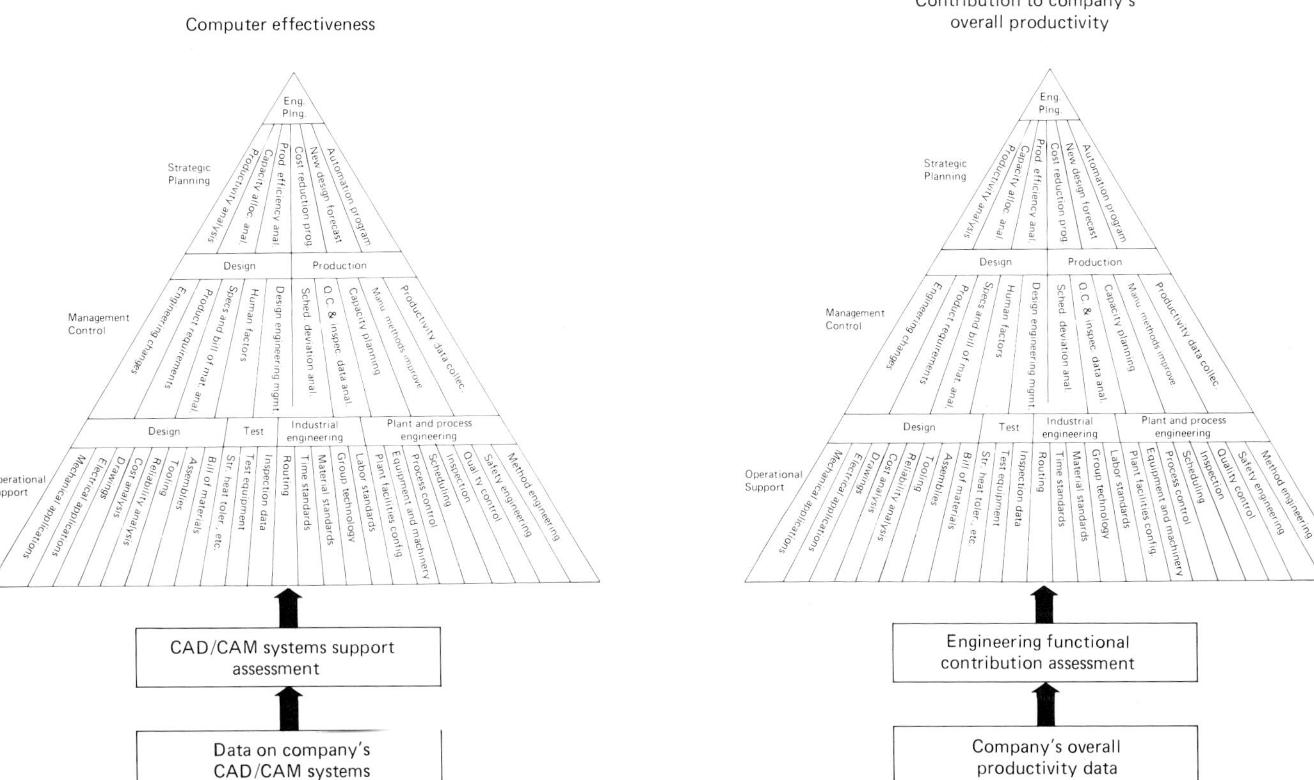

Figure 3.19 Productivity improvement assessment procedures.

company can document the data and interface requirements associated with the operational nature of the function, current automation status, and potential contribution to improvement of certain manufacturing performance characteristics. By using structured functional analysis techniques (e.g., flowcharts, SADT, and IDEF), it is possible to map the current information flow of the functional operations management structure for major manufacturing functions independently of their organizational boundaries. Operations may be performed by an individual, a team, or a computer system. Workshops and interviews with technical users' management have proved efficient information-gathering tools. These interviews should emphasize three important questions:

- What are the objectives of your job?
- What functions do you and your personnel perform (manually or with a computer) in order to accomplish those objectives?
- What information do you need in order to perform your job more efficiently? When do you need it and in what form?

The information should be identified in terms of data elements, and it should emphasize what is needed to accomplish the function better than that which is done by the present system. The current and potential workload of all functions should be analyzed from an information standpoint (e.g., repetitive vs. new and retrieving, data storage, and representation requirements). A final top-down flowchart will provide an efficient tool for identifying critical functions that cause information bottlenecks through

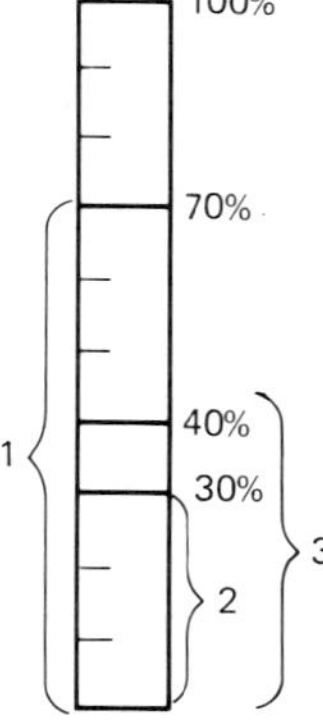

1. Engineer's desired CAD/CAM support
2. Current CAD/CAM support
3. Beneficial CAD/CAM support for company's productivity improvement

Figure 3.20 The results of final analysis of the assessments for a mechanical design function.

their internal or interface inefficiency. Only after these tasks are completed can the actual efficiency of current manufacturing automation support be computed (calculated on a companywide scale) for major manufacturing functions. Figure 3.20 shows that in spite of mechanical engineering's demand for a 40 percent increase in current CAD/CAM support (currently 30 percent), only the first 10 percent would really improve the company's overall engineering productivity. That is because this operation has limited importance to the company's overall engineering business; the current CAD/CAM system is not used efficiently (95 percent capacity for drafting only), and its configuration does not interface with manufacturing engineering for quick engineering changes review.

The final integration of the assessment of all functions of the manufacturing automation portfolio can be presented graphically as a map of opportunities [20, 21, 22] for manufacturing automation support. Figure 3.21 shows the results of comparative evaluation of the CAD/CAM investment opportunity for an overall engineering productivity improvement in automotive component manufacture. This map measures the effectiveness of current CAD/CAM systems—the shaded area inside the circle. The degree of contribution to a company's business improvement for each function is keyed by the overall size of the circles.

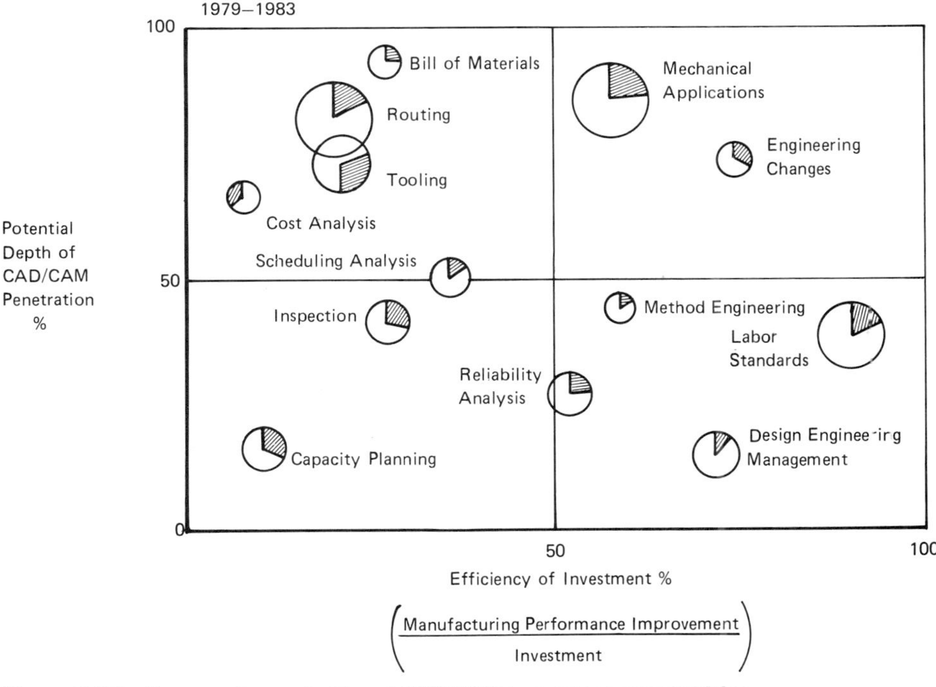

Figure 3.21 Comparative evaluation of CAD/CAM investment opportunities.

3.6.3 Step 3. Develop an organization form and management control mechanism

In order to determine an appropriate organizational form for a CIM program and an effective control mechanism to manage allocated manufacturing automation resources, it is essential to assess the current status of technology planning, administrative management of automation application development, the role and structure of steering committees, and the manufacturing data resource management process.

Observation of a number of implementation efforts shows that there are four major organizational approaches:

- Electronic data processing (EDP) project
- User project
- Matrix project
- Department project

By the EDP project approach, CIM development, purchase, and installation are delegated to the data processing or management information system (DP or MIS) department, which then provides CIM computer service to the engineering and production departments and users. The advantages of this approach are:

- The personnel involved have a thorough knowledge of computers and software development and installation.
- It offers the potential for integration with MIS.

The disadvantages of the approach are:

- Members of this department lack engineering knowledge.
- The approach may result in poor interaction with engineering users.
- It is service-oriented rather than user-oriented.
- It may cultivate empire building.
- It relegates CIM to the role of an orphaned child.

By the user project approach, CIM development, purchase, and installation are placed under a single entity (e.g., a plant or division) both managerially and physically. They are totally independent of DP–MIS and other users. The advantages of this approach are:

- Program managers have a detailed understanding of user needs and requirements.
- It offers total user involvement.
- It enjoys implementation and management flexibility.

The disadvantages are:

- Program managers lack computer knowledge.
- Data may be duplicated.
- The approach can lead to companywide software and hardware standards chaos.
- It offers poor potential for integration or communication with MIS.
- It limits the capacity of the database.

Using the matrix project approach, CIM development, purchase, and installation are placed under the vice president of manufacturing, technology or engineering, with the direct users' development groups involvement. A users CIM steering committee is established for budget scheduling and project management and control. Service is provided and shared by the users themselves. The advantages are that it:

- Represents a companywide effort
- Offers the potential for integrated user database
- Uses unified software and hardware
- Is user-oriented
- Is democratic
- Is flexible to user needs

The disadvantages are:

- Managers lack computer knowledge.
- The CIM group may compete with DP/MIS.
- The approach can lead to dispute over priorities.
- It involves users only part time.
- It relies on "consortium" budgets.
- It suffers from implementation and management inflexibility.

Using the CIM department approach, CIM development, purchase, and installation fall under a "specially appointed" manager of a newly formed department. A task force can be hired from outside or recruited from users. Service is provided to all users in a "technical DP" manner. The advantages are that it:

- Has a single budget
- Enjoys strong management control
- Offers companywide standardization

- Brings in new personnel with outside experience
- Offers system integration
- Enjoys management flexibility

The disadvantages are that it:

- Is service-oriented rather than user-oriented
- Offers poor interaction with users
- Encourages empire building

The analysis of the current and potential balance of power among senior operations management, DP management, manufacturing automation, and users taking the coalition paradigm approach (Fig. 3.6) can identify a workable organizational form for a CIM program.

3.6.4 Step 4. Design custom-made CIM configuration

"Rewiring" current information flow pattern to eliminate bottlenecks of critical operations and functions should be performed on the basis of the potential for competitive manufacturing performance improvement of the functions as identified on the map of opportunities. The information pattern includes configuration, data representation, data storage and distribution, origins of data suppliers and users, and list of procedures and systems for manual and computer-aided information generating, processing, and distribution. Usually, there are several ways to accommodate this tradeoff exercise. As a result, the first part of designing a CIM system configuration involves developing several alternatives and their associated budgets for manufacturing technology modification, acquisition, and integration, personnel hiring and training, data resource management, and overall program supervision and control.

It is important (but often overlooked) that a description of the CIM configuration be translated from system, data, and communications terms into the form of a scenario of the company's operations in the new CIM environment. This top-down scenario, written for managers of the company's major operations, should provide a full understanding of changes in the way of doing business: operations management, implied technical, personnel, and management requirements, and expected performance improvement benefits and their measurement.

3.6.5 Step 5. Design CIM plan

The evolving process of CIM plan design and modification will be efficient only if future management and technical CIM users participate actively.

Thus, during the complex process of establishing a CIM program, it is essential to provide efficient communications among all layers of management. One of the most important communications tools involves documenting the findings and recommendations of the preceding CIM design steps. In addition, the form of the CIM plan documentation should be acceptable to all participants. The form should allow for alternative plans and trade-offs. The plan itself should outline long-term CIM strategies (usually 5 to 7 years) and also specify tactical development and acquisition actions on an annual basis. A model CIM plan structure follows [23]:

CIM plan structure

Management summary:

1. CIM growth requirement to support the business plan
2. CIM alternatives
3. Management, technology, and personnel implications and requirements
4. Feasibility conclusions
5. Selected CIM program

Current status of company's manufacturing automation:

1. Manufacturing automation portfolio: structure and automation penetration
2. Investment history and resources inventory (technology, personnel, and data)
3. System integration and information management
4. Benefits control
5. Value-added analysis for major manufacturing functions and their operations interfaces
6. Critical portion of automation portfolio

Manufacturing competitive position:

1. Business plan summary
2. Manufacturing competitive status
3. Critical CIM users
4. CIM growth requirement to support the business plan
5. Additional CIM benefits
6. Map of CIM investment opportunities

CIM directions and goals:

1. CIM operations management configuration vis-à-vis current operations structure
2. CIM alternatives scenarios
3. Management, technology, personnel, and budget implications and requirements
4. Feasibility conclusion

Selected CIM program:

1. CIM organization and leadership
2. Assessment of the alternatives (technical, business, and implementation)
3. CIM scenario of selected program
4. Structure of selected CIM program
   - Organization
   - Projects and schedule
   - Budget and expected benefits
   - Risk assessment
   - Critical success factors

Implementation strategy:

1. Implementation schedule (annual basis)
   - Implementation outline (projects)
   - CIM penetration into manufacturing automation portfolio
   - Automation resources buildup and management
   - Acquisitions and internal development
   - Installations and training
   - Milestones: integration and training
2. Progress monitoring and control
   - Expenditures and project completion control
   - Benefits milestones
3. Micro-CIM
   - Relation to overall business
   - Degree of extension

3.6.6 Step 6. Test micro-CIM

Selecting an appropriate manufacturing area for micro-CIM implementation is an important technical and business decision. Several guidelines should be followed: (1) The selected product line or its component should be suitable for conversion into full-scale CIM. (2) The manufacturing operation should involve extensive interaction between design and production. (3) Failure to achieve planned competitive benefits from micro-CIM implementation should not damage a company's overall business. (4) Risk assessment analysis for a micro-CIM project under consideration should guarantee a high probability of success. One suggested approach for risk assessment is three-dimensional project description (Fig. 3.22):

1. Magnitude of the project
   - Percentage of overall business involved and potential impact on that business
   - Company activities and personnel involved
   - Project budget
2. Potential for extension to full-scale CIM
   - Supports full-scale CIM concept
   - Results and experience applicability
   - Systems' integration extendability
3. Technology
   - Systems availability and compatibility (hardware and software)
   - Availability of sufficient development resources
   - Capability to integrate the system and develop additional software

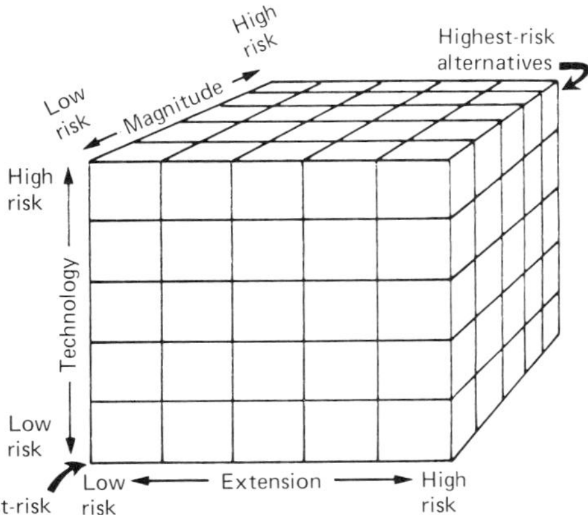

Figure 3.22 Risk assessment of micro-CIM alternatives.

3.6.7 Step 7. Build up resources and awareness

During phases 2 to 4, one of the most important CIM success factors is efficient buildup of CIM human resources and companywide CIM awareness. Tables 3.1 and 3.2 describe typical problems that companies experience and improvement opportunities associated with this step.

3.6.8 Step 8. Direct full-scale implementation

The implementation results of the micro-CIM program should be reviewed and analyzed from a technical, management, and economics standpoint vis-à-vis the rest of the company's manufacturing business assisted by conventional manufacturing automation. This analysis should focus on factors that contributed to the success of the micro-CIM project. It is important to evaluate how manufacturing automation technology has evolved since the start of micro-CIM implementation and whether the micro-CIM integration concept is viable for full-scale CIM. Possibly one more (probably larger) micro-CIM effort will be required to confirm that the company is ready to proceed with full-scale implementation.

3.7 Conclusions

Manufacturing automation technology has unexplored potential in spite of the fact that it is becoming more integrated into the mainstream of manufacturing business. Many factors influence the automation technology capital investment decision. The experience of a number of companies demonstrates that even significant investment in automation technology (e.g., CAD/CAM) does not always increase overall manufacturing performance to the point where it contributes the ultimate business goals of a

TABLE 3.1 Improving CIM Awareness

Lack of awareness of	Ways to improve awareness
General concept	Conduct education programs on CIM: • Concept • Strategy • Tools and integration • Benefits • Problems Increase exposure to ongoing CAD/CAM efforts
State of the art	Publish CIM newsletter Make outside presentations
Company's approach	Develop CIM scenario Encourage high micro-CIM exposure

TABLE 3.2 Human Resources

Problem	Improvement opportunities
CIM leadership	CIM policy and program Assign motivated top manager
CIM operations people CIM people development	Overall CIM human resource development program: ▪ Conduct skill inventory ▪ Obtain critical personnel ▪ Develop CIM scenario ▪ Conduct training programs ▪ Hire from a company with an advanced CIM program ▪ Develop an incentive program ▪ Hire or develop full-time CIM personnel

company. Often the immediate benefits achieved by single operations do not improve the company's competitiveness as expected. In these cases, large-scale manufacturing automation programs have not been designed to address the company's problems of poor competitive manufacturing performance [24, 25, 26].

The required improvements in experience curves of critical operations often have not been fully achieved because automation investment programs are unfocused and unintegrated. Thus, the typical approach—"let's get started and we'll plan later"—becomes very expensive in the long run. "Quick" implementation of randomly selected automation tools, without prior competitive and operations management analyses, promises only "quick" return on investment for a few manufacturing operations. However, this approach will not achieve the critical return on investment: sustained improvement of manufacturing performance so the company can compete efficiently in the marketplace.

Successful development of a custom-made, companywide CIM environment requires a long-term commitment, the leadership of senior business management, and an integrated effort by the various manufacturing operations. Business-driven, long-range CIM planning is essential for success.

ACKNOWLEDGMENTS

The author would like to thank Mr. Tom Gunn and the late Dr. J. Harrington (Arthur D. Little, Inc.), Messrs. R. Hermon-Taylor and T. Hout (Boston Consulting Group), Prof. W. Skinner (Harvard Business School), Dr. R. Nolan (Nolan, Norton Co.), Dr. Y. Mizuno (Nippon Electric Corporation), Messrs. E. Hashimoto and R. Hidaka (Fujitsu, Ltd.), Mr. D. Edison (Westinghouse Corp.), Mr. W. Carter (CAM-I, Inc.), and the many Japanese executives and manufacturing managers who attended numerous seminars delivered by the author as an honorary adviser to the Japan Management Association for their helpful discussions, his wife, Evgenya, for her inspiration and encouragement, and his daughter Michelle for her patience.

REFERENCES

1. "Factories of the Future as a Management Challenge," *New England Business,* May 17, 1982.
2. L. Lipchin, "CAD/CAM Program as a Strategic Tool for Manufacturing Business Planning," *Proceedings,* CAMP '83, March 1983, Berlin, W. Germany.
3. Joseph J. Harrington, *Computer Integrated Manufacturing.* Industrial Press, New York, 1973.
4. Joseph J. Harrington, *Understanding the Manufacturing Process.* Marcel Dekker, Inc., New York, 1984.
5. L. Lipchin, "The Role of Strategic Planning for Productivity Improvement through Computer-Aided Engineering," *Proceedings,* Compcon '81, IEEE, Washington, D.C., September 1981.
6. L. Lipchin, "Strategies for Integration of CAD, CAM, GT and Robotics into Factory Automation," *Nikkei Mechanical,* 12–19, December 1983, Japan.
7. L. Lipchin, "CAD/CAM Users Can Learn from Data Processing's Mistakes," *Computer Graphics News,* May/June 1982.
8. "Factories of the Future as a Management Challenge," *New England Business,* May 17, 1982.
9. L. Lipchin, "Managerial and Strategic Planning Aspect of CAD/CAM Technology Implementation," *Proceedings,* CAD '82, International Conference, March 1982, Brighton, U.K.
10. L. Lipchin, "CAD/CAM Program as a Strategic Tool for Manufacturing Business Planning," *Proceedings,* CAMP '83, March 1983, Berlin, W. Germany.
11. L. Lipchin, "Managerial and Strategic Planning Aspect of CAD/CAM Technology Implementation," *Proceedings,* CAD '82, International Conference, March 1982, Brighton, U.K.
12. L. Lipchin, "Corporate Strategies for CAD/CAM Implementation," *Proceedings,* International Manufacturing Systems Conference '82, July 1982, Buffalo, NY.
13. L. Lipchin, "CAD/CAM Program as a Strategic Tool for Manufacturing Business Planning," *Proceedings,* CAMP '83, March 1983, Berlin, W. Germany.
14. L. Lipchin, "Strategies for Integration of CAD, CAM, GT and Robotics into Factory Automation," *Nikkei Mechanical,* 12–19, December 1983, Japan.
15. D. Abell and J. Hammond, *Strategic Planning,* Prentice-Hall, Englewood Cliffs, N.J., 1979.
16. L. Lipchin, "The Role of Strategic Planning for Productivity Improvement through Computer-Aided Engineering," *Proceedings,* Compcon '81, IEEE, Washington, D.C., September 1981.
17. L. Lipchin, "Managerial and Strategic Planning Aspect of CAD/CAM Technology Implementation," *Proceedings,* CAD '82, International Conference, March 1982, Brighton, U.K.
18. L. Lipchin, "CAD/CAM Program as a Strategic Tool for Manufacturing Business Planning," *Proceedings,* CAMP '83, March 1983, Berlin, W. Germany.
19. L. Lipchin, ICAM "Integrated Decision Support System," Interim Report (Economic Model Chapter), HOS, Inc., March 1980, Cambridge, Mass.
20. L. Lipchin, "The Role of Strategic Planning for Productivity Improvement Through Computer-Aided Engineering," *Proceedings,* Compcon '81, IEEE, Washington, D.C., September 1981.

21. L. Lipchin, "Managerial and Strategic Planning Aspect of CAD/CAM Technology Implementation," *Proceedings,* CAD '82, International Conference, March 1982, Brighton, U.K.
22. L. Lipchin, "CAD/CAM Program as a Strategic Tool for Manufacturing Business Planning," *Proceedings,* CAMP '83, March 1983, Berlin, W. Germany.
23. L. Lipchin, "Corporate Strategies for CAD/CAM Implementation," *Proceedings,* International Manufacturing Systems Conference '82, July 1982, Buffalo, N.Y.
24. L. Lipchin, "Secure Business through CAD/CAM," *Appliance Manufacturer,* September 1982.
25. "Consultant Offers on Strategic Caveat to Manufacturers," *New England Business,* January 6, 1986.
26. L. Lipchin, "Competitive Business Strategies and CIM," *Nikkei Mechanical*, 10–21, October 1985.

PART 4

Implementation and Management

CHAPTER

1

An Approach to CAD/CAM Integration

Robert E. Fulton

1.1 Introduction

For the United States to remain competitive in the world market, improvements in industrial productivity are essential. A key element to improved productivity is the advancement and effective use of computer-aided design and manufacturing (CAD/CAM) technology. To stimulate advances in CAD/CAM technology, a unique joint government and industry project, denoted Integrated Programs for Aerospace-Vehicle Design (IPAD), was carried out from 1971 to 1984 (Fig. 1.1). The project goal was to raise aerospace industry productivity through advances in technology to integrate and manage information involved in the design and manufacturing process. IPAD was carried out under the joint sponsorship of NASA's Office of Aeronautics and Space Technology and the Navy's Materiel Command. The program complemented traditional NASA/DOD research to develop aerospace design technology and the Air Force Integrated Computer-Aided Manufacturing (ICAM) program to advance CAM technology. Work under the IPAD project was done principally through a prime contract with the Boeing Commercial Airplane Company under the guidance of an Industry Technical Advisory Board (ITAB) composed of members of aerospace and Navy contractors and computer companies (Fig. 1.2). ITAB reviews provided a regular forum for over 100 engineering and computer organizations to hold in-depth discussions of critical CAD/CAM issues which directed IPAD research and spurred internal company efforts. IPAD had unprecedented industry support and involvement and provided a unique approach to government-industry cooperation in the development and transfer of CAD/CAM technology. This chapter summarizes the IPAD program from the perspective of a prototype approach to industry involvement, development, implementation, and technology transfer of CAD/CAM integration.

1.2 Issues in CAD/CAM Integration

CAD/CAM integration is a highly complex problem which includes integration of three different technical and management activities—design, manufacturing, and computing—each of which may reside in different management components of a company. Some of the questions which need to be answered before commencing a major CAD/CAM effort include:

1. What are the engineering needs?
2. What are the manufacturing needs?
3. What are the combined needs?
4. What is the long-term (5- to 10-year) goal for CAD/CAM integration: reduce development costs, speed development time, achieve product reproducibility, or solve in-house management issues?
5. What are a set of reasonable short-term goals (1- to 3-year)?

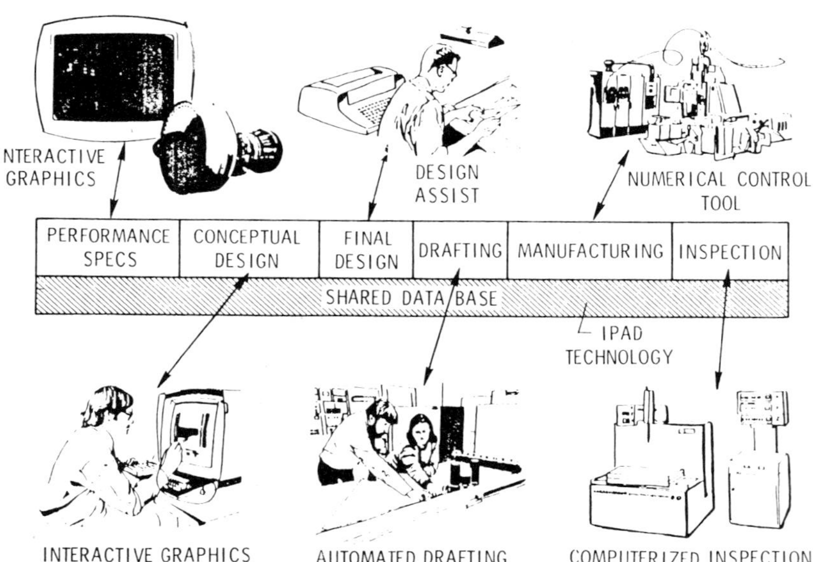

Figure 1.1 Joint government and industry research program to develop technology to manage CAD/CAM information.

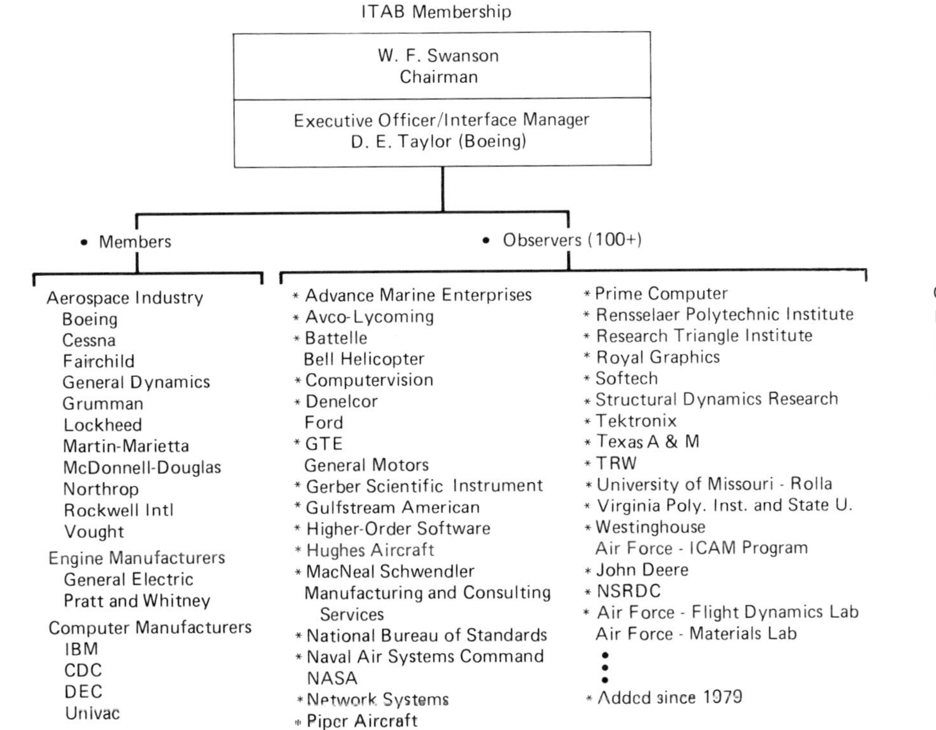

ITAB Activities

Guide development to meet industry needs
Review/critique ongoing work
Evaluate prototype software
Use IPAD technology and products to spur in-house planning and development

Figure 1.2 IPAD industry technical advisory board (ITAB).

6. What are the current status and capabilities of the organization?
7. What resources are available in terms of funds, people, and equipment?
8. How does management deal with failure, overruns, and/or slippages in the high-risk effort?
9. Are the key organizations of design, manufacturing, and computing committed to CAD/CAM integration?
10. Who has responsibility for the integration effort?
11. How much in-house capability is available and is committed to CAD/CAM integration?
12. What external help is available through consultants and contractual support?
13. Is appropriate and competent leadership available to provide a focal point for CAD/CAM integration? Is that leadership acceptable to design, manufacturing, and computing?
14. How long before the company must see a significant return on investment?

Prior to committing to CAD/CAM integration, a long-term development plan should be prepared. It should be developed by a team which represents the CAD, CAM, and computing groups. The plan should have long-term (10-year) and short-term (2- to 3-year) objectives. It should show pertinent ongoing activities in the three separate efforts and how they relate to the integration thrust. Integration products should be defined for the periods of 1- to 3-year, 5-year, and 10-year time frames. The basic plan should be designed to accommodate change, and it should be updated annually. The key development phases include the following:

1. *Define a baseline development process.* Use one or more company products as a baseline and document how development of the product or products has taken place and is expected to take place. Define the tasks which *are* automatable and those which should be automatable. Focus heavily on the information flow through the process and the interfaces which occur.

2. *Develop CAD/CAM integration requirements.* Using the baseline development process as a guide, develop a list of functional requirements for a future integrated CAD/CAM system. These requirements should be jointly developed by engineering, manufacturing, and computing with their representatives participating in the process. The requirements should be reviewed and approved by management from each of the three areas. Ways to test and/or evaluate each requirement should be identified.

3. *Conduct an integrated system preliminary design.* A technical team should develop the preliminary design of a future integrated system which

meets most if not all requirements. The system preliminary design should not be constrained by resources or technology and should be carried down to three or four levels of detail to identify key hardware and software components, interface issues, system requirements, current and projected vendor products, and gross resource requirements.

4. *Implement a phased development.* Implementation of a subset of the total system should be carried out as a phased development toward the total system. This should be guided by a set of planned and increasingly complex demonstration scenarios to test, validate, and demonstrate capabilities and enhancements. It should be based primarily on available commercial products, with the company effort focusing on integration. The implementation strategy should be modular and planned for change through module enhancement and/or replacement. A key ingredient for successful integration is a flexible database management approach for a distributed heterogeneous computing facility which integrates several modular applications tasks. Assessments of system performances should be made a priori and should be incorporated in the test scenarios.

5. *Conduct application tests and evaluations.* The evolving system implementation should be tested on selected company applications while still in the prototype phase. The applications should be of engineering-manufacturing value and be sufficient to demonstrate future system benefits. When key functions are not implemented, they should be modeled or simulated so that a total integrated task can be perceived by reviewing technical and management staff. Assessments of human and system performances should be portrayed together with potential speedups when a full implementation is achieved.

6. *Provide technology transfer and training.* Training and technology transfer for potential users should begin at the outset and continue throughout the program. It should include (1) limited review of goals and approaches, (2) review and critique of requirements and design approaches, (3) training and testbed use of prototype software for system testing, and (4) formal training when useful systems are ready for limited production use.

The above steps provide a guideline for a development plan. The following section gives details on how these steps were carried out on the IPAD project.

1.3 IPAD Approach to CAD/CAM Integration

IPAD was an engineering and computer science R&D effort to use computer technology to improve the integration of engineering analysis, design, and manufacturing processes. The technical approach included several phases, each of which was the basis for subsequent developments. The IPAD technical approach described here was not only useful in addressing integration issues; it also could serve as a guide for other technology programs associated with engineering software development. The basic stages are summarized below together with a brief description of results. (See Fig. 1.3 for an approximate schedule.) The basic strategy was to dissect the aerospace design process, both from an engineering system viewpoint and for specific engineering scenarios, establish functional requirements for a future CAD/CAM system, conduct the preliminary design of a future system (denoted "full IPAD") which best meets all requirements, and develop prototype communication and data management software to meet some of those requirements (Figs. 1.3 and 1.4). Close coordination was maintained with CAM technology being developed in the Air Force ICAM program.

1.3.1 Planning, feasibility, and definition phase, 1971–1976

This phase included two contractual studies to investigate the potential need, approach, and benefits from use of computers to support companywide integration of design (Fig. 1.5). One study (General Dynamics) investigated a spectrum of design tasks from the viewpoint of individual disciplines. The other study (Boeing) developed a systems viewpoint of the total work process for product development. The two studies provided a gross characterization of the computer-aided design capabilities needed to support future product developments as well as the benefits of such a capability (Fig. 1.6). Such benefits included improved designer productivity, improved work environment, reduced design cost, and reduced downstream manufacturing rework cost. Also considered was why NASA should lead such an effort (Fig. 1.7). Industry reviews of these and other IPAD activities (Fig. 1.8) concurred with the results, recommended that NASA

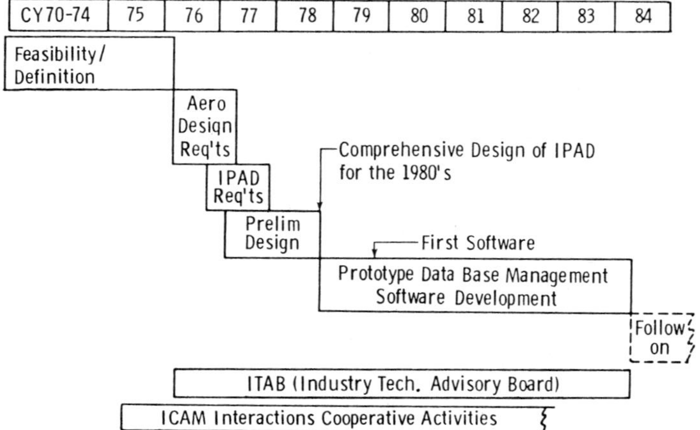

Figure 1.3 IPAD project schedule.

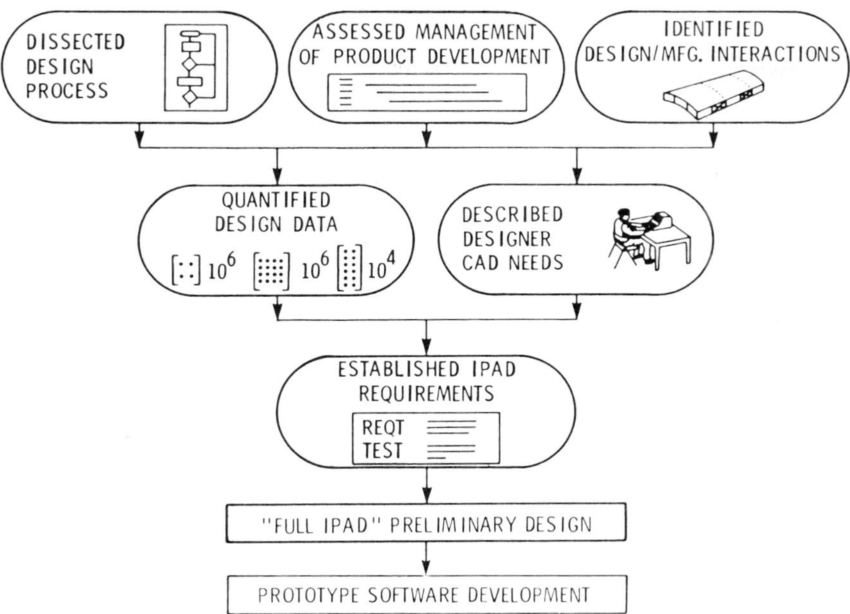

Figure 1.4 IPAD development approach.

proceed with IPAD, and expressed willingness to aid development (Fig. 1.9). This phase culminated in the establishment of a program plan and the selection of a single prime contractor, the Boeing Company, to lead subsequent phases.

1.3.2 Design process definition phase, 1976–1977

The initial goal of the IPAD project was to develop a future integrated CAD system. Since development of such a system was unprecedented, not only the requirements but also the approach to defining the requirements were unknown. The approach taken was to organize a multidisciplinary design team to conduct a systems analysis of the development process for a recent

Boeing Commercial Airplane Company
General Dynamics/Convair Aerospace Division
17 months, $611,000

- Dissected design process
- Examined designer needs
- Defined software elements
- Assessed computer requirements
- Assessed IPAD feasibility/benefits

Figure 1.5 Initial IPAD feasibility and definition studies.

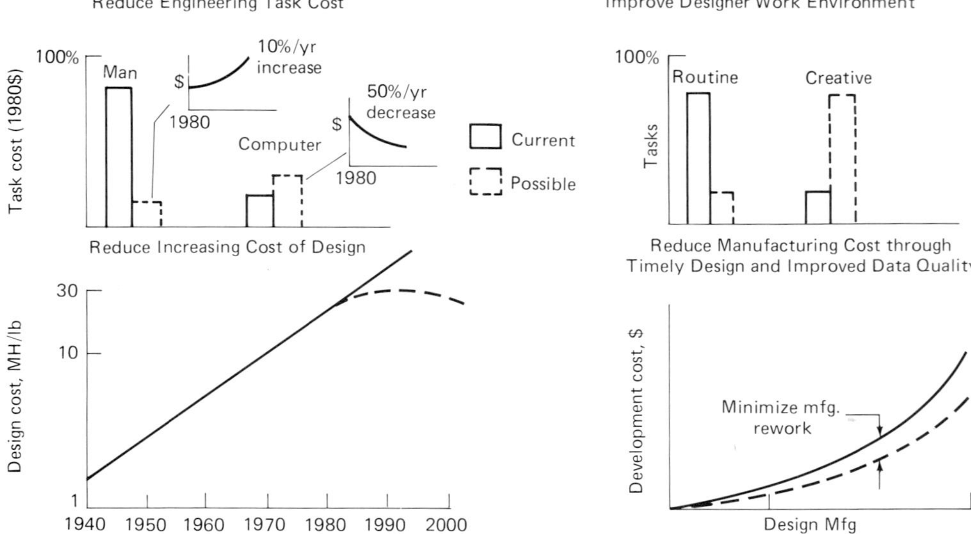

Figure 1.6 Benefits from CAD/CAM technology.

aerospace product, a Boeing 747 commercial air transport. The team prepared detailed logic charts of all stages of design characterizing design levels, the major engineering steps within the levels, interfaces among steps, design iterations, and analysis and design computer programs used to support this work (Fig. 1.10). The 747 study was subsequently expanded to consider other vehicles including a future advanced aircraft (SST), military aircraft (fighter), and a nonaircraft vehicle (hydrofoil).

The system analysis work was also broadened to investigate design in-

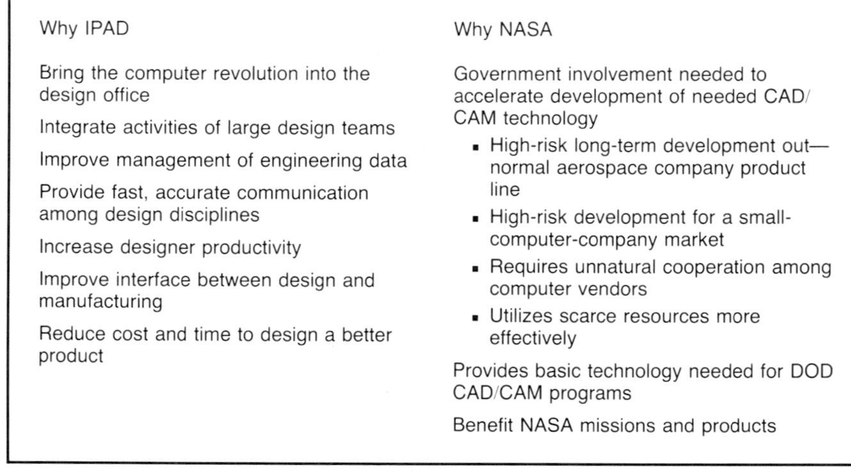

Why IPAD

Bring the computer revolution into the design office

Integrate activities of large design teams

Improve management of engineering data

Provide fast, accurate communication among design disciplines

Increase designer productivity

Improve interface between design and manufacturing

Reduce cost and time to design a better product

Why NASA

Government involvement needed to accelerate development of needed CAD/CAM technology

- High-risk long-term development out—normal aerospace company product line
- High-risk development for a small-computer-company market
- Requires unnatural cooperation among computer vendors
- Utilizes scarce resources more effectively

Provides basic technology needed for DOD CAD/CAM programs

Benefit NASA missions and products

Figure 1.7 Reasons for NASA development of IPAD technology.

- Four oral reports of study contract results (74–187 attendees)
- Aerospace management review (January 30, 31, 1974)
  NASA, ASD, AFFDL, Boeing, Grumman, MDAC, DAC, P+W, RI, Lockheed, Vought
- Funded technical critiques (Oct. 19, 1973–March 19, 1974)
  MDAC, Lockheed, Grumman, RI, CDC, IBM, UNIVAC
- Numerous informal visits and meeting presentations
- Prospectus and survey of 41 aerospace executives (February 1975) sought commitment to:
  - Become member of ITAB
  - Evaluate IPAD software
  - If IPAD is useful, support user organization for maintenance and/or improvement

Figure 1.8 Industry scrutiny of IPAD feasibility definition studies.

terfaces to manufacturing and where major data exchanges take place (Fig. 1.11). This aspect of the study was limited to design and manufacturing interfaces, since the ICAM program was well underway to investigate technology associated with the details of manufacturing processes. The studies did, however, include an investigation of how an engineering development was managed through schedule and resource control, tracking, and assessment processes. The above work provided a system-level description of a representative aerospace design process, the first such description available in the literature.

The aerospace design process description was used to conduct an investigation of the detailed information flow through the design process. Charts were prepared for defining in detail the quantities and types of data flowing among the various design stages. Figures 1.12 and 1.13, for example, show the data flow associated with preliminary design of an aircraft fuselage frame. Also, a complementary study was conducted to identify the individual tasks engineers wanted computers to do and how engineers would like to interface with the computer. This work characterized and quantified the data flow through a design process, quantified the daily information work load on a company having several products under simultaneous development, and established a list of user interface capabilities needed by engineers to control a future integrated design process.

- IPAD development should proceed
- Government should develop core and leave technical modules to industry
- Provide early user involvement
- Cost estimates probably low
- Aim toward "working prototype" system

Figure 1.9 Industry advice from IPAD critique.

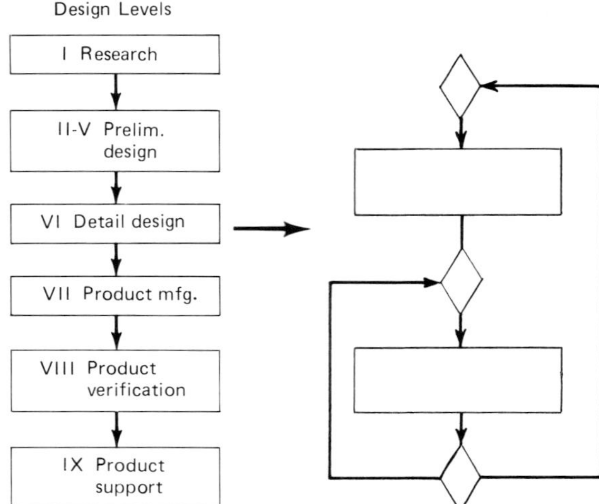

Figure 1.10 Dissection of the design process for representative vehicles.

1.3.3 Requirements definition phase, 1976–1977

The design process definition phase provided the basis for development of a list of functional requirements for a future integrated CAD system. These requirements were compiled by a joint engineering and computer science team through a very demanding review and assessment process. In developing the requirement list, the effort focused on user need rather than on the feasibility of satisfying a need. Great difficulty occurred in bridging the communication gap between the two different technical cultures of engineering and computer science. To help bridge it, each requirement was supported by an abstract of an acceptance test to determine if the requirement was satisfied. No requirement was included if an acceptance test could not be conceived.

The requirements document was subjected to intensive review by ITAB and NASA staff, and several major reviews were held. These reviews often caused major reworks to consider areas overlooked, correct deficiencies, and clarify issues. The final requirements document was prepared and edited by a four-person technical team composed of a senior engineer and computer analysts from Boeing and NASA working together on site for approximately one month. Figure 1.14 shows the number of requirements according to various categories. User interface and information access, transfer, and management needs made up the bulk of the requirements. For example, the management of data required a demanding set of functions greatly exceeding existing software capabilities (Fig. 1.15), and a typical company would require rapid access to trillions of words of design data (Fig. 1.16).

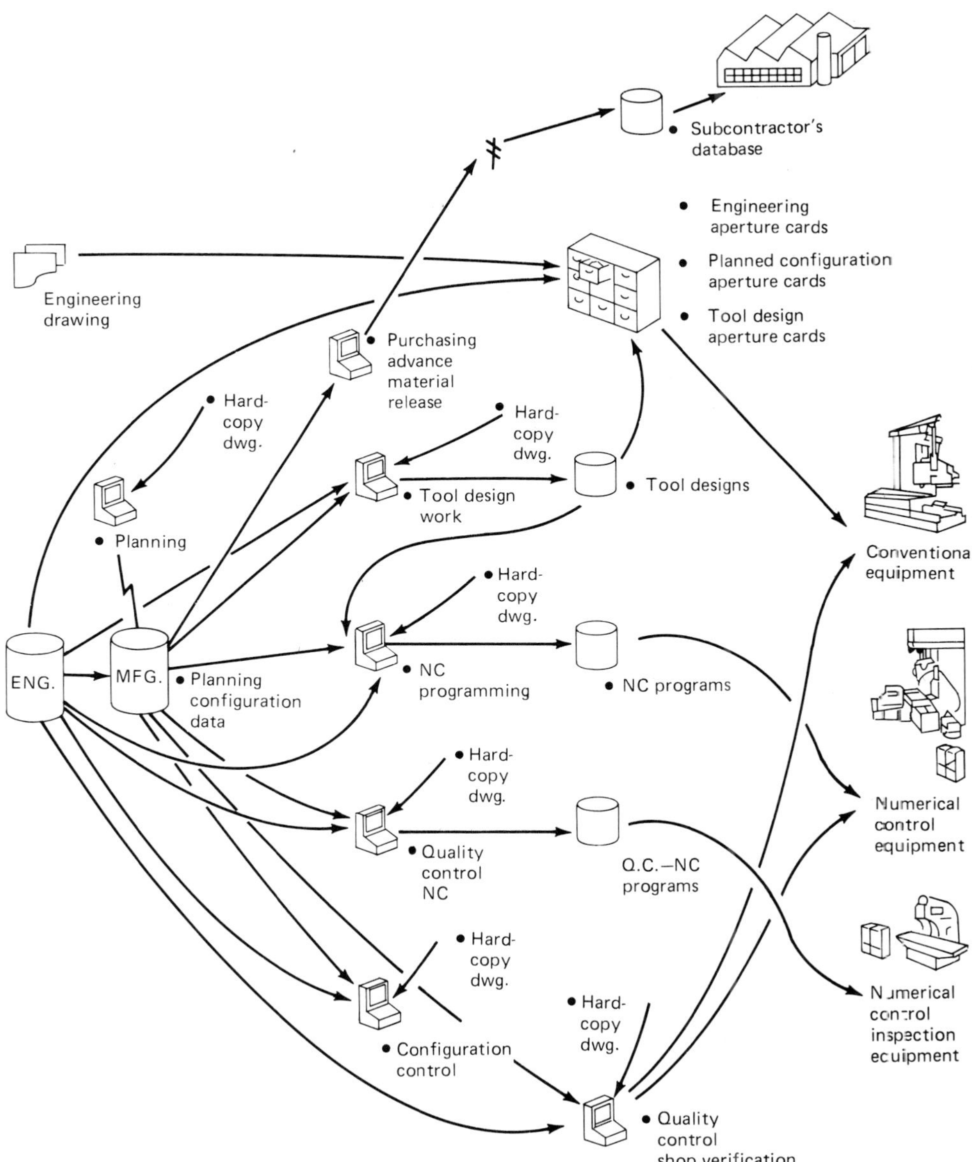

Figure 1.11 Flow of data from engineering to manufacturing.

1.3.4 Preliminary design of a future integrated CAD system, 1977–1978

Preliminary design of a future IPAD system was developed to meet the IPAD requirements. Software design was carried out through a top-down approach coupling requirements to system design. This coupling was not always possible, since there was not a one-to-one connection between specific requirements and software functions and there was heavy reliance on distributing operating system technology not available. Ultimately, a level 3 preliminary design of this future advanced system was developed with varying levels of further details for each component. The resulting system, denoted "full IPAD," could be viewed as a general-purpose interactive integrated computer-aided design system developed to support engineering design processes. It would be built on the operating systems of a distributed, heterogeneous computing complex. Its primary function would be to handle engineering data associated with the design process. Full-IPAD software would be installed by each company on its computers and used in a manner similar to vendor-supplied operating system software. The

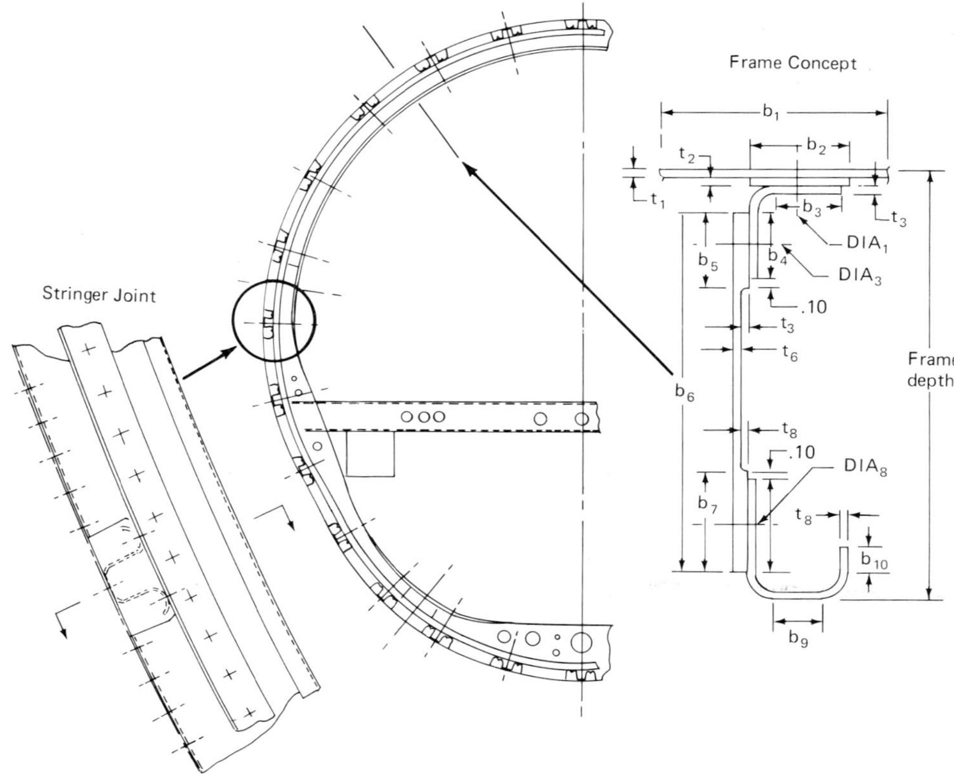

Figure 1.12 Geometry data for aircraft fuselage frame.

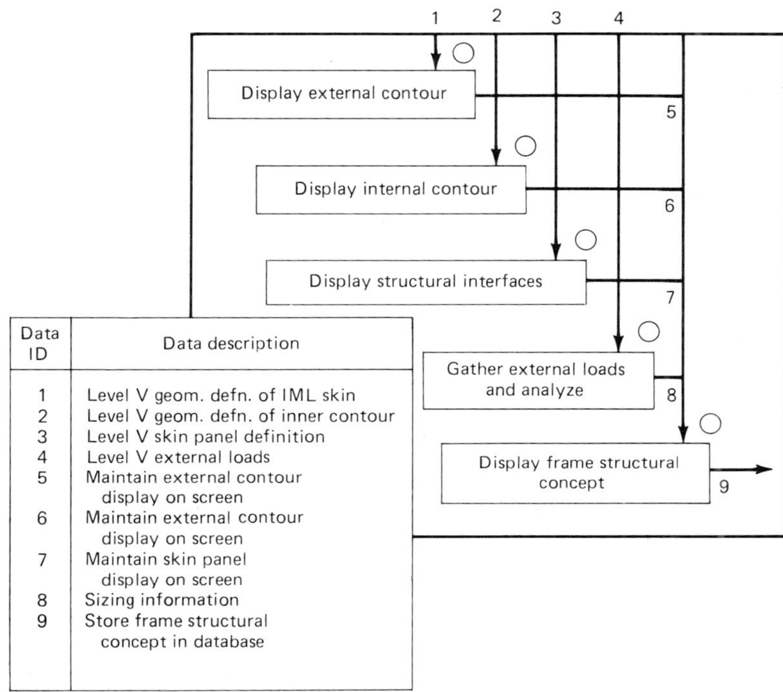

Figure 1.13 Typical data flow for preliminary design of aircraft fuselage frame.

full-IPAD software would augment, rather than replace, existing operating system software. It would support the continuous design activities of a typical company mix of multiple development projects. The full-IPAD system would serve management and engineering staffs at all levels of design (conceptual, preliminary, and final) and aid in the assembly and organization of design data for manufacturing processes.

The full-IPAD system would support generation, storage, and management of large quantities of data. Its capacity would be limited only by the

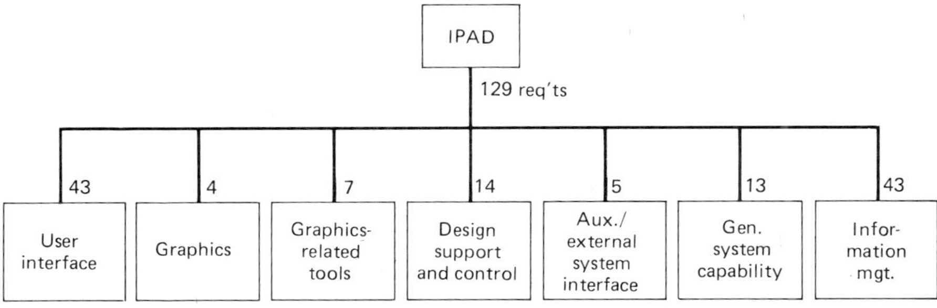

Figure 1.14 IPAD requirements in various categories.

- Accommodate multiple views of data
- Allow multiple levels of data description
- Permit changes and extensions in data definition
- Distribute data over computer networks
- Manage geometry data
- Provide configuration management capabilities
- Manage information about data

Figure 1.15 CAD/CAM data management systems requirements.

computer hardware configurations selected by each company. The system would be used in a distributed computing environment having one or more central host computing systems and many remote computing systems. One such arrangement of full-IPAD components is given in Fig. 1.17. The number of terminals might be several hundred or more and be distributed across the host and remote systems. The full-IPAD software would function on the computer complexes in use today by aerospace corporations, but it would achieve its full potential on the computers of the next decade.

The functions of the three major IPAD software components illustrated in Fig. 1.17 are (1) *executive software* (IPEX) to control user-directed processes through interactive interfaces with a large number of terminals in simultaneous use by engineering and management personnel and to provide communications between computer hardware within and outside the IPAD distributed computing system, (2) *data management software* (IPIP) to provide a comprehensive, versatile capability for efficiently storing, tracking, protecting, and retrieving exceptionally large quantities of data maintained on multiple storage devices, and (3) *geometry and graphics utility software* to provide a wide range of capabilities for information and geometry creation, manipulation, and display functions including design and drafting and interactive and display graphics.

500 to 1000 simultaneous user terminals

Large, rapid-access data volumes to support vehicle design

	Online data, 10^9 words	10-min availability, 10^9 words
10 preliminary designs	1.7	—
10 sustaining designs	7.8	15.7
2 detail designs	2.5	3.2
	12.0	18.9

Figure 1.16 Key IPAD performance requirements driving software package.

Libraries within the databases might include analysis and design computer programs utilized by various disciplinary specialists and extensive quantities of data. The analysis and design computer programs would not be part of the full-IPAD system; they would be provided by each company to form the complete design-software system. Selected publicly available technical programs might be included in IPAD releases to demonstrate capabilities. The data in the database would include all official project information defining the characteristics of current baselines and alternative designs and their performance, as well as archival handbook information forming the technology base for company designs. Simultaneous access to the same baseline design information by all disciplinary groups would thus be possible. Temporary storage for design information being actively used by individuals or teams would also be provided.

A full-IPAD system would not be a hands-off "automated design" system and would not constrain company design methods. The quality of future aerospace designs generated in an IPAD environment would depend on the same primary factors as in today's design environment: creativity of designers, quality of technical staff, quality of analysis tools and design data, and coordination of design and manufacturing information. IPAD should also be a tool to improve manufacturing direct access to engineering data. Although support for the manufacturing process was not a specific requirement for the full-IPAD system, it is believed that many manufacturing needs are met by the resulting system design.

A preliminary design of each of the above components was carried out to understand key elements, logic interfaces, and scope required for de-

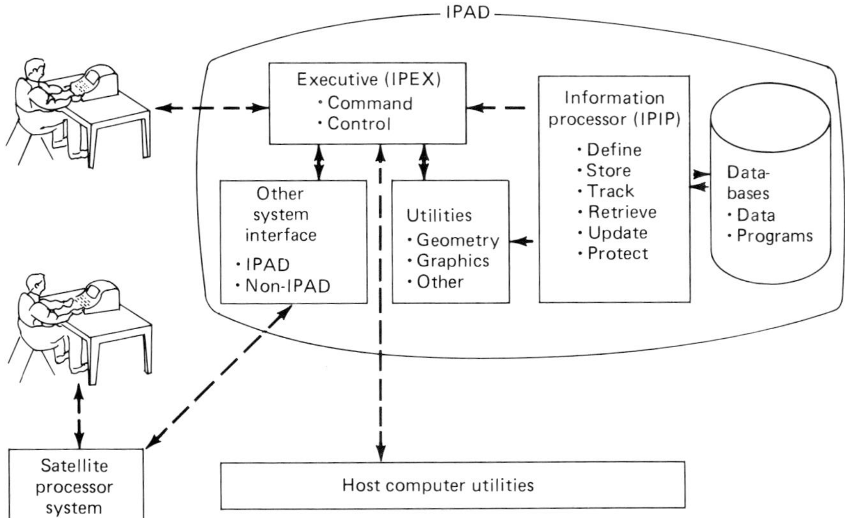

Figure 1.17 Key software components of a full IPAD system.

velopment and/or technology advance. Development of the total full-IPAD system was estimated in 1978 to be a high-risk effort costing in excess of $100 million and well outside the scope of an R&D effort. Two critical components which required major technology advances were IPIP and IPEX. After extensive reviews by ITAB, it was decided to focus the next phase on development of prototype software to manage engineering data. Although there was strong sympathy for IPEX, a distributed engineering operating system, it was felt that computer vendors, with greater resources, would continue to address this issue and that there could be a higher payoff to engineering for the IPAD project to concentrate on advancing data management technology (Fig. 1.18).

1.3.5 Prototype software development, 1979–1984

Under the guidance of ITAB, the IPAD project developed prototype computer software to meet many CAD/CAM information management requirements. Some of the basic requirements driving CAD/CAM systems development are identified in Fig. 1.15. The IPAD approach taken was to conduct appropriate research and to develop prototype software for a future network of computers. Data structure models considered and their status in 1978

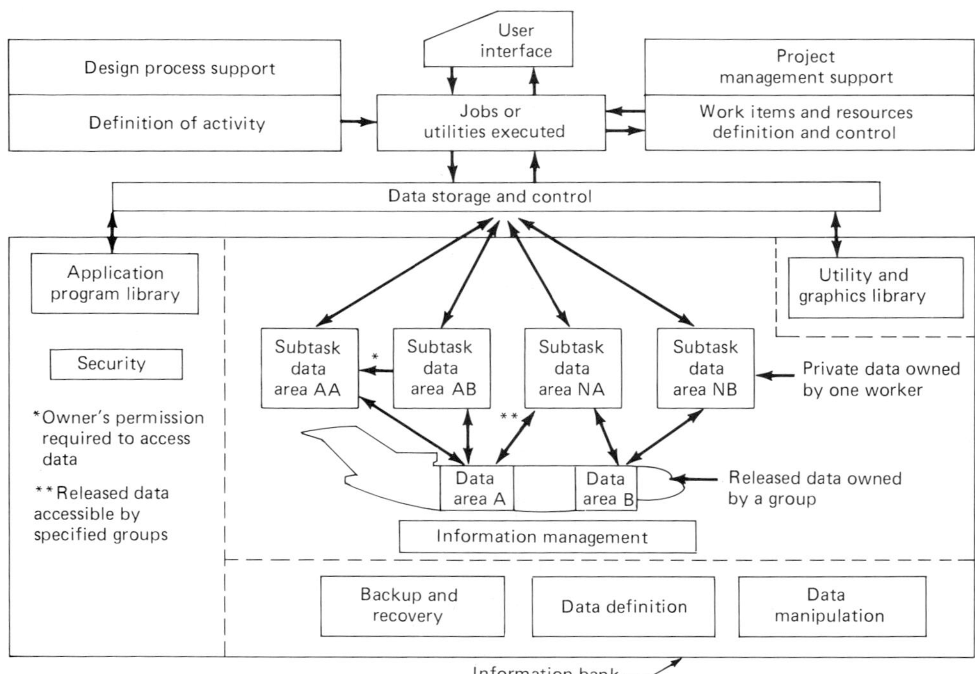

Figure 1.18 Engineering use of integrated CAD/CAM data management system.

included hierarchical, network, relational, and three-schema (Fig. 1.19). To provide the required CAD/CAM functionality, and yet meet software performance requirements, database management would be staged at two or more levels with different software capabilities needed for both the local (user) level and global (project) level (Fig. 1.20). With such a tiered database management approach, current inconvenient file-oriented procedures (Fig. 1.21) would be replaced in an IPAD environment by future procedures (Fig. 1.22) in which convenient user languages would efficiently create, store, manipulate, access, and control information in accordance with CAD/CAM requirements.

Prototype software was developed under the IPAD project at both the local and global levels (Fig. 1.20). A system denoted relational information management (RIM) was developed for local-level data management. RIM is based on the highly flexible relational model which organizes and manages engineering and scientific information according to tables and relations among tables. Its features include interactive queries, report writer,

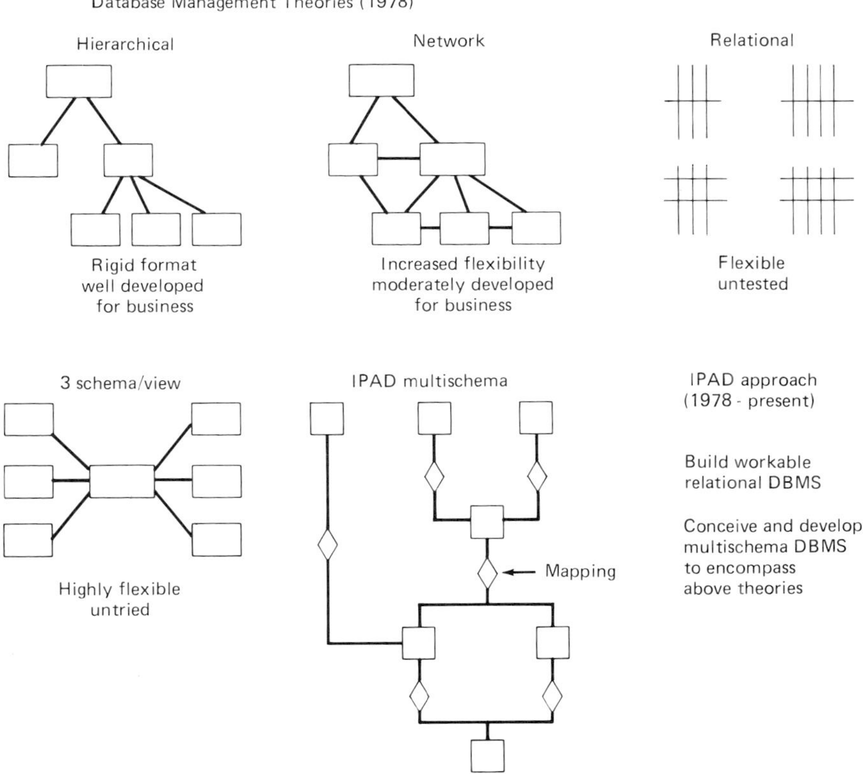

Figure 1.19 Alternative database management models considered by IPAD.

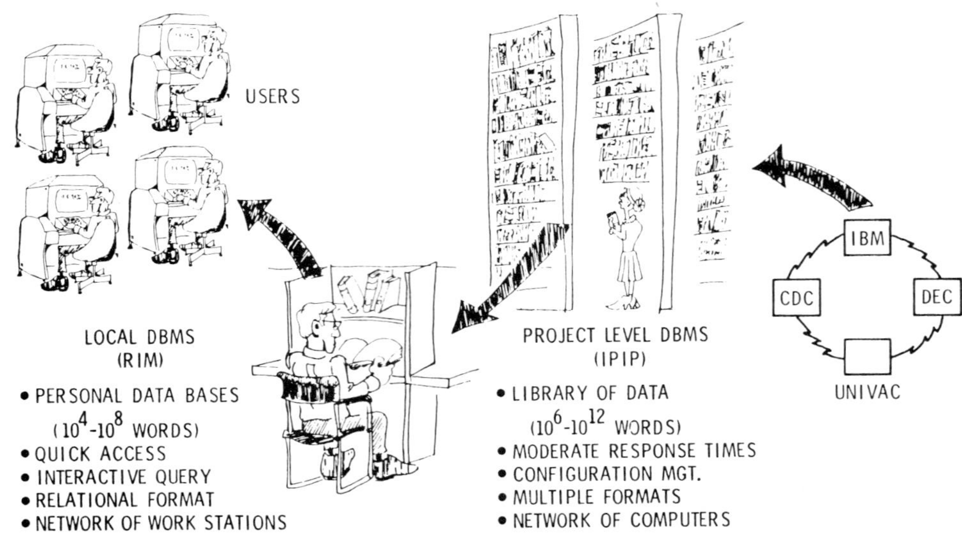

Figure 1.20 IPAD multilevel approach to engineering database management.

and Fortran interface. RIM was first operational in 1979 and is now a mature system (Fig. 1.23). In 1981, it served as a critical information management capability to support NASA investigations of the integrity of 30,000 tiles on the space shuttle (Fig. 1.24).

The success of RIM in such evaluations led to its continued development and enhancement by government and industry. Over 300 copies of RIM were distributed by the IPAD project, and a public version 5.0 is available from COSMIC* for CDC, IBM, DEC, UNIVAC, PRIME, and Harris computers. Commercial organizations have continued to enhance RIM and now provide compatible RIM derivative software (e.g., BCS/RIM and MicroRIM) and associated maintenance and support for such software operational on a wide range of computers (from personal computers to supercomputers). Commercial versions of RIM are being used extensively by industry, and one version has been adopted by the Naval Sea Systems Command for use in its early-stage ship design integration process. NASA used RIM as the common data management system (Fig. 1.25) to integrate several engineering disciplines through development and application of the Prototype Integrated Design System (PRIDE).

IPAD research on development of a global database management system (DBMS) denoted IPAD Information Processor (IPIP) has continued. The approach taken in IPIP is to provide the capability within one system to

*Computer Software Management and Information Center, 112 Barrow Hall, University of Georgia, Athens, GA 30602.

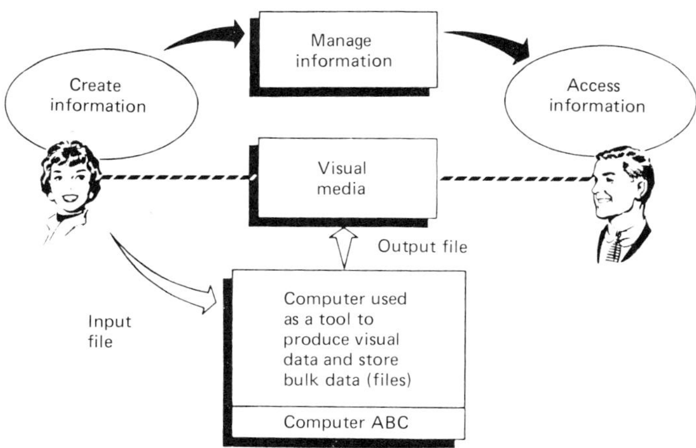

Figure 1.21 Current file-oriented approach for CAD/CAM information.

manage information composed of a wide variety of data structures including hierarchical, network, relational, and geometric. The IPIP approach is to use multiple levels of information formats (schemata) to permit unlimited reorganization of information as work progresses (Fig. 1.26).

Each IPIP schema is connected to other schemata via a general-purpose mapping capability (language). It provides a new concept called structure processing (Fig. 1.27), in which an unlimited number of layers of data structures appropriate for different uses can be created on top of the basic database. At the time of writing, IPIP was a new concept, still in test and

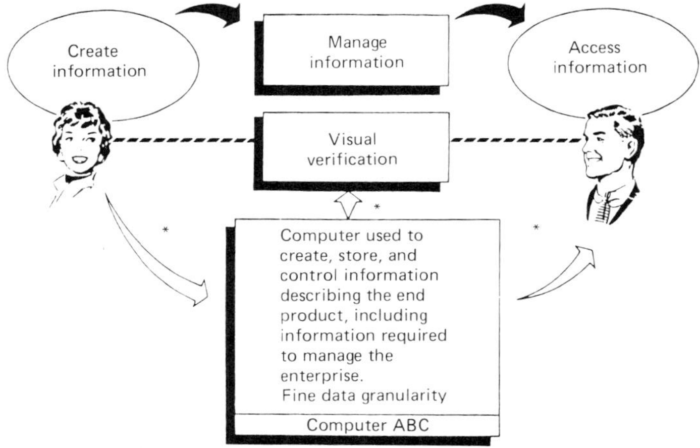

Figure 1.22 Future database management approach for CAD/CAM information.

- Written in fortran
- Code highly portable (CDC, VAX, Prime, IBM, Univac, Cray, Micros, etc.)
- Based on relational algebra
- Flexible query language
- Application program interface capabilities
- Rim-to-rim communications file
- Schema can be modified as needed
- Variable-length attributes
- Commercial vendors supporting and/or expanding rim

Figure 1.23 Features of RIM relational information management system.

evaluation phases, and was operational only on a CDC computer. Its approach to management of geometric data together with nongeometric data is a unique concept which could be very important to future integration of design and manufacturing such as the application illustrated in Fig. 1.28. A critical technical challenge in IPIP development has been to provide a

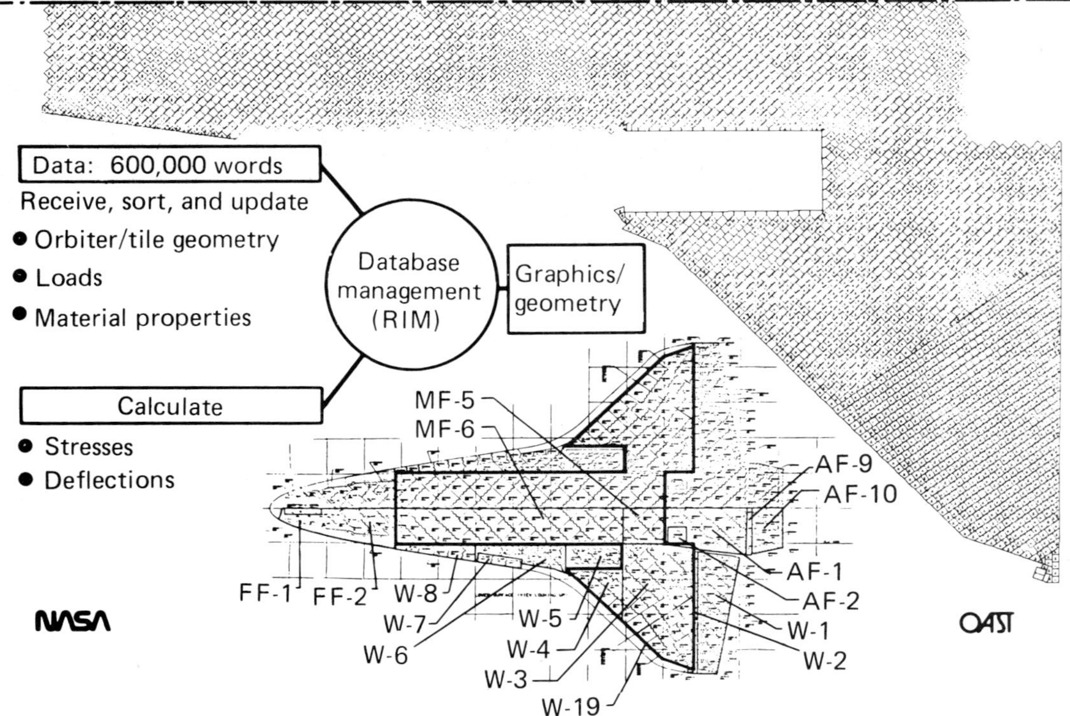

Figure 1.24 Use of IPAD/RIM data manager to support investigation of space shuttle tile analysis.

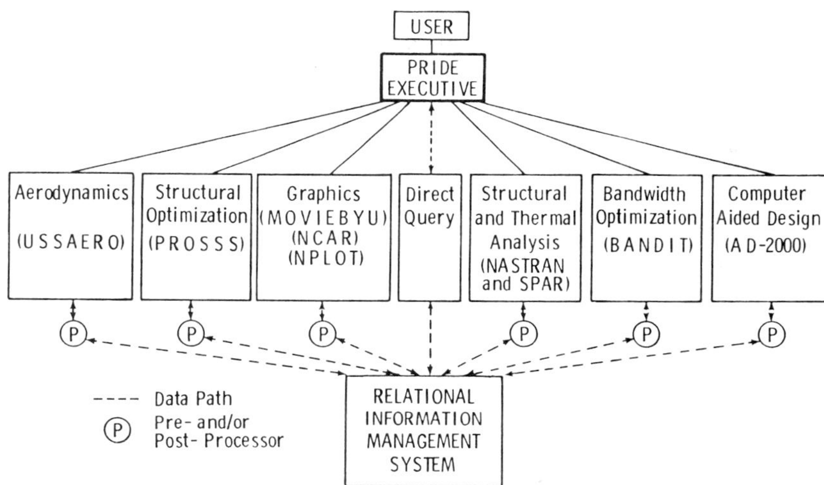

Figure 1.25 Organization of PRIDE prototype integrated design system.

high degree of engineering user flexibility and yet achieve acceptable response times. In late 1983, a test system (denoted version 5.0) which has user responses for test problems of less than 0.5s per data manipulation language was provided to selected ITAB organizations to support evaluations such as that illustrated in Fig. 1.29.

One vendor (CDC) established IPIP as an as-is commercial product and then planned limited support for its installation and evaluation. IPAD results to date in defining CAD/CAM data management requirements and in developing prototype software have helped stimulate development of

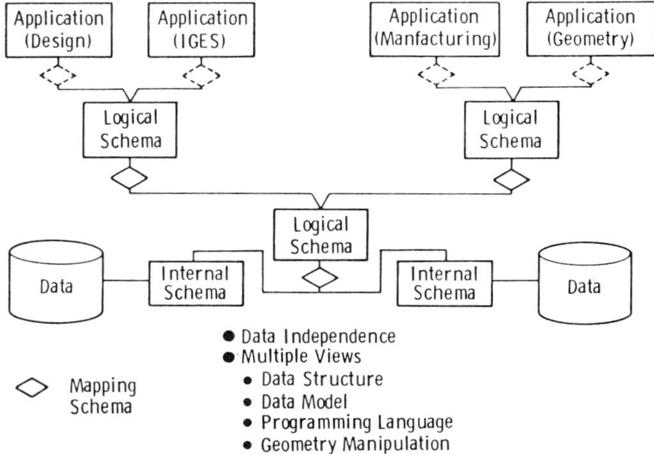

Figure 1.26 Typical arrangement of IPIP data schemata (formats) to connect application schemata to storage schemata.

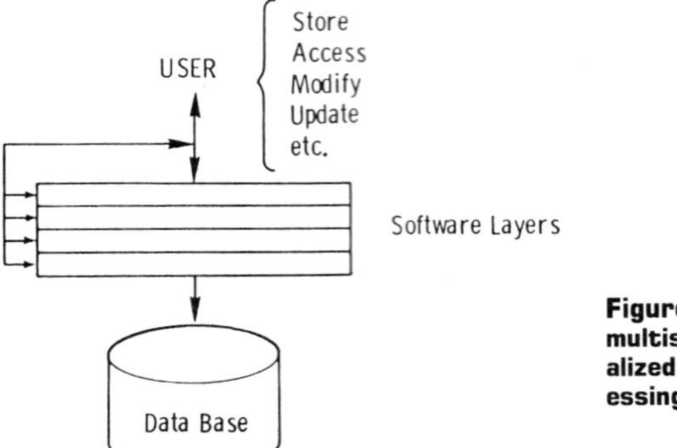

Figure 1.27 Application of IPIP multischema capability to generalized information structure processing.

commercial CAD/CAM data management software, and several computer vendors planned release in 1984–1985 of relational-type data management systems which address many of the CAD/CAM requirements identified in IPAD research. IPAD results have also helped stimulate infusion of database management technology into university engineering research.

A critical CAD/CAM requirement not yet contained within any available or planned commercial data management system is the ability to efficiently manage geometry information as data in concert with other engineering data (Fig. 1.30). Through use of its multischema capability, and the structure processing concepts, IPIP provided the first approach to management of geometry information within a data management system. The IPIP approach provides software capability to create, on top of the basic geometric data, an information structure having an unlimited number of geometric descriptions (schemata). One geometry schema includes the evolving geometry-graphics standard, Initial Graphics Exchange Specifications (IGES). This IPIP information structure concept opens the door for convenient integration of geometric information with other types of information asso-

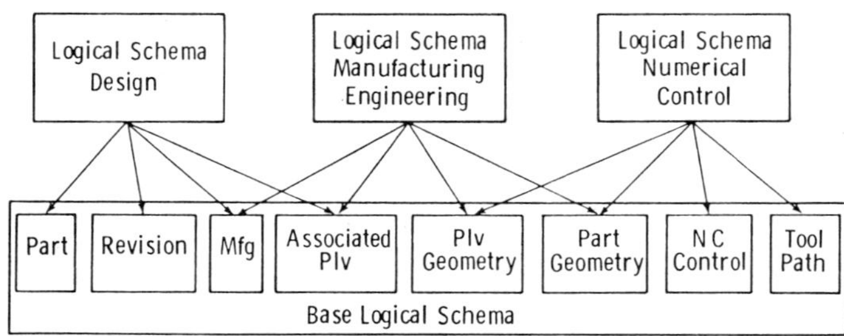

Integration of Design/Manufacturing Information

Figure 1.28 Application of IPIP multischema capability to integration of design and manufacturing information.

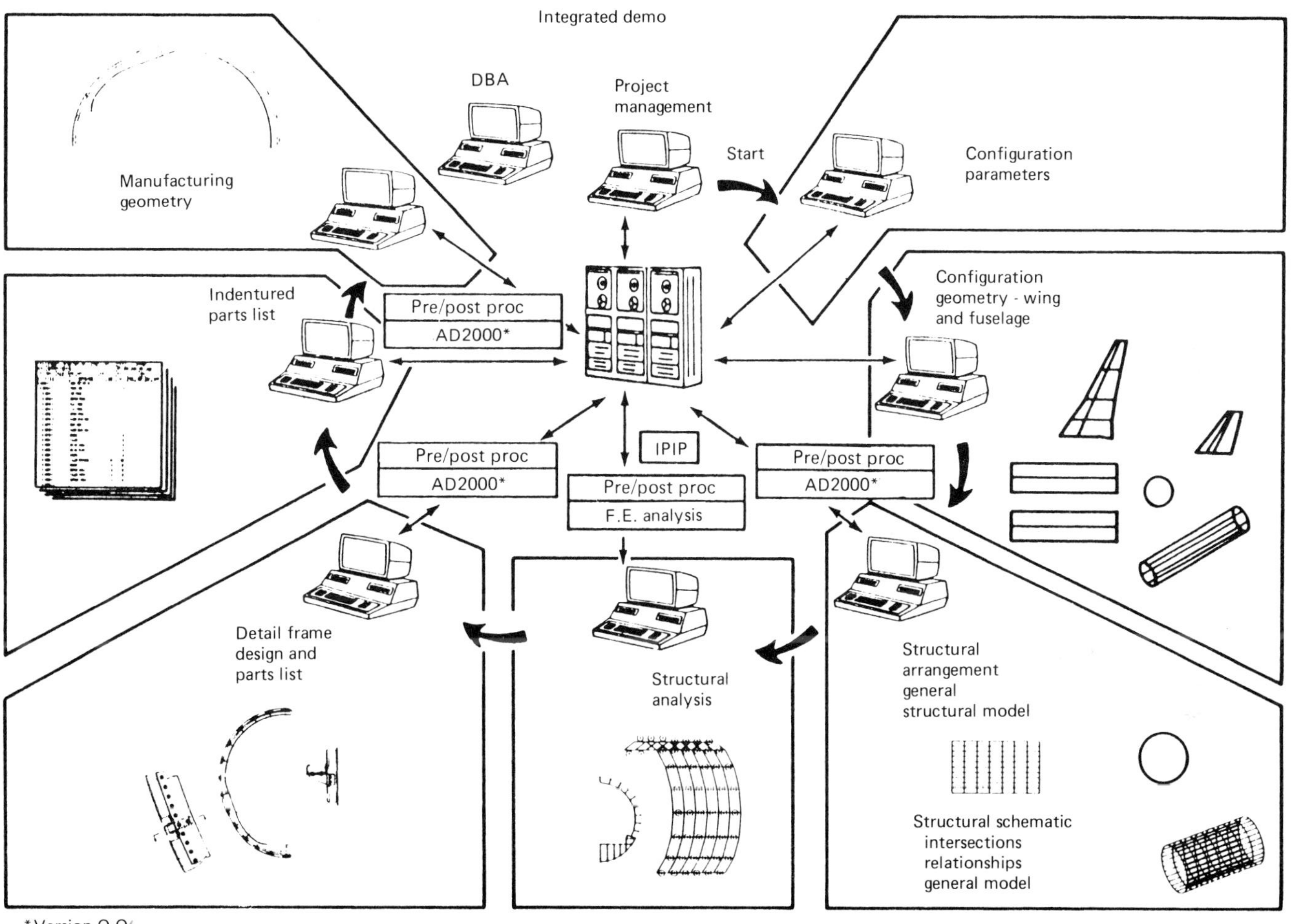

*Version O.O/.

Figure 1.29 Planned test demonstrations to evaluate IPAD/IPIP for CAD/CAM use.

ciated with a CAD/CAM development process (Fig. 1.31). At the time of writing, an evaluation of the IPIP geometry concept was underway, and comparisons were being made with other approaches in which management of the geometric data takes place as a "geometry engine" outside a basic relational database manager.

1.3.6 CAM data management requirements phase, 1982–1984

The IPAD requirements were continually revised as the subsequent development phases progressed, but a major revisit to requirements technology occurred in 1982. It was triggered by Navy cosponsorship of the IPAD project as well as by a growing awareness of the need to address CAM data management requirements. This CAM requirements effort work built on earlier studies of manufacturing interaction with design (Fig. 1.11), a limited study of the information flow of a sheet metal part (Fig. 1.32) during fabrication, comparable studies by Grumman for three other parts, several ICAM studies, and a joint IPAD/ICAM workshop, in 1980, focused on CAM data management issues. The results of those efforts were integrated into an assessment of the database management requirements to support CAM

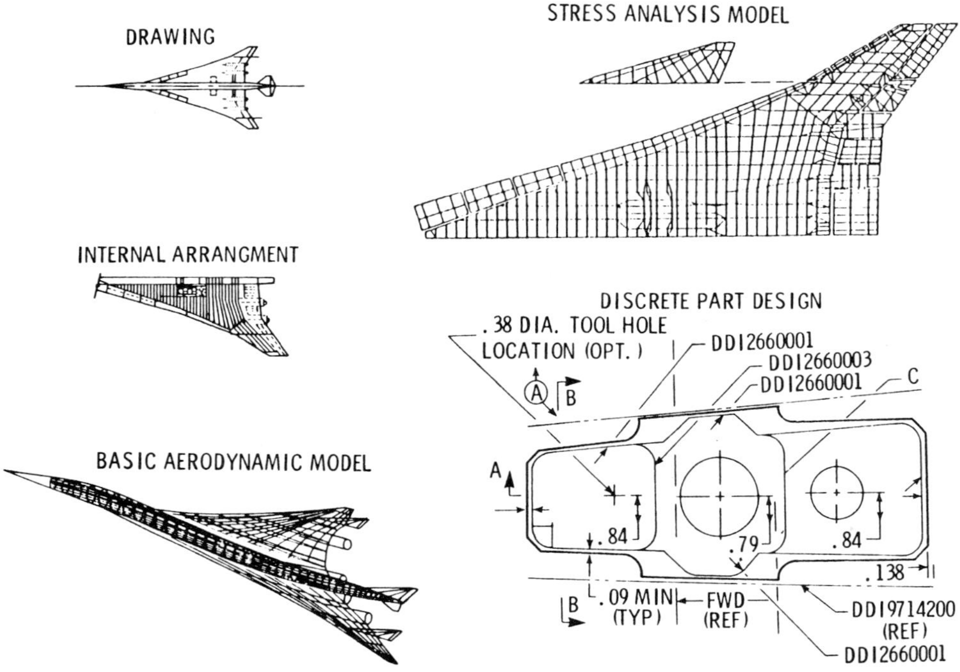

Figure 1.30 Critical role of geometry information in CAD/CAM.

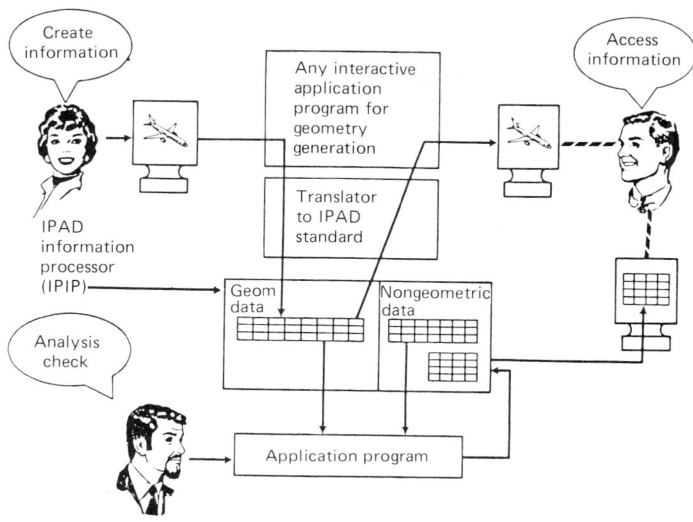

Figure 1.31 IPAD approach to unified management of combined geometric and nongeometric CAD/CAM data.

(Fig. 1.33). At the time of writing, a Navy-supported follow-on effort with the remaining IPAD team was continuing work to define CAM data management requirements and to develop prototype software and applications to test the CAM requirements. Close coordination was also continuing between Boeing and ICAM studies such as the Product Data Definition Interface by McDonnell-Douglas and the Integrated Information Support System by General Electric. One aspect of the continuing assessment of CAM data management requirements is to be a correlation with IPIP functionality.

1.4 Technology Transfer Mechanisms

"Technology transfer" is used here in the broadest possible context of ensuring that the appropriate technology is transferred to those who need it. The IPAD project in its totality was a major technology transfer mechanism, and this technology transfer showed up in all facets of the project. There were, however, some major elements that should be highlighted as drivers to make the technology transfer process a success. IPAD research, for example, provided ideas and focused attention on critical CAD/CAM integrations issues, which stimulated technology transfer among the triad of the IPAD team, computer vendors, and aerospace computer users. A limited number of universities were also involved in the process. Thus IPAD research not only provided new technology but was also the catalyst for

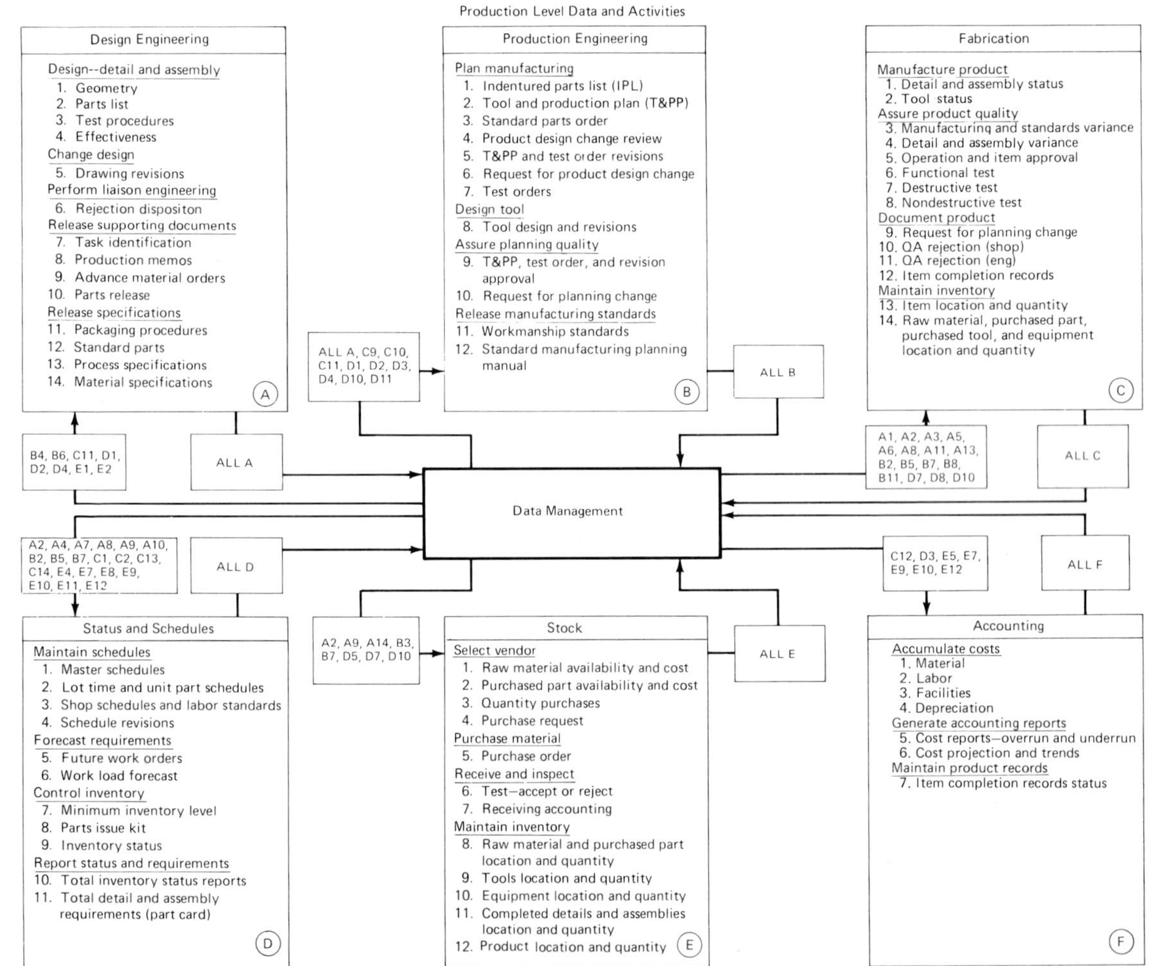

Figure 1.32 Approach to defining initial CAM data management requirements.

technology transfer between the users and producers of CAD/CAM integration technology.

The major focus for technology transfer and industry involvement in the IPAD project was through the Industry Technical Advisory Board (ITAB). ITAB was formed by the development contractor soon after contract initiation to afford industry the maximum opportunity for influencing the course of IPAD development. NASA invited ITAB companies to participate, and the development contractor established the board composition subject to NASA approval. Board members were typically high-level executives having overview responsibility for company CAD/CAM-related work. They were supported by senior technical staff when appropriate.

ITAB was deliberately limited to 20 members, which was felt to be a manageable but representative group. The number of observers eventually grew to over 80. The total board, representing major U.S. aerospace and computer companies, met at approximately 3-month intervals. ITAB activities included review of planning and technical documents, critique of key development decisions, prioritization of IPAD activities, identification of demonstration programs, sponsorship of symposia, and evaluation of prototype software. As the IPAD software was released, ITAB member companies and other potential IPAD users aided in its evaluation and use.

The ITAB chairman was selected from industry and was supported by

Data integration across heterogeneous computer systems
Transaction processing data
Real-time interactions
Distributed processing and data
Multiple views of data
Levels of security; data integrity
Multiuser networks across dispersed work environment
Local and global data distribution
Geometry management
Versioning
Backup and recovery
Good user interface and knowledge base support
Integrated manufacturing and material management control
Quality assurance integration
Configuration control
3-D electronic product definition data
Incremental implementation; organized to evolve; expandable

Figure 1.33 Functions needed in a CAM information management capability.

an ITAB executive officer on the contractor's staff who was independent of and detached from the development group. The relations among ITAB, the contractor (Boeing), and NASA are shown in Fig. 1.34. This arrangement ensured direct communication between ITAB and the contractor for ongoing work while maintaining strong communication with NASA at both the IPAD Project Office and NASA Headquarters.

Operationally, a typical ITAB meeting was convened at different ITAB member sites, consisted of various status reports, and resulted in a series of recommendations to the contractor. ITAB recommendations were drafted by an ITAB Control Board, finalized by ITAB members, and meticulously tracked at subsequent meetings. Special ITAB audit committees were often established to conduct on-site technical reviews of Boeing work at specific major milestones. The work of the audit committees often provided the background for ITAB recommendations. Each meeting usually included extensive technical discussions and varied opinions which provided an excellent forum for education, information exchange, and clarification of issues. The resulting recommendations provided program direction, and recommendations were tracked to ensure contractor response.

ITAB members were a highly participating group, and they assumed many responsibilities which significantly aided the program. All member organizations and several observers actively participated in hosting meetings, technical exchanges, and software or document evaluations. A few companies contributed staff and/or products to support specific research or development tasks. Government funding was provided to support ITAB travel, but the funds were never fully used by ITAB. The total ITAB effort was so pervasive and extensive that, if contracted, it would likely have equaled the government funds contracted for the IPAD technology development.

To aid technology transfer, NASA contractually required that ITAB and Boeing work together as a team without NASA serving as mediator. NASA

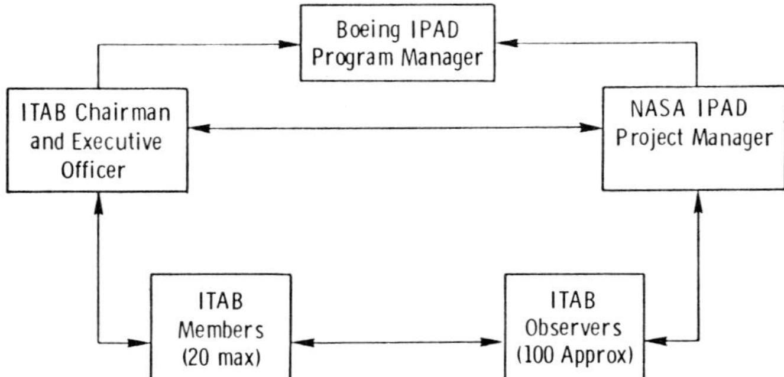

Figure 1.34 IPAD organization and communication channel.

usually limited technical direction to situations in which Boeing and ITAB could not resolve an issue. The success of the ITAB-Boeing-NASA triad was due partly to the contractual structure of the concept and partly to the commitment of the leaders of the three groups. It was also facilitated by key Boeing staff, including technical submanagers, the ITAB executive officer, and the IPAD program support manager, who provided administrative and technical support to ITAB. After approximately one year of operation, ITAB began to assume ownership of the IPAD project and its technology thrusts; this ownership lasted until the program's conclusion. The number of ITAB members and observers quadrupled over the project life, and the group became active enthusiasts to facilitate CAD/CAM technology transfer through a variety of formal and informal mechanisms. ITAB strongly supported continuing the IPAD program because of its success as a technology transfer mechanism and because there were still many issues to be resolved.

1.5 IPAD Products and Their Uses

Over its lifetime the IPAD project produced a large number of CAD/CAM integration products which were widely distributed to ITAB members, observers, and other organizations. These products included technology reports, prototype software, and technology exchange processes (Fig. 1.35). Because of the nature of the products and the early release distribution process, it is difficult to document their extensive use. Informal discussions indicate the IPAD technology permeated deeply into numerous organizations. Several unique concepts developed in the IPAD project were critically

Technology reports (13,000 documents)
- Requirements and design of integrated CAD/CAM system
- Data management, geometry, and networking
- Data communications standards

Prototype software (600 software packages to 325 organizations)
- Data management
- Networking
- Graphic design/drafting (purchased)

CAD/CAM coordination mechanism (60 ITAB meetings, workshops, working groups, seminars, and symposia)
- Information exchange
- Concept and software evaluation
- Technology guidance
- Focus technology attention

Figure 1.35 IPAD products.

important to achieving a future integrated CAD/CAM capability. Figure 1.36 shows the approximate timeline for completion of some of the products. Some of the IPAD products and their representative uses include:

1. Description of an aerospace design process and its interfaces to manufacturing

   USE: The aerospace design process description has been used by several companies including Lockheed Georgia (Fig. 1.37) and Rockwell.

2. Requirements and methodology for defining a future CAD/CAM system

   Preliminary design of a future integrated CAD system based on a unified data management approach

   USE: The requirements definition and full IPAD preliminary design work have been used by many computer vendors as bases for their CAD/CAM products. For example, IBM acknowledges significant use of the IPAD requirements in definition of its CAD/CAM products and CDC's major CAD/CAM product ICEM (Integrated Computer Aided Engineering and Manufacturing) is based on IPAD requirements and the full-IPAD software preliminary design.

3. Relational information management (RIM) system, which demonstrated the need for relational data management for engineering use

   USE: RIM is used at hundreds of organizations and is becoming a national data management standard. It is embedded in company products at Mentor Graphics, ICARUS, EXXON, EXECUCOM; it has spawned two new companies, MICRO-RIM and RIM Technology, to offer RIM-based products and services; and it has formed the basis for a commercial software product BCS-RIM marketed by Boeing Computer Services.

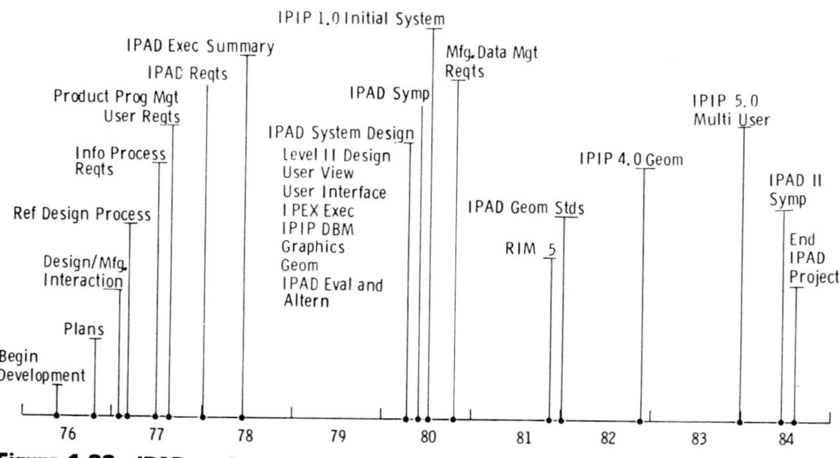

Figure 1.36 IPAD products and events.

4. Prototype Integrated Design System (PRIDE), which demonstrated the approach to and benefit of a relational data management system and executive to integrate geometry, graphics, and several disciplines to achieve a multidisciplinary computer aided engineering capability

   USE: The PRIDE system has served as the guide for several aerospace organizations in the development of integrated engineering analysis systems based on RIM and other relational DBMSs.

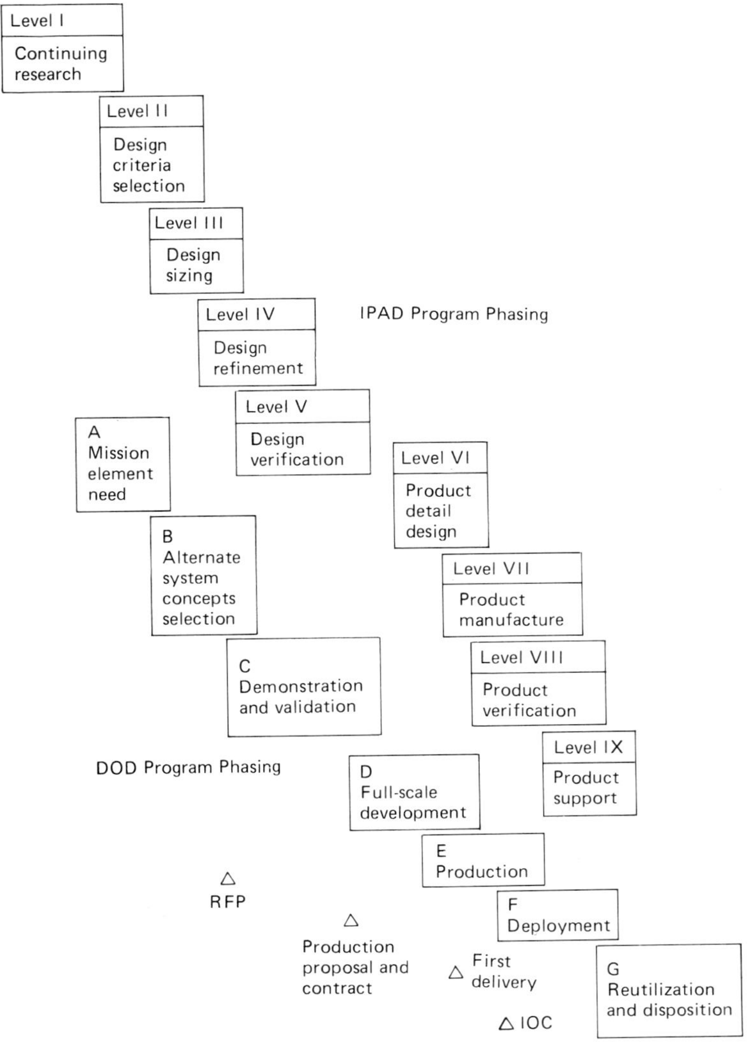

Figure 1.37 Commercial vs. military program phasing. *(Courtesy of Lockheed Georgia)*

5. Prototype future multischema engineering data management system, which encompassed in one system hierarchical, network, relational, and geometry data structures

   USE: IPIP has served as a baseline for advanced data management approaches and is being offered by CDC as a product denoted CDC-IPIP. Several vendors plan the release in 1984–1986 of engineering database management systems based on relational and/or three-schema approaches. A variety of approaches to management of geometry data is also under study.

6. Data networking approach for communication of data among heterogeneous computers

   USE: Networking software has been installed at several aerospace companies, and it was a prototype for several commercial networking products including Network Systems NETEX and by CDC's LCN.

7. CAD/CAM standard activities to facilitate development of data management standards

   USE: IPAD has played a significant role in several national standards efforts including cosponsorship of the IGES effort to establish an Initial Graphics Exchange Specification. IPIP Logical Schema Languages and Data Manipulation Languages were based on 1978 CODASYL Data Description and Data Manipulation Languages, and RIM is becoming a standard for communication of engineering data among heterogeneous computers.

8. Unique technology transfer approach to speed use of high technology by industry

   USE: The ITAB industry involvement mechanism has been used as a model for several major programs. For example, ITAB was specifically included in a November 1983 RFP for development of "Guidelines for Applying Computer Aided Engineering Systems to Generate Plant Designs" by the Electric Power Research Institute and a new Air Force program for engineering data management for logistics needs (Integrated Design Support System) plans to establish an ITAB-like advisory board to guide its activities.

1.6 Priorities for Future CAD/CAM Technology Development

There are many areas of technology which need attention to achieve the level of database management required for integration of CAD/CAM activities in a major aerospace company. Some major priorities are noted here. A key data management requirement not yet commercially available is the ability to manage unified CAD/CAM information distributed across computers of different manufacture with the user flexibility provided by software such as RIM and IPIP. A typical company may have several different computers to support its combined engineering manufacturing activity; the addition of subcontractors introduces even more heterogeneity in computers. Examples of recent high-technology developments include the NASA space shuttle (Fig. 1.38) and Navy advanced aircraft (e.g., Fig. 1.39) wherein major components were developed by many widely dispersed companies, each having different computer complexes. To support such developments in the future, ITAB, NASA, and the Navy identified critical needs for CAD/CAM-related data management research over the next few years to include:

1. Refinement and assessment of manufacturing data management requirements
2. Development of executive software to control information management over a network of heterogeneous computers
3. Development of distributed data management software capabilities
4. Extension and evaluation of geometry data management software capabilities
5. Development of multidisciplinary analysis and design data management approaches for sequential and concurrent processing computers
6. Development of data management approaches to support expert engineering systems
7. Definition of the elements of expert systems technology to be utilized internally by advanced data management systems

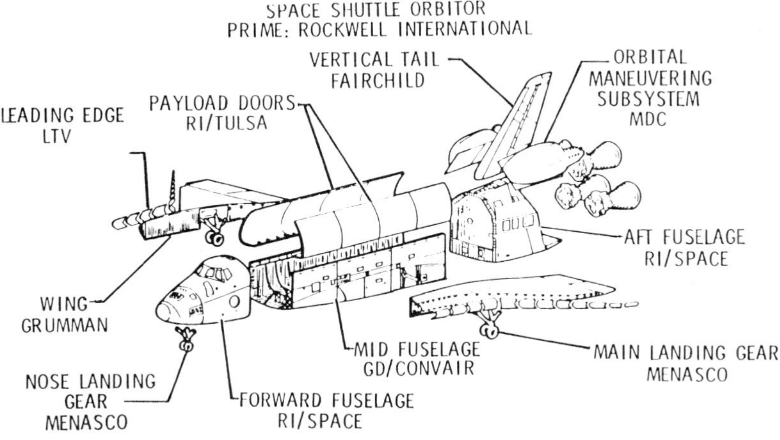

Figure 1.38 Typical multicompany development approach for aerospace products.

A National Research Council report states:

> The use of computers in design and manufacturing offers the potential of an integrated information system that encompasses product planning, designing, manufactural engineering, purchasing, materials requirements planning, manufacturing, quality assurance, and customer acceptance. A single product definition database containing an electronic description of the designed products that are being constructed or manufactured is a keystone to the successful utilization of CAD/CAM technology. The NASA/Navy IPAD project under the guidance of ITAB has helped focus unified government and industry technology development on this important national productivity need.

1.7 Management Issues

Although there are many technological issues critical to CAD/CAM integration, the success of an integrated CAD/CAM system is heavily dependent on management support. The availability of good computer software and hardware does not ensure a productive system. Special issues that require management attention include:

- Procedures to ensure long-term management planning and commitment
- Ways to enhance communication, coordination, and planning among design, manufacturing, and computing
- Engineers with vision and perspective in both design and manufacturing

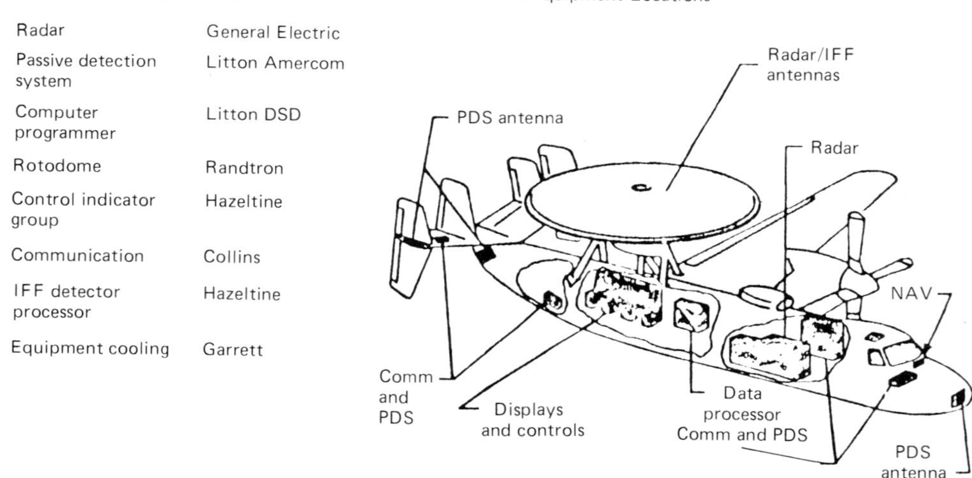

Figure 1.39 Major avionics suppliers for advanced navy aircraft E2C. *(Courtesy Grumman Aerospace Co.)*

- Good computer systems and applications programming staff
- A group of catalysts who can get things going and sustain the effort
- A strategy which is based on evolutionary (not revolutionary) changes in the work environment

The Industry Technical Advisory Board (ITAB), which guided the IPAD development, served as an excellent arena in which to clarify concepts and stimulate CAD/CAM developments within member and observer companies. ITAB meetings, work sessions, project critiques, and software evaluation activities served not only to improve the IPAD project results but, equally important, to help clarify and resolve management aspects of CAD/CAM planning and acceptance. In some respects, the IPAD project was an ongoing case study for industry managers on the critical technical and management issues associated with developing and implementing an integrated CAD/CAM capability. Such issues are often difficult to understand, because of the complex coupling of traditional engineering needs and approaches with the rapidly changing computer science technology. Resolution of issues often required unprecedented cooperation and frankness among the participating ITAB companies as they used the IPAD baseline approach to address mutual problems. The explicit products of the IPAD project, such as reports and prototype software, provided important and needed advances in CAD/CAM technology; yet the implicit benefits of the ITAB/IPAD process, such as use of the IPAD development effort to clarify management issues, reduce risk, and resolve company commitments, may be even more important to the long-term advancement of CAD/CAM technology and to the improved productivity critically needed by industry.

1.8 Concluding Comments

A unique joint government and industry project designated Integrated Programs for Aerospace-Vehicle Design (IPAD) was carried out from 1971 to 1984 with the goal of raising aerospace industry productivity through advances in computer-based technology to integrate and manage information involved in the design and manufacturing process. IPAD research was guided by an Industry Technical Advisory Board (ITAB) composed of over 100 representatives from aerospace and computer companies. The project complemented traditional NASA/DOD research to develop aerospace design technology and the Air Force's Integrated Computer-Aided Manufacturing (ICAM) program to advance CAM technology. IPAD had unprecedented industry support and involvement and served as a unique approach to government-industry cooperation in the development and transfer of advanced technology.

The IPAD project has made several contributions to CAD/CAM technology such as (1) development of methodology to integrate engineering activities, (2) definition and design of a future integrated CAD/CAM system, (3) demonstration of the need for relational database management for engineering, and (4) development and demonstration of new concepts for distributed database management. It also demonstrated a unique and highly successful approach to joint industry and government cooperation in achieving technology transfer in an advanced engineering computer sciences research program.

The lessons learned from the project suggest the basic IPAD program plan was appropriate and successful for the combined goals of technology development and technology transfer. A retrospective assessment identified issues which should be addressed in future IPAD-like projects: They include the:

1. Natural conflict between engineering and computer science disciplines
2. High turnover rate in computer-related technical staff
3. Conflict between research-oriented and useful software products
4. Benefit of an involved user group from a program's inception
5. Need for a clear image of a software program throughout its lifespan
6. Importance of small modular useful products completed at regular intervals throughout a program's life
7. Critical need for and cost of technology transfer as a planned part of an engineering software research program

CHAPTER 2

Considerations for Successful Implementation: Controls, Feedback, Benchmarking

David C. Penning

2.1 Introduction

The preceding chapters have illustrated some of the many ways in which computer-integrated manufacturing (CIM) technologies provide new opportunities. The factories of the future will be far more efficient than those we have had. If successful, manufacturing enterprises in the twenty-first century will both appear and operate very differently from such enterprises of the past.

Although the technology and the need for CIM are clear, the implementation has proved difficult. Some automated equipment markets, such as robotics, computer vision systems, and automated machine tools, have grown more slowly than was expected, especially in the United States. CIM implementation difficulties occur because:

- Investment in CIM is expensive and hard to justify by traditional direct labor savings accounting.

- CIM tends to be organizationally disruptive, since both information and operational responsibilities require sharing among formerly autonomous departments.
- CIM is resisted by personnel who fear that they may not be able to cope with new methods or technologies.
- Rapidly changing technologies, supply sources, and prices tend to confuse implementation decisions.
- CIM implementation requires development of functional and information flow-planning models, but the models can be controversial when they expose weaknesses in current methods.

In spite of the above inhibiting factors, implementation of CIM is necessary if manufacturing organizations are to remain competitive. The benefits that the pioneers have achieved not only give them competitive edges but also provide the basis for further CIM development through reinvestment of the savings. This adds further pressure upon manufacturers who have yet to implement significant CIM technologies.

This chapter describes how some pioneering firms are successfully implementing CIM concepts. They have benchmarked, controlled, and documented the results of their efforts so that the experience of CIM implementation can be used to obtain even more benefits in the future. Table 2.1 lists the basic steps that should be considered.

2.2 Defining Objectives

Implementation of CIM is a *strategic* issue because:

- The focus of the factory must be on carrying out the competitive business strategy of the company.
- CIM must compete with other needs of available resources for investment in time and dollars.
- The CIM planning horizon of 5 to 10 years exceeds ordinary tactical planning limits.
- CIM involves the total organization, and so it requires integration of all functional departments into a single business system.
- Implementation of CIM includes assessment of factors outside the business, such as technology developments, competitors' activities, and economic and socio-political trends.

Strategic objectives of the business can be used to determine which of two basic approaches to CIM implementation is most appropriate. When a new product line is being developed, there may be a unique opportunity

TABLE 2.1 CIM Implementation Steps

- Define objectives
- Set up CIM organization
- Determine status
- Identify needs
- Determine evaluation criteria
- Prioritize needs
- Estimate cost and schedule
- Document plans
- Get approvals
- Define projects
- Document system requirements
- Prepare procurement specifications
- Purchase software and hardware and services
- Install and train
- Report progress
- Analyze results
- Develop new concepts
- Recycle implementation steps

to set up a new facility or a separate portion of an existing facility that will achieve an accelerated, major advance into CIM. Considerations such as equipment arrangements to take advantage of grouping similar parts processing together (group technology), floor layout to accommodate automated storage and retrieval systems (AS/RS), automated guided vehicle (AGV) materials handling, and other CIM concepts can be integrated as a single system. This opportunity eliminates the problems caused by disruption of existing operations, relocation of equipment, piecemeal development of islands of automation, and changes to current operating procedures and organization structures.

Even more important, the new facility can be focused to achieve the strategic objectives of the business. The system can be tailored for production of small lots of a wide variety of parts or for a high-volume, low-variation product line, depending upon competitive market opportunities. Utilizing CIM concepts to provide a strategic competitive advantage is an essential part of the implementation of CIM because it gets the whole organization's interest and support behind the activity.

Unfortunately, the opportunity to build a new facility for CIM implementation occurs infrequently to many organizations in mature industries, yet those are the very organizations that most need to find a way to use CIM to gain competitive advantages. The approach that is taken in these situations can be described as a 5- to 10-year evolutionary development as defined by an overall strategic CIM plan. Within the existing facility, islands of automation are created and are gradually integrated into the total CIM system as defined in the CIM plan.

A common problem with the incremental approach is the failure of top management to establish long-term strategic business objectives that can be guidelines for setting CIM objectives. One way to force the resolution of this issue is to require that all CIM project funding approvals be contingent upon evidence that the project clearly supports the business objectives.

A compromise between the new-facility, high-risk approach and the slow, evolutionary approach is to develop a "factory within the factory." In this case, only a portion of the factory is used to prove out CIM technologies for specific product lines. Once the objectives have been accomplished in the pilot area, implementation experience provides guidance for development of the rest of the factory. A variation of this approach in multidivisional corporations is to establish one facility to be the lead factory in one technology and another in a second area. These separate factories then share their complementary CIM experiences during subsequent implementations.

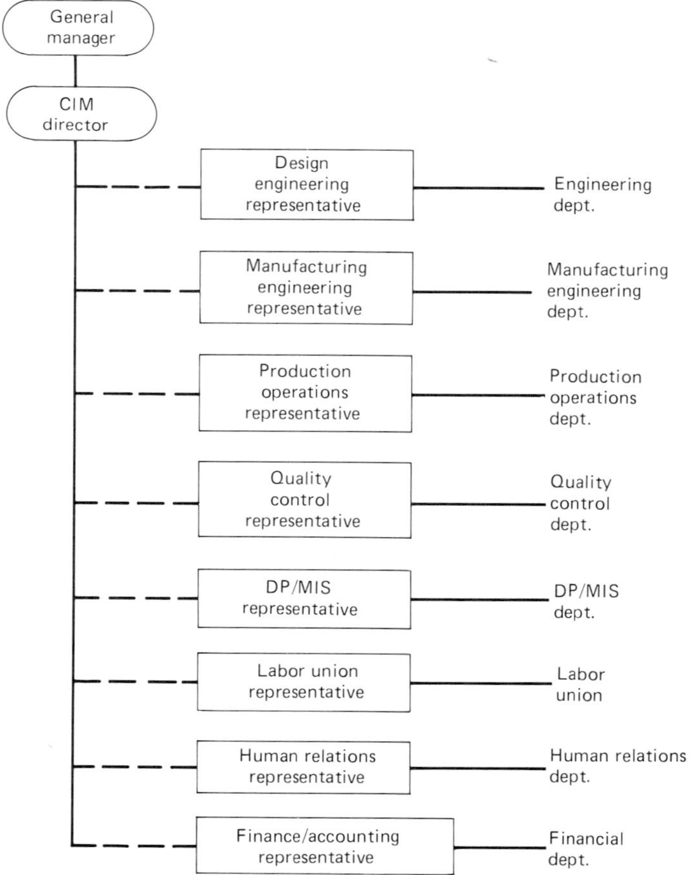

Figure 2.1 Organizing for CIM within the factory.

2.3 Organizing for CIM Development

Implementation of CIM requires the integration of functions and information flows that traditionally have been separate. Organizations have discovered that there are vast misunderstandings in what functional departments think other departments do. In manufacturing, no one person accurately understands the entire operation. Successful implementation requires that the barriers between organizations be removed. For that reason a multidisciplinary approach to CIM development is necessary.

Figure 2.1 shows a typical CIM implementation organization. Each major functional department is represented by one person who has knowledge of the department and authority to make CIM decisions on behalf of the organization. That person reports, on a project basis, to the CIM director, who reports directly to the factory general manager. Note that if the factory has a union, a representative of that group will help gain the viewpoints and support of workers for changes in work rules and organization structures. Union or no union, however, CIM implementation is guaranteed to fail without informed worker support.

The role and mission of the CIM implementation organization must be clear and be actively supported by the general manager. The tangible evidence of this support will be setting up an organization of the best representative members and providing an adequate budget for its operation.

Most organizations have found that the CIM director position is one of great opportunity for the selected individual. Who else, in the future, will know more about the operations of the factory than the person who was responsible for planning and implementing the integrated manufacturing system? Yet the position is fraught with problems such as resentment by individuals in functional departments of invasion of their turfs, general opposition to change, and a lack of full support by the general manager, who is caught in the squeeze between meeting quarterly goals and spending resources for CIM implementation that has a payoff sometime in the future.

The CIM director must act as the CIM "champion," be the catalyst that pulls all of the diverse, and adverse, activities into solid implementation results, and resolve the functional differences within the CIM organizational structure. When the director cannot do those things, the general manager must settle conflicts. The latter has a full-time job that requires a broad understanding of the needs of the company, knowledge about the opportunities of CIM technologies, and ability to replace heterogeneous chaos with an organized system. Clearly, few individuals can meet those qualifications.

The objectives of the CIM organization as a whole are to:

- Identify, prioritize, and recommend CIM implementation projects for each functional department

- Integrate individual projects into an architecture that supports overall CIM strategic goals
- Prepare an annual CIM development plan that is a part of the factory's total business plan
- Monitor and report progress of CIM implementation projects regularly to the general manager

In addition to the factorywide CIM implementation organization, there must be individual CIM project groups that are formed for the purpose of implementing specific CIM technologies, usually within a single functional department. For example, there may be one group that will be responsible for expanding the CAD system, another group that is installing a flexible manufacturing system (FMS), a third group to evaluate group technology software packages. Each group reports not only to its functional department manager, but, on a matrix basis, to the factory CIM implementation organization for its project.

In large multidivisional organizations, a corporate CIM committee has been shown to be valuable for implementation. The committee consists of a corporate director of CIM development and each of the individual factory CIM directors, plus both internal and external consultants as required. The committee's role is to:

- Provide a means for corporate sharing of CIM experience and resources among individual factories
- Integrate individual factory CIM plans into a total corporate plan for executive review and decisions
- Assist in setting CIM standards for the total corporation when that is desirable
- Provide specific project guidance for individual factories' CIM projects as requested

Because the factory of the future will operate as an integrated system, coordination of functional departments and the sharing of manufacturing information are necessary. That is why CIM implementation groups that cross functional and organizational boundaries are required. The experience of nearly every corporation has been that establishing these groups goes against the grain of traditional manufacturing organizations. The result has been that effective CIM implementation has often been thwarted by advocates of status quo, especially at middle-management levels. But if that opposition is anticipated and is planned for as a normal part of the implementation effort, it can be overcome through timely communications and education at all levels.

2.4 Identifying CIM Needs

A bewildering array of CIM technologies are available, and more are being developed. Figure 2.2 shows some of the opportunities. But a key issue is this: Which of these technologies are appropriate for a specific factory at a given time? The factory CIM development organization should determine which CIM technologies are needed. It can do so by first assessing the status of automation that already exists in the factory and then relating the as-is position to the business objectives for the future. Where are the bottlenecks? What are the critical success factors that are needed to gain a competitive edge? What would the ideal factory be capable of doing 5 to 10 years in the future?

Assessing the existing factory status nearly always produces some surprising results—mostly negative. For example, even though numerical control (NC) machine tool technology has been available for at least 30 years, few of today's factories have equipped even 50 percent of their machines with NC controls. CAD systems routinely show engineering savings many times the investment costs, yet manual drafting continues at many factories. Other obviously illogical and outmoded practices will usually be "discovered" during the initial automation status assessment.

How should the status assessment be made? One simple way is to prepare a checklist of available CIM systems, methods, and technologies. Table 2.2 illustrates a sample list for a metalworking factory. Each of the CIM development group representatives should perform the assessment within a department, and the results should be audited by the CIM group acting together. The audit should be documented, reviewed with the responsible department manager to ensure accuracy and concurrence (often with great reluctance), and then consolidated into a factory automation status report to be reviewed and approved by the factory general manager.

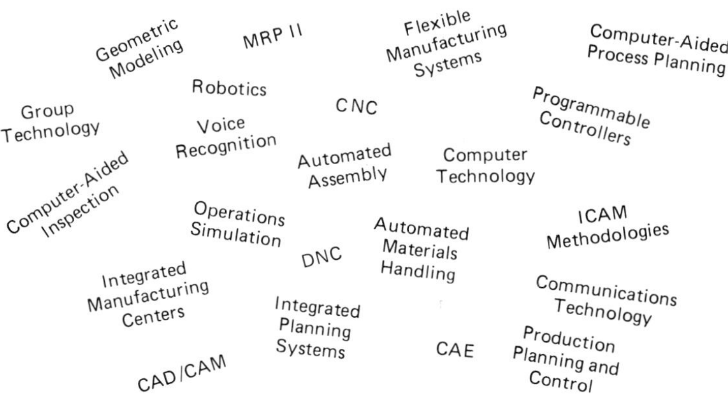

Figure 2.2 Factory-of-the-future building blocks.

The status of automation at the factory having been accurately determined, the needs for CIM development, in the light of the objectives of the business, come into better focus. Typically there are far more needs and CIM opportunities than can be reasonably satisfied simultaneously, so it is necessary to determine which of the CIM developments offer the greatest opportunities. The determination can be made by establishing a standard set of evaluation criteria against which all of the opportunities can be measured.

Development of CIM needs evaluation criteria is based upon the business goals for the factory. For example, what levels of quality, capacity utili-

TABLE 2.2 CIM Technology Audit

Functional Department ______________________ Date __________

CIM technologies	Estbl. sys.	Being implm.	Under study	Future reqmt.	N/A
CAD/CAE					
Mech. design/drafting					
Elect. design/drafting					
Other design/drafting					
Mechanical analysis					
Electronic analysis					
Engineering workstations					
Printers and plotters					
Digitizers					
Engineering data management					
CAM					
CNC machine tool control					
DNC machine system					
Flexible machining system					
Automated storage/retrieval					
Automated warehousing					
Automated conveyor system					
Guided-vehicle system					
Robotics: materials handling					
Robotics: assembly					
Robotics: other					
Inspection & test: mechanical					
Inspection & test: electronic					
Manufacturing programming					
Quality lab automation					
Production support					
Factory data collection					
Production monitoring					
Local area network					
Manufacturing database mgmt.					
Group technology					
Expert systems					
Process planning					
Material reqmt. planning					
Production control					
Production planning					
Material/tooling control					
Maintenance management					
Facilities management					
Procurement management					

TABLE 2.3 CIM Project Priority Assessment

Assessment criterion (1)	Score (2)	Weight (3)	Weighted total (2 × 3)
Quality improvement		10	
Customer requirement		9	
Greater capacity		7	
Reduced inventory		7	
Reduced direct labor		4	
Support other CIM projects		2	
CIM development experience		1	
Total project value			

zation, inventory, new product developments, and so on, are desired? What resources in funding, time, technical expertise, facilities, and equipment are available? Determination of what is important for reaching the business goals requires top-management input to development of the needs evaluation criteria. Top-down direction is essential for providing guidelines for CIM requirements evaluation.

Selection of which CIM needs should be implemented should be based upon an evaluation of all alternative CIM projects. Each need is evaluated in the light of the standard criteria for all needs. Needs are then scored and ranked. Table 2.3 is an example of CIM needs evaluation structure. In this example, the degree to which the CIM need supports any criterion is represented by a value from 0 (no support) to 3 (high support). Each criterion has a weighting factor to account for the differing importance of the factors. Each need score is multiplied by the criterion weight to arrive at a value. All values are summed for each need to arrive at a total need value. All proposed CIM needs can be ranked based upon their total value.

The ranked CIM needs list provides the basis for determining which CIM projects are most important. It provides the basis allocating limited funds to the most valuable CIM projects. Setting factorywide CIM priorities forces development of clear objectives and integration of various implementation activities.

To assist in the determination of automation status and the setting of CIM goals for the future, structured modeling techniques have been developed. Models of the manufacturing environment are graphic diagrams that help to simplify and clarify the complex relationships that exist in every factory. These models typically describe four views of the factory:

1. Current functional relationship status (as-is)
2. Functional relationships as they are envisioned 5 to 10 years in the future (to-be)

3. Information flows associated with the functions as the flows presently exist (as-is)
4. Information flows that are expected for future functions 5 to 10 years ahead (to-be)

Various methodologies have been developed for generating the factory functional and information flow models. The key concept is that the model has a structure that permits the complexities of factory operations to be decomposed into hierarchies of simple diagrams. The diagrams are carried to whatever level is required to adequately define, analyze, and communicate the manufacturing environment both as it exists and as it is intended to be in the factory of the future. Because every factory is a unique system, each model will be unique, but the methodology for representing that is a standard that helps break down communications barriers during CIM implementation.

Analysis of the models assists in establishing CIM needs evaluation criteria, as well as determining individual CIM project implementation priorities. For example, if a proposed implementation project does not logically support the development of the to-be manufacturing models, its value relative to the values of other projects that clearly support the model of the future is lower.

Although difficult to develop, manufacturing models are worth the effort. They not only assist in setting implementation priorities but permit what-if development scenario analyses. Table 2.4 lists some of the many benefits that can be derived from models of the factory. Figures 2.3 to 2.5 illustrate examples of models that have been developed for U.S. Air Force factory modernization (ICAM) projects. This methodology is called IDEF, meaning *ICAM definition language*. Further information about this CIM system design tool can be obtained from the U.S. Air Force Material Laboratory, WPAFB, Dayton, OH 45435.

2.5 CIM Project Management

Once CIM requirements have been identified and implementation priorities have been set up, specific projects can be established. A project leader for

TABLE 2.4 Benefits of Factory Models

- Reduction of complexity
- Common understanding
- Basis for logical analysis
- Integration of functions
- Reduction of omission and duplication
- Tool for simulation
- Promotes system optimization
- Identification of CIM needs
- Validation of CIM priorities
- Structures incremental implementation

each project is chosen. This individual is usually one who has experience in the functional area that is most affected by the new technology. He or she will be responsible for defining the tasks that will be required to implement the CIM development, will set budgets and milestone schedules for each task, and will assign the responsibility for accomplishing the tasks on time and within budget. All of the above will be documented in a project plan.

Figure 2.6 shows the elements of CIM project plan. Each plan should be reviewed by the factory CIM implementation organization as defined above. In addition, the plan must have the approval of the department managers who will be impacted by the implementation development.

The factory CIM director will assemble all of the individual project plans in order to establish total factory CIM development budgets and schedules. This total factory CIM plan must be integrated with the factory's annual business plan. It obviously will be reviewed, and usually revised, several times before the factory general manager will approve the total CIM plan.

There are many reasons for incorporating the CIM plan into the factory's business plan. On the one hand, CIM implementation will consume resources that then will not be available for other factory developments, such as new products and office automation. CIM expenditures must be considered in the context of available factory resources. Trade-offs with other non-CIM needs will be required.

On the other hand, implementation of CIM technologies may be a major part of a business strategy for the factory. For example, implementation of manufacturing resource planning software system modules may be the cornerstone of reduction of inventory costs or late product deliveries.

The individual CIM project plans not only support the total factory CIM development and business plans but also provide the basis for measuring the impact of changes in the original goals. Changes during CIM implementation are to be expected. Change is made the rule rather than the exception by the dynamic nature of CIM technologies and hardware and software systems that support the technologies. Further, the systems interactions among functions that once were uncoupled but are now inte-

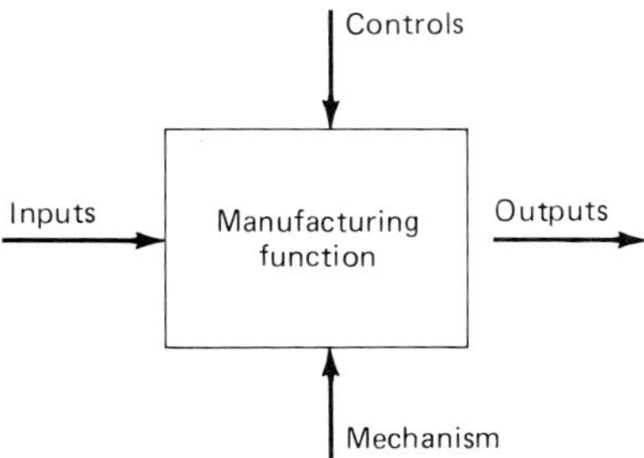

Figure 2.3 IDEF functional model structure.

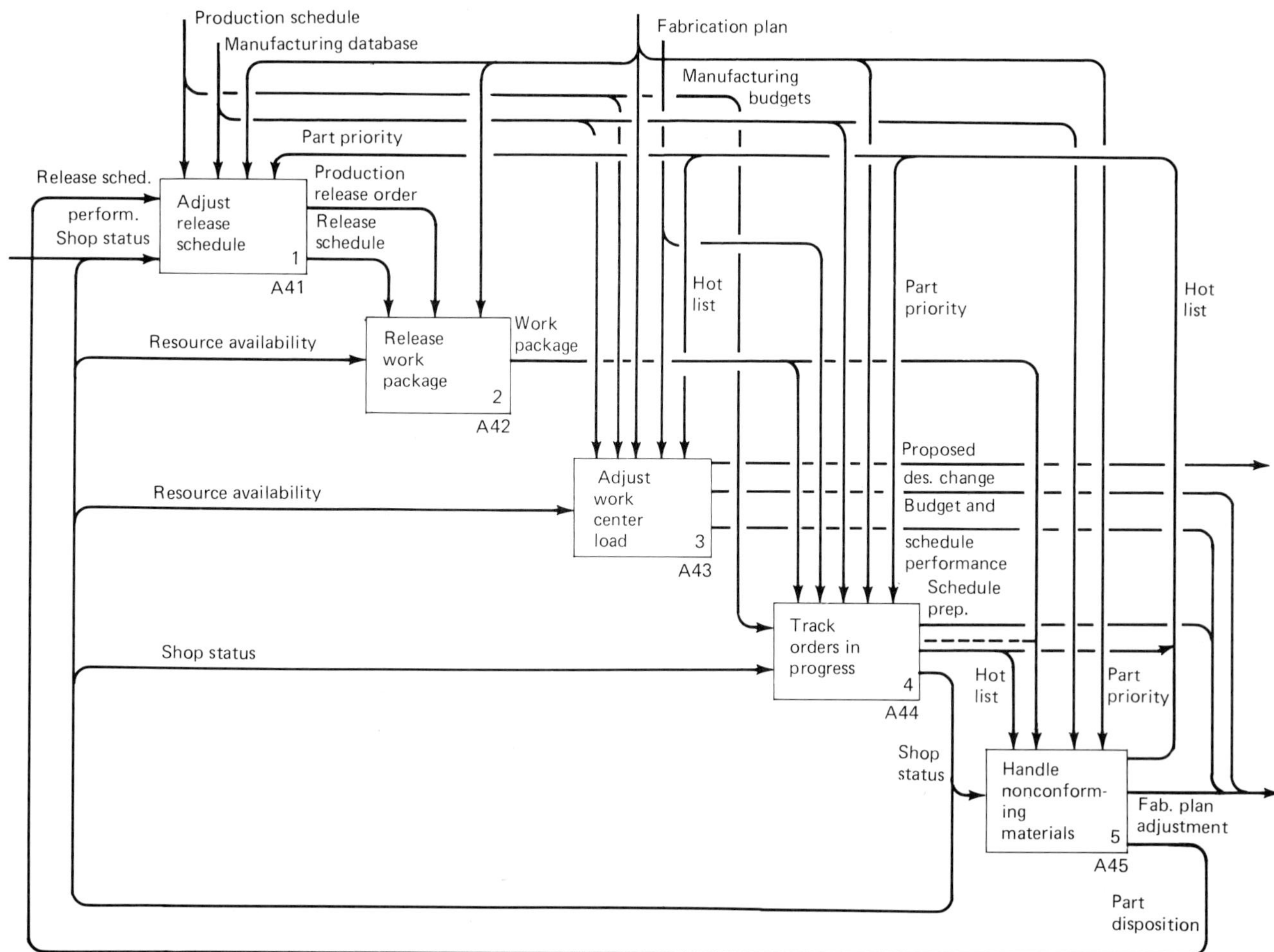

Figure 2.4 Example of U.S. Air Force IDEF structure: model of production control functions.

grated into one system can force changes in individual projects. Often the impact and change requirements cannot be foreseen.

But changes may frustrate efforts to complete projects on time and within budget. Delayed decisions, or no decisions, caused by changes in plans often result in "analysis paralysis." It is the responsibility of the CIM director of development to weigh the benefits of changes against the costs and delays. The director must decide, on the basis of overall knowledge of ongoing CIM activities, whether to approve proposed changes, must be certain that the impacts of one change are reflected in all other impacted CIM implementation projects as required, and must also see to it that the to-be manufacturing models accurately reflect approved project changes.

One of the first tasks in every CIM project is to define the system performance requirements. Seven categories of requirements should be examined for project application:

- Information resources
- System function definition

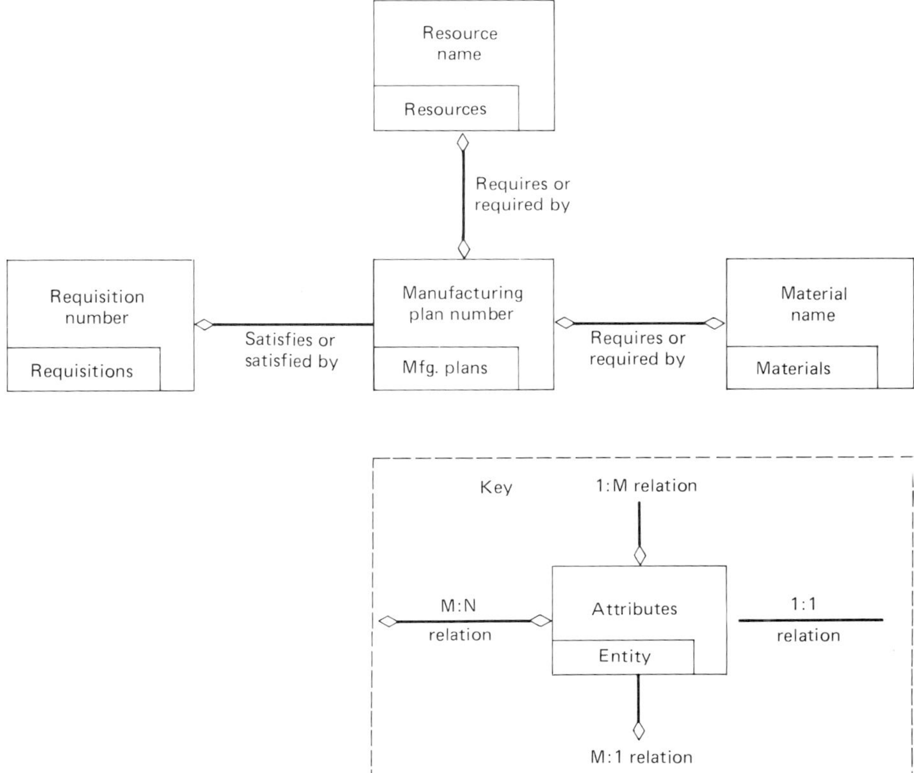

Figure 2.5 A simple IDEF model of one area of manufacturing information.

Project Name ______________________ Department ______________

Date ______________ Manuf. Model Reference ______________________

Project Leader ______________________ Revision No. ______________

Project Description ____

____

Project Objectives ____

____

Implementation Milestones:

Feasibility ______________ System delivery ______________

Vendor evaluation ______________ Training ______________

Vendor selection ______________ Installation ______________

Purchase approval ______________ Startup ______________

Order acceptance ______________ Project completion ______________

Cost Estimates:

Fiscal year	Hardware	Software	Labor	Other	Total	Source of funding
198__						
198__						
198__						
198__						
198__						
Total						

Remarks:

Figure 2.6 CIM project planning summary.

- Management control
- Quality assurance
- Human resources
- Financial and schedule targets
- Material resources

Completion of the system requirements will provide the basis for generating procurement specifications. A very serious concern of the CIM project leader is to determine who can provide the equipment and services to do the job. Although there are many vendors who will sell hardware or software, few are willing to take responsibility for integrating a tailored system into a specific factory environment. The end user will find that CIM project implementation cannot be delegated to outside vendors. At best, adopting new manufacturing technology will require a cooperative partnership of the hardware and/or software suppliers and the user.

The relations with the vendors of CIM hardware and/or software are critical to the success of the implementation project. Suppliers expect the end user to provide knowledge of unique requirements. These will be reflected in the procurement specifications. Since the supplier will provide the knowledge of the technology being applied, the specifications should define the performance requirements of the system rather than how these results should be obtained. However, the project leader must understand the technology well enough to select the best supplier approach from among various bidders. Also, the end user must have enough understanding of the technology to be able to judge the limitations of the selected supplier's equipment.

The procurement specification must be written at a level of detail sufficient to be the basis for a request for proposal (RFP) from prospective CIM system suppliers. The RFP should, in addition to specifying the system requirements, identify the response time window and background information on the specific factory and product or process background.

When the proposals have been received, they should be evaluated against the functional procurement specifications. Cost should be separated from system capabilities during this initial evaluation. In the scoring and ranking of the vendors' proposals, the same analysis structure concepts that are used for ranking CIM projects can be utilized. The specification line items become the evaluation criteria against which the proposals are scored.

The CIM project leader, on the basis of the RFP evaluation analysis, can select the top two or three vendor system configurations. At this time, benchmarking of all hardware and software components can be done. Then if the requirements that are essential to the success of the system have been identified, consideration can be given to making changes in operations that could simplify the application. This may make the vendors' proposed systems more feasible or more cost-effective.

It is essential that benchmark tests be well conceived. Samples of typical parts or processes should be provided to the vendors if possible. Actual benchmark tests should be designed to evaluate only the critical functions of the system and should be performed in the presence of the end-user evaluation team. The same tests should be required of each prospective vendor, and they should not last more than one day per vendor. The vendor should not be provided with complete benchmark test descriptions in advance, because the element of surprise may be required to get a true assessment of each system's limitations as well as its capabilities.

The capabilities of the vendor's proposed system to meet the benchmark tests should be scored by the end-user project team members. The results of all of the vendors should be collectively analyzed and ranked by degree of capability to meet the procurement functional performance specification. At this point, relative costs of the competing systems, as well as other important factors should be included in the vendor selection process.

Some of the other criteria include the experience of the vendor with the end-user's application, the number of existing installations, evaluations of the vendor's capability by other end users, and whether the vendors support integration standards that may exist for the end-user's corporation. Together with these other considerations, the benchmark tests will determine which vendor is recommended by the project team.

Selection of the winning vendor marks the beginning of a close working relationship between the CIM project team and the vendor. Installation plans and schedules should be developed with the vendor; they include site preparation, operator training, equipment delivery, installation procedures, start-up requirements, and acceptance criteria.

All employees who will be involved in CIM system implementation and operational usage should be fully informed of the system well in advance. It is a truism in manufacturing that the only thing faster than electronic communication is the grapevine! The natural fear of change can be converted into strong employee support by timely, open communications with affected workers. Research done by the author has shown that, instead of being a problem, involved workers are some of the most highly motivated employees in CIM development. That is because they can see that the investment in CIM represents a commitment by management in the future of their factory—hence job security for them. Involved workers frequently are an excellent source of ideas for system improvements and resolution of the myriad of unforeseen glitches in CIM system implementation.

2.6 Implementation Control and Feedback

The key to successful CIM project implementation is an adequate plan. The tasks must be broken down into work packages that are logical units but which also are of predetermined time frames and budgets. These limits vary with the nature and overall scope of the project. Work packages of one month's duration and costing not over $100,000, have frequently been used. The size of a work package reflects the level of risk in progress toward the plan that management is willing to accept.

In addition to breaking the tasks down into manageable packages, the interdependence of the tasks must be recognized. This can be depicted graphically in two ways: a hierarchical work breakdown structure and a PERT chart showing the time dependencies of all the work packages. Validation of the project budget and schedule is obtained by summing assigned resources of each of the work packages and then comparing the sum in dollars and worker-days to that for the total project. A plan is not realistic until resources balance both schedules and budgets.

The essence of effective CIM project control is that the method used for management must be simple to use. Required changes in the plan must be made without replanning the entire project. One way that helps to ensure simplicity, as well as consistency, is to use the same document that outlined the key project goals as the summary reporting document. If further explanatory documentation regarding the progress of implementation is required, additional sheets can be added to the basic report format.

Figure 2.6 can be used as a combined planning and reporting document. This level of documentation may seem too superficial for proper feedback of implementation progress, but experience has shown that overly bureaucratic reporting systems will not be used. Recently many project management software packages for use on desktop personal computers have become available. These may be useful, as long as project management does not become a project by itself.

Routine feedback of actual progress against planned implementation milestones is essential to keep a project under control. Fast recognition that a milestone has not been met will usually trigger corrective action while the problem is still small enough to permit recovery within the overall project budget and schedule.

Implementation progress feedback should not be for only the project leader; the progress should also be tracked by the factory CIM director. Ultimately, the factory general manager must be kept aware of CIM implementation on at least a quarterly basis. Most factory managers tend to follow CIM developments much more closely than reviewing quarterly summary reports.

Since CIM implementation is often a cornerstone in the successful accomplishment of overall business strategic goals, the factory manager's view of CIM status tends to be less oriented toward feats of technological competency and more toward how CIM projects impact business goals such as quality, delivery, and productivity. The factory CIM development manager will need to report CIM progress to the general manager in terms of support for the critical success factors for the business. To the degree that the manager is adept in making this connection, he or she will garner support from the general manager to assist in overcoming the inevitable obstacles to CIM implementation.

Relating the CIM implementation to critical business success factors as perceived by the factory general manager will:

- Help ensure that the general manager will focus significant attention on CIM progress regularly and carefully
- Force the general manager to actively seek good measures for CIM progress and also seek reports that measure accomplishment effectively
- Provide the CIM director with guidelines to the amount and frequency of reporting information that is necessary to satisfy the general manager
- Eliminate the trap of building CIM reports around data that is easy to collect, rather than on what is significant to the general manager

Critical success factors ensure the successful accomplishment of the overall factory goals. Reporting CIM implementation progress against these measures will assist CIM development in several major ways. First, knowledge of these critical factors helps identify CIM projects that should have enhanced value to the factory because they clearly support accomplishment of goals.

Second, support of the general manager is essential for the backup authority that the CIM development director needs in order to resolve organizational impasses and/or uncooperative line managers. Experience has proved that not everyone in a manufacturing organization will think that CIM implementation is important or even desirable. Unfortunately, this is the real world of CIM today. However, the factory general manager who is motivated to use CIM to achieve corporate goals can take whatever steps are required to overcome organizational opposition.

A well-conceived and documented CIM project plan will include clear implementation milestones. As the system installation proceeds, progress against the milestones can be determined objectively. For example, a progress measurement might be whether a predetermined number of system operators have completed a given number of days of vendor training. Objective progress milestones include associated completion dates, expenditure levels, and task accomplishments. Definition of the milestones should be based upon four sources:

1. In-house engineering development feasibility studies (the factory floor is no place for R&D)
2. Experience of other end users who have installed similar systems
3. Consultants who have participated in comparable installation activities
4. The hardware, software, and services vendor(s)

Progress against milestones should usually be reported on a weekly basis to the factory CIM development director, who usually reports progress of each project quarterly to the general manager. Milestones that are missed, or unreported, should be investigated promptly by the CIM director. Together, the director and the affected project leader should determine the reason for the slippage. They can determine when the milestone will be met, the impacts upon other milestones, and how recovery can be made. If recovery is not possible, replanning will be necessary.

Should the cause of the slippage turn out to be a continuing problem, the CIM director should request assistance from the CIM development organization and/or the involved line management. If the recovery still appears to be unresolved, the CIM director must appeal to the factory general manager for help. Obviously this last resort is an admission of failure by the CIM organization. Weekly monitoring of progress and verification of milestone accomplishment, together with rapid resolutions of problems before they grow large, will go a long way toward ensuring that CIM implementation targets will be met as planned.

2.7 Benchmarking CIM Success

Full development of the opportunities of CIM is being accomplished by an evolutionary process. Factories start with conventional installed equipment and use existing methods and operating experience. Figure 2.7 shows the anticipated evolution of conventional machine tools into an automated flexible manufacturing system.

The strategy that is usually followed is to implement a series of viable, economically justified CIM project steps. Each individual step has the potential for sufficient economic return to justify it by itself. If the project is successful, the return from it will generate savings that can be used to support the development and implementation of the next CIM step.

Each CIM project should be viewed in the context of the 5- to 10-year CIM to-be architectures. This long-range system view of each CIM project ensures that as the islands of automation and the successive steps toward the factory of the future evolve, the islands will gradually come together into a single, optimized, computer-based system. This system will be focused upon providing the most efficient manufacturing technologies that will provide a strategic business competitive edge.

Since justification for continuous CIM evolution on a project-by-project basis often depends upon verification of accomplished benefits, a key implementation issue involves benchmarking CIM success. Several research studies concerning how major U.S. corporations have justified their CIM projects in robotics, machine vision, and CAD systems were conducted by this writer. The studies revealed that less than one-third of the CIM projects had been formally evaluated 6 months to 1 year after installation by any of the 45 corporations surveyed. Reasons for the lack of reviews ranged from "we knew the results were good, so we didn't try to measure the benefits," to "we tried hard to establish the return on the investment, but found it too hard to do—everything was changing too fast."

The few CIM projects that were evaluated after implementation were nearly always more beneficial than had been anticipated. However, the benefits were usually not "hard" savings, i.e., in direct labor savings, but rather in "soft" benefits. The latter were often such things as improved employee morale, better on-time product deliveries, enhanced image to customers, and greater safety. Frequently those benefits had not been the primary justification objective before the installation, but they turned out to be the greatest benefits in hindsight.

The research study suggests that a broader definition of CIM benefits than has traditionally been used must be developed. Table 2.5 lists some of the common hard (tangible) and soft (intangible) benefits that end users have experienced. Financial analysts should structure CIM benefits analyses so that nontraditional soft benefits, as well as tangible benefits, are

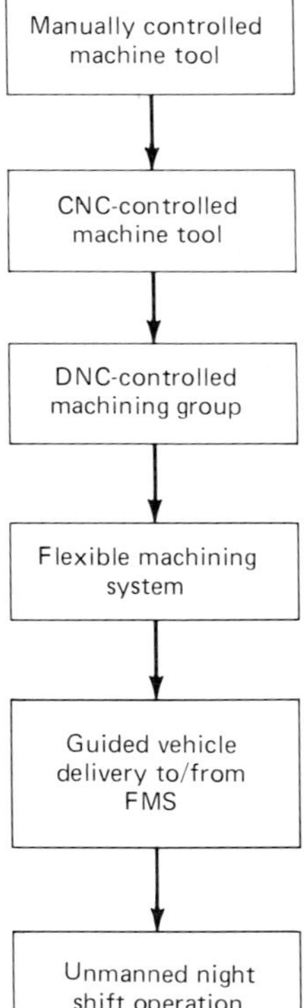

Figure 2.7 Evolution of conventional machine tools into a flexible manufacturing system.

anticipated in advance. That will help ensure that CIM project priorities are based upon all benefit measures, and not just on labor savings.

It should be the accounting department's function to use the CIM financial evaluation structure to collect data and measure both hard and soft benefits of CIM implementation. The measurement of results is not for the purpose of proving that the original estimates were correct, because the investment has already been made. The purpose of the evaluation is to learn from experience so that future CIM planning and implementation can become more effective than they were before. For that reason, knowing what went wrong may be more valuable than learning what went as expected.

Even though reviews of completed CIM projects seldom produce exact investment returns calculations, enough can be learned from hindsight that the attempt will be worth the effort. Further, assessing the risks of too much or not enough CIM investment levels, based upon actual experience within the particular factory environment, is more easily accomplished.

CIM success must not only be measured against internal experience; it must also be evaluated with regard to external opportunities and threats. One of the functions of the CIM director should be to monitor technical developments. This can be done informally by reading professional publications, attending conventions such as AutoFact, and belonging to professional organizations such as the Society for Manufacturing Engineers, Institute of Industrial Engineers, and the National Computer Graphics Association.

A more formal approach would involve setting up a routine library search of online databases, such as DIALOG monitoring an array of selected pub-

TABLE 2.5 Examples of CIM Benefits

Tangible
• Higher profits • Less direct labor • Increased machine utilization • Reduced scrap and rework • Increased factory capacity • Reduced inventory • Shortened new product development time • Fewer missed delivery dates • Decreased warranty costs
Intangible
• Higher employee morale • Safer working environment • Improved customer image • Greater scheduling flexibility • Greater ease in recruiting new employees • Increased job security • More opportunities for upgrading skills

lications, and subscribing to CIM services offered by consulting and market research firms. Monthly summaries should be sent to CIM development organization members. Communications ties should be established with other organizations so that technical research can be shared.

Just as the technology should be monitored, so the wise CIM director will also monitor the activities of competitors. The methods and sources of competitive evaluation are beyond the scope of this chapter. Suffice it to say that no one organization has all the answers. Since competitors often have similar CIM needs, an evaluation of their CIM implementation activities should be an excellent means of measuring the value as well as potential threats that CIM holds out to the factory.

2.8 Conclusion

CIM is changing every aspect of the factory environment from developing new products to shipping the finished current products. Further, the rate of change is accelerating. The issue is not whether to use CIM but how CIM can best be implemented.

Not only is CIM development giving the pioneering factories a competitive edge, but the experience gained from implementation is a source of new concepts for future development. Many factories are implementing CIM not so much to solve immediate problems as to provide the basis for integrating CIM technology in future factories.

Some factory managers, when faced with a business downturn, curtail their CIM implementation activities, saying that they must cut back on expenditures. But other managers realize that slack times, when surplus resources are available and capacity is not stretched to the limit, are the best times in which to implement CIM projects. The renaissance of Detroit automakers was due in large measure to the investment in CIM even though the industry was in a recession.

CIM implementation is both the major present problem and the key to future success in manufacturing. Research by this writer has shown that the more a factory becomes involved in CIM, the better it gets at successful implementation. Those who fail to begin installing automation systems are in danger of falling so far behind the innovators that they may be unable to catch up.

One thing is clear: CIM opportunities are limited only by one's imagination. Implementation of CIM technology *is* the factory of the future.

CHAPTER 3

The Economics of CIM*

John L. Canada

Robert Edwards

3.1 Introduction

3.1.1 What is this chapter; what is CIM?

This chapter was extracted from the authors' how-to manual for considering the intangible and economic factors which are important in the consideration of modern manufacturing systems which incorporate high technology.* Normally such systems are characterized by use of computers for product and/or process design, manufacturing controls, and/or decision support systems. Figure 3.1 gives a further description of what are normally considered the component functions of the broad field of computer-integrated manufacturing (CIM).

We loosely refer to CIM as any computer-oriented equipment or system which aids in, or achieves, the automation of a manufacturing enterprise and which is planned to increase, if not eventually complete, integration of the enterprise. The CIM concept of manufacturing is a continuous flow process beginning with product design and ending with shipment or cus-

---

***Should We Automate Now? Evaluation of Computer-Integrated Manufacturing Systems: An Applications Manual with Handy Forms*. (The book is available from Industrial Extension Service, North Carolina State University, Box 7902, Raleigh, NC 27695-7902, for $39.) Copyright to the material belongs to North Carolina State University. The editors of this handbook thank the University for permission to use the material.

tomer service. The CIM acronym is an expression of the ultimate goal and not necessarily the approach to achieving that goal. Integration connotes tying together adjacent operations and overall control systems.

3.1.2 For whom is the chapter intended?

The manual from which this material is taken is intended for investment-planning decision makers and analysts. Although the forms may be filled out and computations performed largely by analysts, the answers and thinking required (particularly for strategic planning) call for very active, thoughtful consideration by higher management.

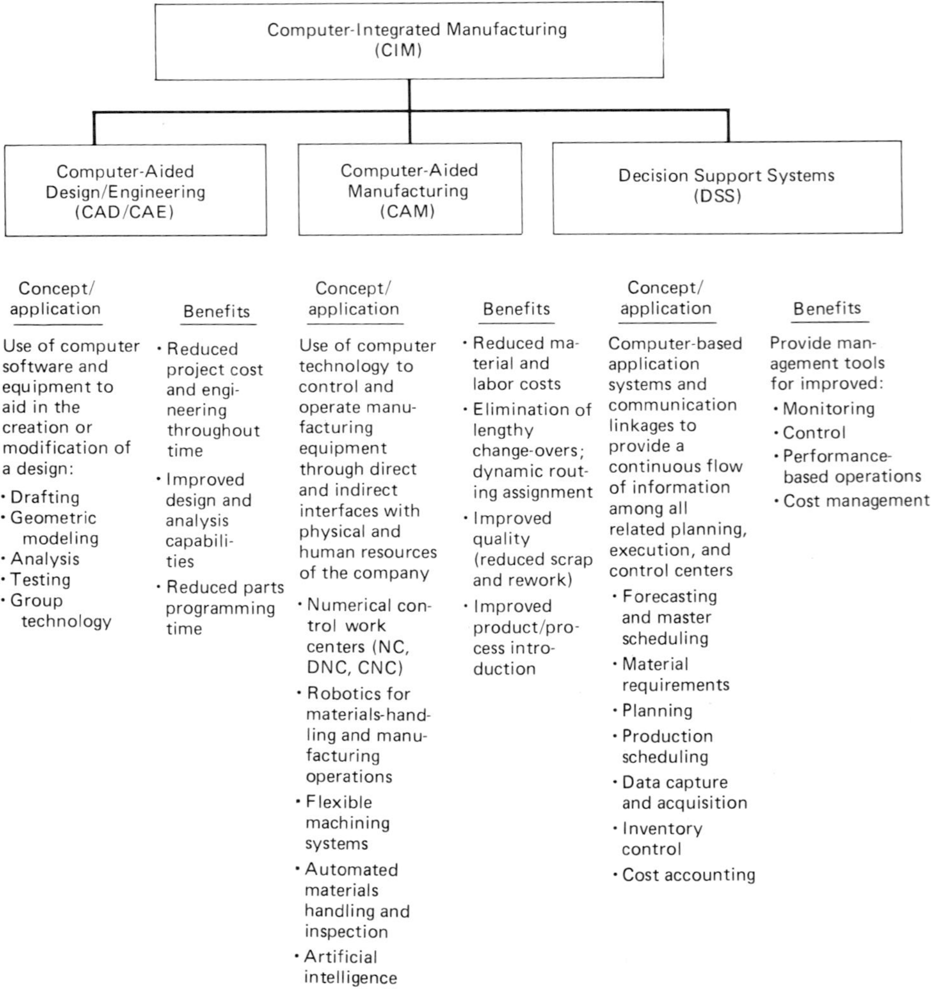

Figure 3.1 The component functions of CIM. (From Michaels et al., Material Handling Engineering, March 1984, p. 97.)

The *strategic planning* process is of such long-term importance—including product and competitive as well as manufacturing concerns—and the technical considerations are often so complex or at least hard to predict that for most firms it is recommended that a group of knowledgeable specialists and managers address the questions. The *tactical planning* process usually would require less detailed consideration by higher management than in the strategic case. Nevertheless, it normally justifies group consideration—usually different but overlapping group(s)—to ensure proper identification of alternatives, evaluation of expected benefits, costs, and risks, and hopefully final consensus which will lead to full backing for implementation of the chosen alternative(s). (*Note*: It is recommended that mechanisms be used to ensure that group thinking will not be unduly dominated by one person. Two ways widely advocated for achieving this are the Delphi technique [Ref. 1] and the nominal group process [Ref. 2].)

3.1.3 Importance

Because of competitive pressures, firms must plan and initiate sound steps toward CIM systems. Some writers advocate bold steps, but most writers recognize the need for phased development of resources (personnel, technological, product, financial, and so on) based on a general master plan. Many writers advocate looking on investments in CIM systems as R&D projects to gain knowledge and experience (options) for the future.

Most persons recognize that a firm cannot expect to maintain the status quo indefinitely. Rapidly advancing technology renders products obsolete and processes noncompetitive. Unless the firm prepares to achieve rapid product design innovation [using computer-aided design and drafting (CADD)] to attain efficient manufacturing process planning execution and control [using computer-aided manufacturing (CAM)], and to utilize sound decision support systems (DDS), all significantly based on computer technology, it can expect to be eventually forced out of business by competition which successfully adapts to the new technology (see Sec. 3.2.1). Hence, when one is considering the economic and other attributes of a CIM system, one should compare it to what will be rather than what now is without the system. In the words of James Baker, General Electric Senior Vice-President: "Automate, emigrate, or evaporate."

3.1.4 The need for nontraditional justification analysis procedures

Manufacturing firms within the United States are increasingly considering computer-based automation equipment and systems in their efforts to increase productivity and regain or maintain a competitive edge. Such systems always require massive capital investments. Many firms have tended to base their investment decisions on traditional discounted cash flow financial justification methodologies best suited to meet short-term profitability criteria rather than long-term strategic goals. The traditional financial justification methodologies, when used in conjunction with the high hurdle (minimum attractive) rates prevalent in today's uncertain and capital-scarce environment, often result in the rejection of proposed high-technology equipment and systems.

The inadequacy of such traditional justification procedures has become apparent. Much of the literature blames the problem on the inability of traditional justification procedures to quantify and formally consider so-called intangible benefits. Other literature points out the shortcomings of contemporary and allegedly outmoded cost accounting systems, which inhibit the use of relevant but often unconventional measures of performance. Some authors even totally disclaim attempts to justify modern automation through traditional capital budgeting evaluation practices on the ground that such decisions should primarily be based on strategic considerations.

Bela Gold, in a landmark article in *Harvard Business Review* [Ref. 3] said:

> The real promise of CAM technology lies not in its use as yet another, perhaps fancier than usual, machine tool located at a single point in an otherwise unchanged production process. CAM's promise lies, by contrast, in the ability to integrate adjacent operations with each other and with overall control systems. Because it offers a systemic—not a "point" capability—neither its purchase nor its performance should be evaluated in the traditional way.

3.1.5 Overview of capital budgeting and the decision-making process

Decisions regarding investments in CIM systems would normally be made as part of a firm's overall capital budgeting process. That process is usually highly specific to the individual firm. Figure 3.2 is a diagram of the somewhat simplified selection process recommended in this chapter. First the representative good investment opportunities are identified. For each opportunity, "representative good" is meant to be one of perhaps several mutually exclusive alternatives which, upon relatively brief screening, seem to be good for further comparison purposes. Figure 3.2 then shows categorization of all such opportunities into three groups:

Group	Consequences	Resource commitment	Examples
Strategic	Long-run	Major	CIM systems New product line
Administrative	Medium-length	Moderate	Replace gen'l-purpose eqpt.
Operational	Short-run	Minor	Fixtures Discretionary repairs

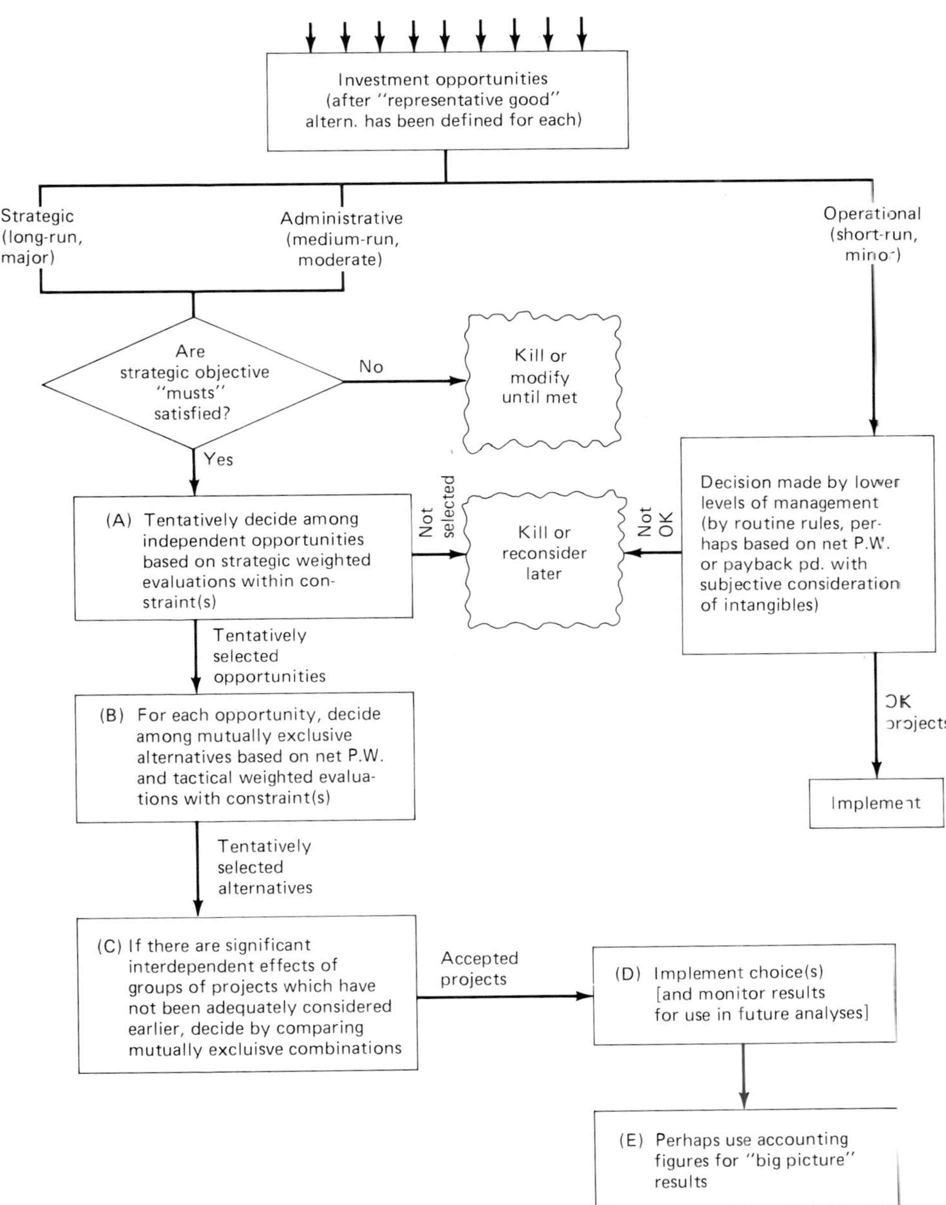

Figure 3.2 Recommended selection process.

Operational decisions are shown as being made by routine decision rules, and they will not be discussed further herein. Investment opportunities in the strategic and administrative categories, because they tend to overlap, are shown to go through the same subsequent steps. They are first screened according to whether strategic objective "musts" are, as the name implies, criteria or questions which must be satisfied before an opportunity will be eligible for further consideration. Examples are: "Does it maintain or advance the firm's technological capabilities?", and "Does it keep us concentrated in the ______ business?", and "Does it risk less than ______ percent of the firm's net worth?".

Opportunities which pass the strategic musts test are then ranked and tentatively selected within whatever constraints exist according to a method called "strategic weighted evaluation." Next, any mutually exclusive alternatives for each tentatively accepted opportunity are considered and tentatively selected by consideration of (1) financial, and (2) probably intangible, factors within whatever constraints apply. The financial measure used is net present worth (Net P.W.), and the intangibles measure method is called the tactical weighted evaluation. If there are significant interdependent effects of combinations of projects, they can be considered by comparing mutually exclusive combinations.

3.1.6 Categories and classification for opportunity selection purposes

Opportunities considered and/or selected are commonly categorized into convenient, meaningful groupings for thinking and presentation purposes. The groupings might be as routine as by plant or department of the organization. Another grouping used in the simplified selection process in Fig. 3.2 is:

1. Strategic
2. Administrative
3. Operational

Yet another possible indicative grouping is:

1. Cost reduction
2. Replacement (routine in kind)
3. Expansion (additional capacity for existing products)
4. Expansion (new product lines)
5. R&D technical capability
6. Other

A CIM system probably would fit into two or more of the above categories, such as numbers 3, 4, and 5.

Another, more detailed indicative grouping [Ref. 4] is:

1. The kinds of scarce resources used, such as cash and key personnel
2. The amount of each of the above resources required
3. The way benefits from the investment are affected by other benefits
4. The form in which benefits are received (attributes aided)
5. Whether incremental benefits are the result of lower cost or increased sales or merely prevent a decline in market share
6. Functional activities most closely related
7. Industry classification
8. Degree of necessity, such as essential, very desirable, or optimal

3.2 Strategic Benefits of CIM and Accounting Measures

In this section we discuss the strategic benefits of CIM. Our complete manual also contains descriptions of strategic and tactical planning processes, an assortment of easy-to-apply graphic and tabular techniques for exploring and evaluating the effect of lack of certainty in planning estimates, extensive checklists and estimating aids, and a complete set of forms. It shows use of a weighted evaluation of the objectives and attributes technique, together with after-tax net present worth analysis for quantifiable benefits and costs. It also includes methodologies for considering sequential investment alternatives and interdependent combinations of projects.

3.2.1 Avoiding competitive wipeout due to costs and prices

Perhaps the most compelling reason for seriously undertaking steps toward the adoption of CIM is *survival*—through avoiding competitive wipeout due to manufacturing costs squeezing profits.

CIM installations commonly increase fixed costs and production capacity but substantially reduce unit variable costs. The result is that firms which adopt CIM tend to reduce prices to gain a greater share of the market and benefit from economies of scale. Competing firms may be reluctant to do likewise because the prospective return on the added investment, particularly at lower market prices, would then seem to be too low—below their common "hurdle rate" or payback period standards.

The results of firms declining or delaying modernizing investments, so that their facilities become obsolete compared to those of the competition, are sometimes descriptively referred to as *disinvestment*. The following describes what can happen [Ref. 5]:

> Companies so often become trapped in this sort of disinvestment spiral—deferred investment leading to reduced profitability, which further reduces the incentive to invest. . . . No company, however, can be sure of recovering lost ground . . . postponed investments may not be reversible; complete recovery may be impossible.

To further illustrate the dynamics of automation (CIM) vs. nonautomation (status quo) decisions, Figs. 3.3 to 3.6 show a series of typical scenarios using breakeven charts for three firms competing on the basis of prices and costs in a given market.

Figure 3.3 shows the base case, in which the firms have differing capacities (2M, 1.5M, and 1M units, respectively) but roughly comparable investments and fixed costs per unit capacity and the same variable costs per unit. Sales per unit of capacity for each are roughly comparable, resulting in the ROI (return on investment, or profit/investment) for each being close to 10 percent. (*Note*: Detailed calculations are not shown herein.) This would be a good example of a stable situation with a limited number of competitors.

However, suppose that, in Fig. 3.3, firm B decides to be more competitively aggressive by installing automation, which will increase fixed costs and decrease variable costs per unit considerably. The dashed lines in Fig. 3.3 show the assumed new fixed and total costs. If firm B kept the same sales, its profit would still be about the same but its ROI would be almost halved (to 5.8 percent) because of the larger investment required. Firm B probably would not be happy with this, and it probably would take steps to increase the sales. The remaining figures in this section assume that total costs (TC) and variable costs (VC) remain the same for each firm and that only firm B has automated.

Figure 3.4 (competitive scenario 1) shows assumed results if firm B reduces the revenue (price) to $90 per unit and firms A and C keep the revenue at $100 per unit. It is further assumed that the total market remains the same at 3350 units but firm B's unit sales double and the sales of firms A and C are proportionally decreased. The results are that firm B makes a very high 20.2 percent ROI but firms A and C have losses. Even though the losses are relatively small, firms A and C will probably try to regain sales through price competition.

Figure 3.5 (competitive scenario 2) shows assumed results if firms A and C match firm B's revenue (price) of $90 per unit and sales for all firms increase 30 percent over the base case in Fig. 3.3. The results are that firm B's ROI decreases to 8.5 percent and firms A and C have ROIs of 12.2 and

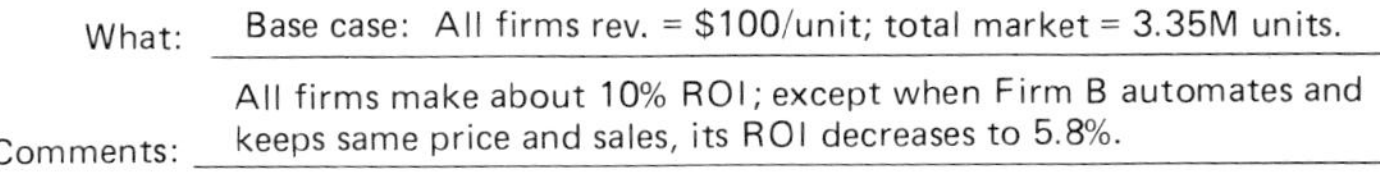

What: Base case: All firms rev. = $100/unit; total market = 3.35M units.

Comments: All firms make about 10% ROI; except when Firm B automates and keeps same price and sales, its ROI decreases to 5.8%.

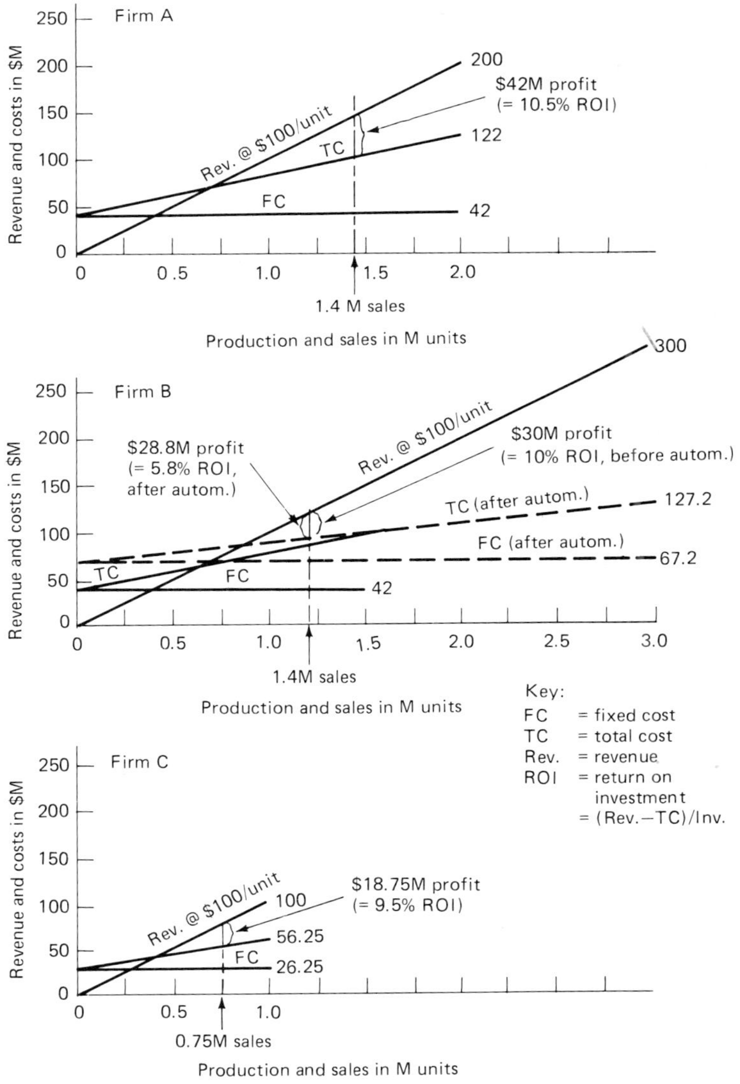

Firm A

Revenue and costs in $M

250
200
150
100
50
0

Rev. @ $100/unit
TC
FC
200
$42M profit
(= 10.5% ROI)
122
42

0 0.5 1.0 1.5 2.0

1.4 M sales

Production and sales in M units

Firm B

Revenue and costs in $M

250
200
150
100
50
0

$28.8M profit
(= 5.8% ROI,
after autom.)
Rev. @ $100/unit
300
$30M profit
(= 10% ROI, before autom.)
TC (after autom.)
127.2
FC (after autom.)
67.2
TC
FC
42

0 0.5 1.0 1.5 2.0 2.5 3.0

1.4M sales

Production and sales in M units

Key:
FC = fixed cost
TC = total cost
Rev. = revenue
ROI = return on investment = (Rev.−TC)/Inv.

Firm C

Revenue and costs in $M

250
200
150
100
50
0

Rev. @ $100/unit
100
$18.75M profit
(= 9.5% ROI)
56.25
FC
26.25

0 0.5 1.0

0.75M sales

Production and sales in M units

Figure 3.3 Breakeven chart 1.

What: Compet. scenario I: Firm B reduces rev. to $90/unit. Firms A and C keep rev. = $100/unit. Total market stays same, but B's sales double to 2.4M and A's and C's decrease to 0.619M and 0.133M, resp.

Comments: B makes 20.2% ROI, but both A and C have small negative ROIs.

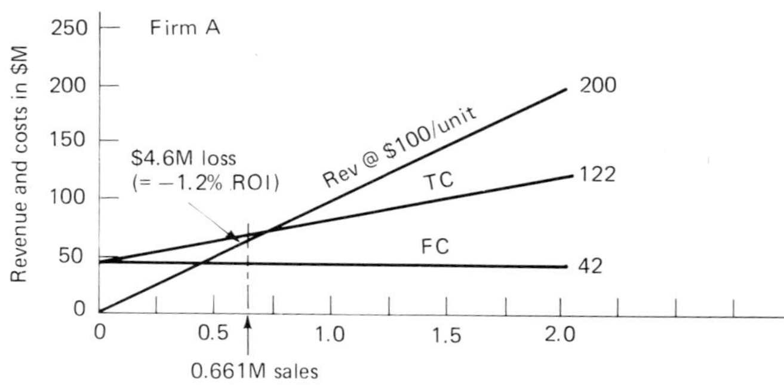

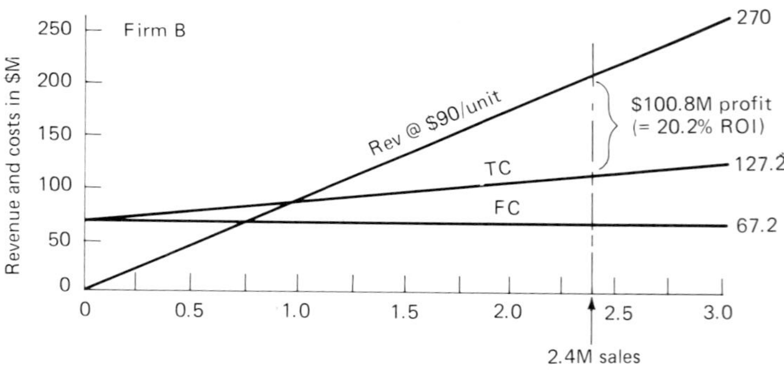

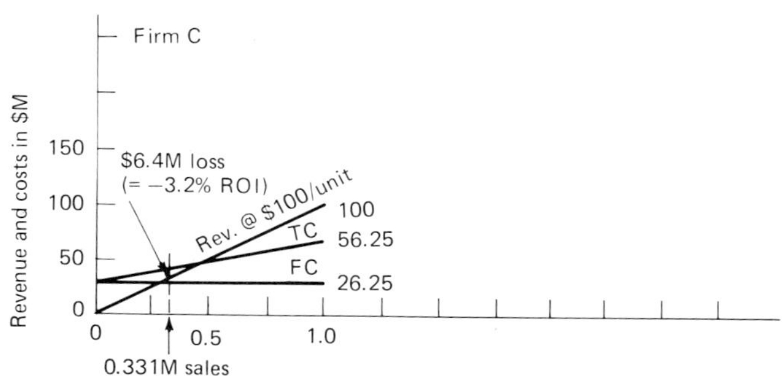

Figure 3.4 Breakeven chart 2.

What: Compet. scenario II: Firm B reduces rev. to $90/unit. Firms A and C also reduce to $90/unit. Total market increases by 30% for each. New sales are A, 1.82M; B, 1.56M; C, 0.975M.

Comments: B's ROI reduces slightly to 8.4%, but A's and C's ROIs increase to somewhat more than 10%.

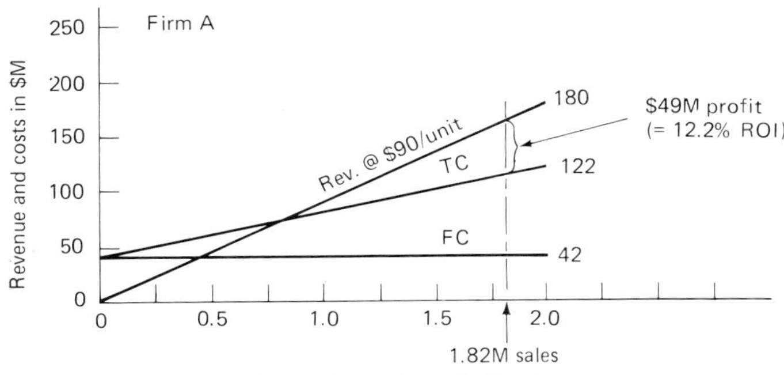

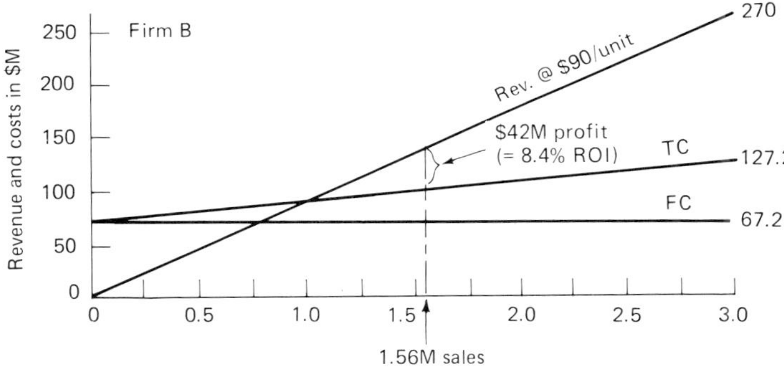

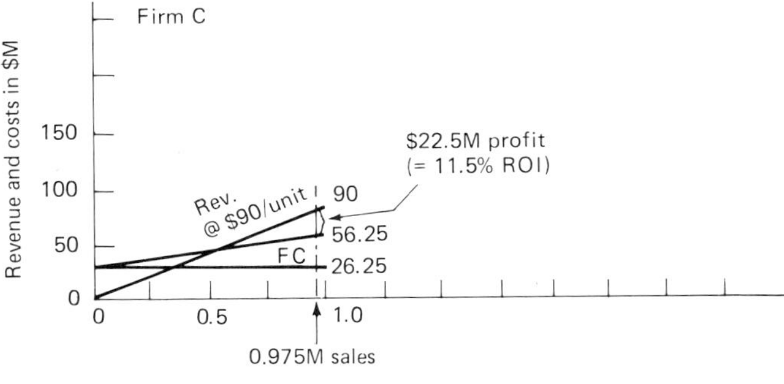

Figure 3.5 Breakeven chart 3.

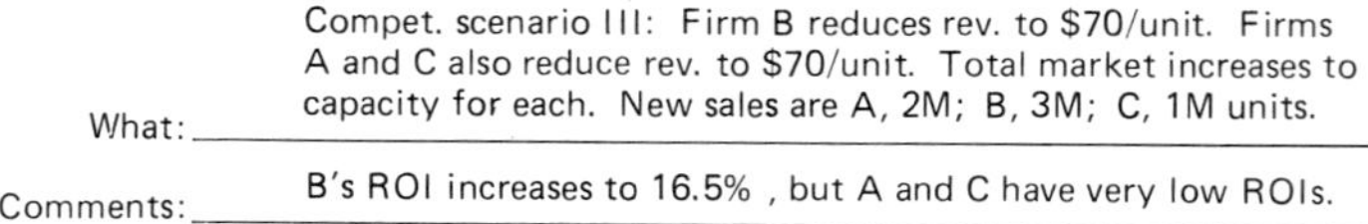
What: Compet. scenario III: Firm B reduces rev. to $70/unit. Firms A and C also reduce rev. to $70/unit. Total market increases to capacity for each. New sales are A, 2M; B, 3M; C, 1M units.

Comments: B's ROI increases to 16.5%, but A and C have very low ROIs.

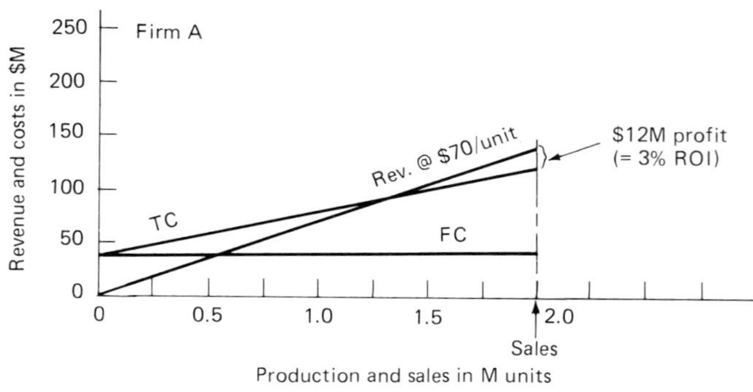

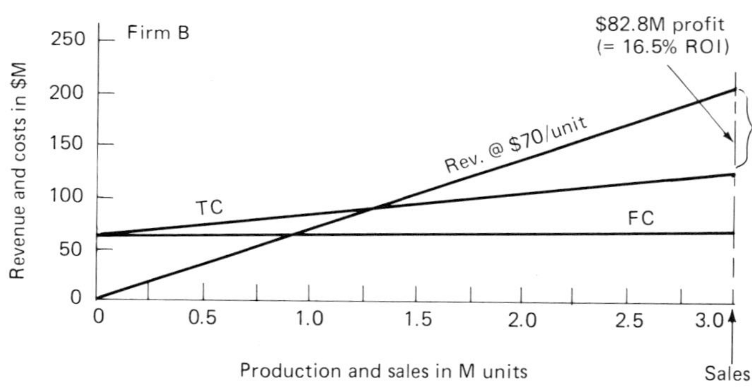

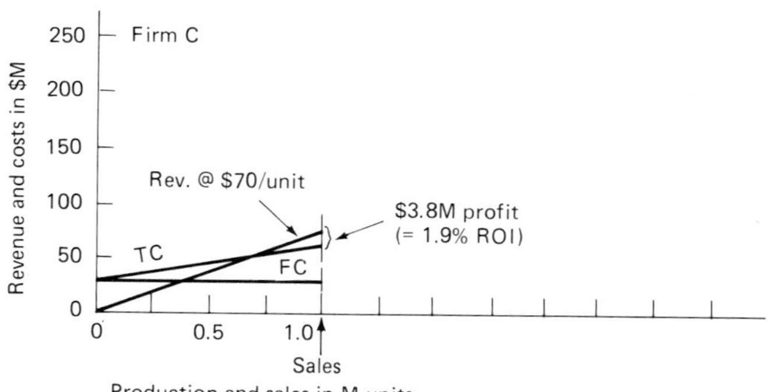

Figure 3.6 Breakeven chart 4.

11.5 percent, respectively. This will probably cause firm B to compete on price even further to increase its sales (and utilization of capacity).

Figure 3.6 (competitive scenario 3) assumes that firm B reduces revenue (price) to $70 per unit and firms A and C match it. This is assumed to result in all firms being able to sell their capacity outputs. The results are that firm B has a healthy 16.5 percent ROI and firms A and C have 3 and 1.9 percent ROIs, respectively. Firm B is now in command of the market—it is making a good profit, and its competitors are doing so poorly they would probably be reluctant or unable to automate themselves to become cost-competitive. Indeed, firm B might even reduce revenue (price) to less than $70 per unit to further squeeze the revenue and profits of its competitors and probably discourage new competitors from entering the market.

3.2.2 Avoiding competitive wipeout due to other factors

Although competitive cost and profit factors are often compelling reasons to invest in CIM, it should be recognized that companies compete on the basis of many other factors which may well be affected (aided) by the adoption of CIM systems compared to conventional (non-CIM) systems.

3.2.3 Difficulty in avoiding and/or reversing decline due to disinvestment; likely results

The irreversibility of investments and actions postponed too long is partly rooted in the dynamics of human organizations. Further, the cost of making up for lost opportunities gets greater. Faced with these circumstances, top management often concludes that a division or product line is unsalvageable and purposely continues the process of disinvestment (i.e., letting assets depreciate rather than updating). Thus, it eventually gets out of the business, often at a substantial loss (at least an opportunity loss compared to the future business it might have had).

The above points have been expanded upon well in several *Harvard Business Review* articles. For example, Hayes and Garvin [Ref. 5] say:

> It is both difficult and costly for a company to extract itself from such a spiral, for usually no single investment can repair the damage. . . . Managers must be willing to reinvest at the very time such action [may] appear least attractive. They must stop pouring funds into refurbishing their images and upgrade their factories instead. . . . Beyond all else capital investment represents an act of faith, a belief that the future

will be as promising as the present, together with a commitment to making that future happen.

3.2.4 Significant problems and limitations in the implementation of CIM

Even though a well-planned and well-executed approach to achieve CIM contributes to improved product quality and increased productivity, it is not a panacea. Installing part or all of a CIM system without first establishing the appropriate manufacturing disciplines may simply give the factory the ability to produce defective products in greater numbers and at higher rates than before. In other words, "don't computer-automate a mess!"

CIM typically requires quite a large capital investment commitment. The mechanical and electronic equipment probably will require expanded engineering and maintenance skills in the organization. There will be need for initial and ongoing training for managers as well as users.

In implementing CIM, there may be need to reorganize the plant and reschedule production. During the start-up period there probably will be productivity and production losses. Even after the system is implemented well, there may be a problem obtaining enough sales to utilize the (normally high fixed cost and low variable cost) system and people economically.

Finally, it should be emphasized that no substantially new system can be expected to work without the enthusiastic and communicated support of management. Plans should be worked out by knowledgeable people in the organization and endorsed and enthusiastically accepted by the people who are going to "make it happen."

Realistically, most firms cannot and should not try to implement a full CIM system without going through stages of investment, absorption, learning, and proving-in. Initially, management should think in terms of "What is the best springboard for the development of integrated systems?". It may well be a good idea to plan on what are classically called "islands of automation," then linkages between those islands, and then integration of the whole. In any case, this should be done with a road map of where the firm is heading to achieve integration eventually, while recognizing probable changes in product and available CIM technology.

3.2.5 Financial (accounting) models and ratios

Accounting results are normally used for after-the-fact analyses of what has happened. However, for at least big-picture strategic planning, it may be very useful to project results in terms of accounting statements and commonly used ratios to reflect roughly the firm's profitability, activity, liquidity, and/or other financial measures in commonly used terms.

Figure 3.7 shows an example of a very useful flow diagram format for

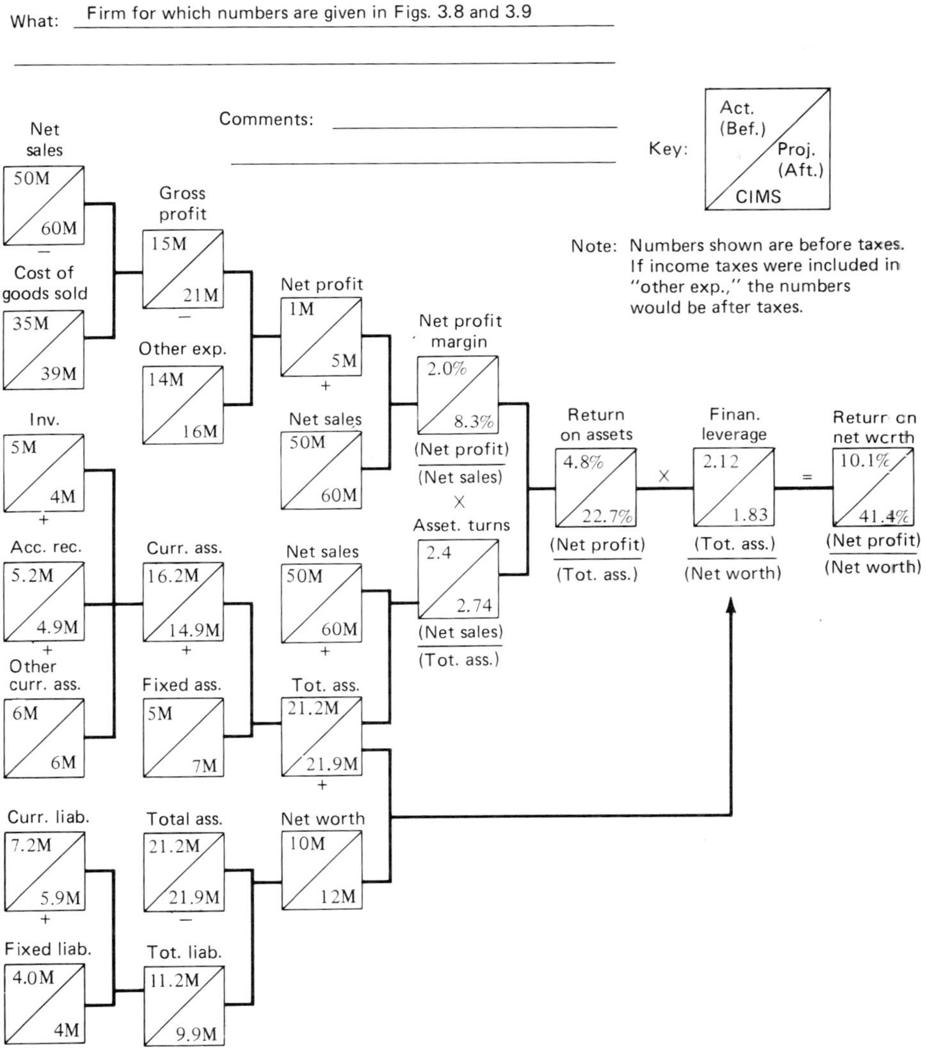

What: Firm for which numbers are given in Figs. 3.8 and 3.9

Comments:

Key: Act. (Bef.) / Proj. (Aft.) / CIMS

Note: Numbers shown are before taxes. If income taxes were included in "other exp.," the numbers would be after taxes.

Net sales 50M / 60M
−
Cost of goods sold 35M / 39M

Gross profit 15M / 21M
−
Other exp. 14M / 16M

Net profit 1M / 5M
÷
Net sales 50M / 60M

Net profit margin 2.0% / 8.3%
(Net profit) / (Net sales)
X
Asset. turns 2.4 / 2.74
(Net sales) / (Tot. ass.)

Inv. 5M / 4M
+
Acc. rec. 5.2M / 4.9M
+
Other curr. ass. 6M / 6M

Curr. ass. 16.2M / 14.9M
+
Fixed ass. 5M / 7M

Net sales 50M / 60M
÷
Tot. ass. 21.2M / 21.9M
÷
Net worth 10M / 12M

Curr. liab. 7.2M / 5.9M
+
Fixed liab. 4.0M / 4M

Total ass. 21.2M / 21.9M
−
Tot. liab. 11.2M / 9.9M

Return on assets 4.8% / 22.7%
(Net profit) / (Tot. ass.)
X
Finan. leverage 2.12 / 1.83
(Tot. ass.) / (Net worth)
=
Return on net worth 10.1% / 41.4%
(Net profit) / (Net worth)

Figure 3.7 Financial flow diagram for comparing actual (before) with projected (after). (*Adapted from John Shewchuk, "Justifying Flexible Automation,"* American Machinist, *October 1984, p. 95.*)

displaying actual (or before) and projected (or after) results. Each of the numbers shown in the upper left-hand corners of the boxes is the "before" for some time period or point in time, and each of the numbers in the lower right-hand corners is the "after" for some comparable later time period or point in time. The numbers are illustrative only; they do not correspond to numbers for firm B in Figs. 3.3 to 3.6. The numbers do show very favorable

	198X (before CIMS)	198X + 2 (projected, after CIMS)
Gross sales	$54.0M	$61.0M
Sales returns and allowances	4.0M	1.0M
Net sales	$50.0M	$60.0M
Cost of goods sold	35.0M	39.0M
Gross profit of sales	$15.0M	$21.0M
Expenses		
Selling	$ 2.0M	$ 2.0M
General and administrative	12.0M	14.0M
(incl. depreciation and interest)		
Net profit before taxes	$ 1.0M	$ 5.0M
Income taxes (at 50 percent)	0.5M	2.5M
Net profit after taxes	0.5M	2.5M
Number of common shares outstanding	1.0M	1.0M
Profit per share after taxes	$ 0.50	$ 2.50

Figure 3.8 Example income (P&L) statement for years ending December 31, ________.

	198X (before CIMS)	198X + 2 (projected, after CIMS)
Assets		
Cash	$ 5.0M	$ 5.0M
Marketable securities	1.0M	1.0M
Accounts receivable	5.2M	4.9M
Inventory	5.0M	4.0M
Total current assets	16.2M	14.9M
Net fixed assets	5.0M	7.0M
Total assets	$21.2M	$21.9M
Liabilities and Net Worth		
Accounts payable	$ 3.0M	$ 3.0M
Short-term bank loan	3.2M	1.9M
Accrued expenses	1.0M	1.0M
Total current liabilities	7.2M	5.9M
Long-term (fixed) liabilities	4.0M	4.0M
Total liabilities	$11.2M	$ 9.9M
Net worth		
Preferred stock	—	—
Common stock	5.0M	5.0M
Retained earnings	5.0M	7.0M
Total net worth	$10.0M	$12.0M
Total liabilities and net worth	$21.2M	$21.9M

Figure 3.9 Example balance sheets as of December 31, ________.

1. *Leverage ratios* indicate the extent to which a firm has financed its investments by borrowing. They reflect the financial structure of a firm and its risk posture.

Ratio		Formula	198X (before CIMS)		198X + 2 (projected, after CIMS)	
a. Debt-equity ratio	=	$\frac{\text{total liabilities}}{\text{net worth}}$	$\frac{\$11.2M}{10.0M}$	= 1.12	$\frac{\$\ 9.9M}{12.0M}$	= 0.83
b. Financial leverage	=	$\frac{\text{total assets}}{\text{net worth}}$	$\frac{21.2M}{10.0M}$	= 2.12	$\frac{21.9M}{12.0M}$	= 1.83

2. *Profitability ratios* describe the profitability of a company in the past. These are the measures most executives are evaluated against. Profitability ratios most often used are:

Ratio		Formula	198X (before CIMS)		198X + 2 (projected, after CIMS)	
a. Profit margin	=	$\frac{\text{net income}}{\text{net sales}}$	$\frac{\$\ 1.0M}{50.0M}$	= 2.0%	$\frac{\$\ 5.0M}{60.0M}$	= 8.3%
b. Return on assets	=	$\frac{\text{net income}}{\text{total assets}}$	$\frac{1.0M}{21.2M}$	= 4.8%	$\frac{5.0M}{21.9M}$	= 22.7%
c. Return on new worth	=	$\frac{\text{net income}}{\text{net worth}}$	$\frac{1.0M}{10.0M}$	= 10%	$\frac{5.0M}{12.0M}$	= 41.4%

3. *Activity ratios* provide clues to how efficiently a firm is managing its assets. Activity ratios used most often are:

Ratio		Formula	198X (before CIMS)		198X + 2 (projected, after CIMS)	
a. Inventory turnover	=	$\frac{\text{net sales}}{\text{inventory}}$	$\frac{\$50.0M}{5.0M}$	= 10	$\frac{\$60.0M}{4.0M}$	= 15
b. Collection period	=	$\frac{\text{accounts receivable(365)}}{\text{net sales}}$	$\frac{6.2M(365)}{50.0M}$	= 45	$\frac{4.9M(365)}{60.0M}$	= 30
c. Fixed assets turnover	=	$\frac{\text{net sales}}{\text{fixed assets}}$	$\frac{50.0M}{5.0M}$	= 10.0	$\frac{60.0M}{7.5M}$	= 8
d. Total assets turnover	=	$\frac{\text{net sales}}{\text{total assets}}$	$\frac{50.0M}{21.2M}$	= 2.4	$\frac{60.0M}{21.9M}$	= 2.7

4. *Liquidity ratios* measure the extent to which a company can meet its current obligations: bills to its suppliers, interest payments, payroll and operating expenses, etc. Commonly used liquidity ratios are:

Ratio		Formula	198X (before CIMS)		198X + 2 (projected, after CIMS)	
a. Current ratio	=	$\frac{\text{current assets}}{\text{current liabilities}}$	$\frac{16.2M}{7.2M}$	= 2.3	$\frac{14.9M}{3.9M}$	= 3.8
b. Quick ratio	=	$\frac{\text{current assets} - \text{inventory}}{\text{current liabilities}}$	$\frac{16.2M - 5.0M}{7.2M}$	= 1.6	$\frac{14.9M - 4.0M}{3.9M}$	= 2.8

Figure 3.10 Commonly used financial ratios (applied to example numbers in Figs. 3.8 and 3.9).

What: 40% incr. in fixed assets (to $7M)

20% decr. in inventory (to $4M)

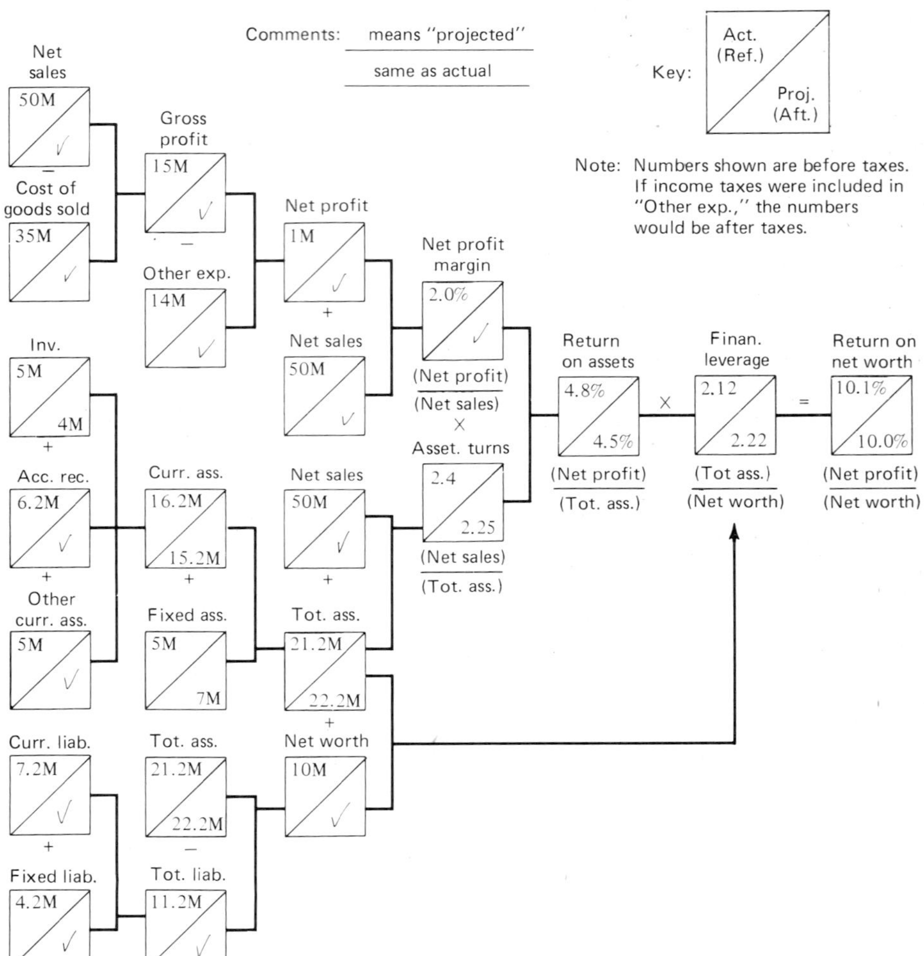

Figure 3.11 Financial flow diagram for comparing actual (before) with projected (after); Case 1. (*Adapted from John Shewchuk, "Justifying Flexible Automation,"* American Machinist, *October 1984, p. 95.*)

improvements, which are mostly due to increased sales with expenses increasing at a moderate rate.

Figures 3.8 and 3.9 show more detail on what was reflected in Fig. 3.7 in the form of income (profit and loss) statements and balance sheets, respectively. Note that both before CIM and after CIM figures are shown for each.

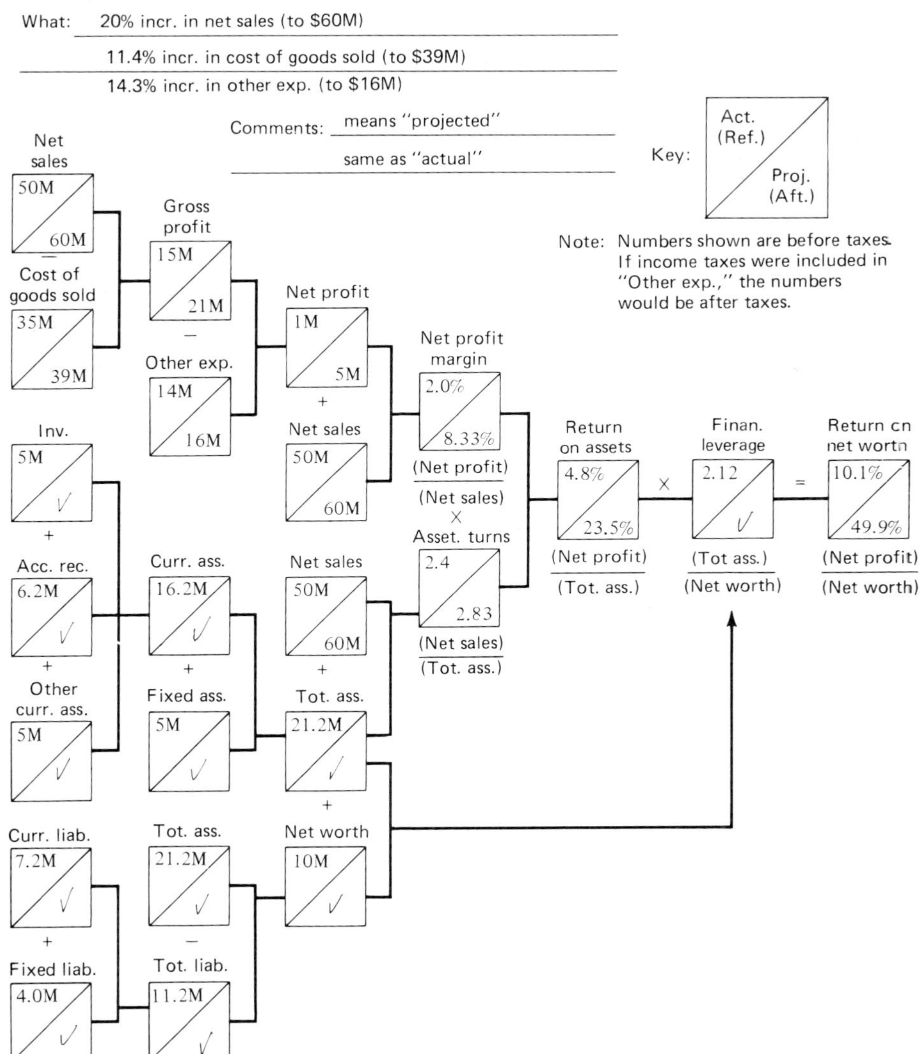

Figure 3.12 Financial flow diagram for comparing actual (before) with projected (after); Case 2. *(Adapted from John Shewchuk, "Justifying Flexible Automation,"* American Machinist, *October 1984, p. 95.)*

Figure 3.10 describes and illustrates some common ratios used to measure a firm's financial health. The ratios are both before CIM and after CIM using the numbers in Figs. 3.8 and 3.9. Analyses of trends or changes in these ratios for the firm compared to other firms in the industry, and/or for the firm itself over time, are considered very useful. Indeed, many of the ratios are also shown in the financial flow diagram in Fig. 3.7.

Management can use any combination of ratios to study what is of greatest interest at a given time. Depending upon the product life cycle, growth stage of the firm, and overall economic condition, the focus of management may shift from liquidity to leverage; from growth to maintenance of profit margins; or from favorable stock price to improvement in operating efficiency.

3.2.6 Use of financial flow diagram for specific or partial changes

The financial flow diagram illustrated in Fig. 3.7 can also be used for simplified or rough analyses of the effects of particular changes. To illustrate, Figs. 3.11 and 3.12 show relevant results for two cases based on the before (actual) figures given in Fig. 3.7 as follows:

Figure 3.11 (case 1) is for 40 percent increase in fixed assets and 20 percent decrease in inventory only. The resulting depreciation charge increase and inventory carrying cost decrease happen to cancel each other out with respect to other expenses. All other amounts are assumed to remain constant, so that the resulting return on net worth stays virtually constant.

Figure 3.12 (case 2) considers the effect of a 20 percent increase in net sales and perhaps plausible increases in expenses, but without any changes in the assets or liabilities of the firm. That results in a nearly fivefold increase in return on net worth.

Other cases or conditions can be studied at will. Figure 3.7 reflects a more realistic case than either of the above because it includes changes in sales, expenses, assets, liabilities, and net worth all considered together.

REFERENCES

1. J. R. Canada and J. A. White, *Capital Investment Decision Analysis for Management and Engineering*, Prentice-Hall, Englewood Cliffs, N.J., 1980, pp. 208–209.
2. D. Scott Sink, "Using the Nominal Group Technique Efficiently," *National Productivity Review*, September 1983, pp. 173–184.
3. Bela Gold, "CAM Sets New Rules for Production," *Harvard Business Review*, November–December 1982, pp. 88–94.
4. H. Bierman and S. Smidt, *The Capital Budgeting Decision*, 4th ed., Macmillan, New York, 1975, pp. 83–84.
5. Robert H. Hayes and D. A. Garvin, "Managing as if Tomorrow Mattered," *Harvard Business Review*, May–June 1982, pp. 71–79.

ANNOTATED BIBLIOGRAPHY

This is a very extensive, but not exhaustive, bibliography for persons interested in the economic and/or noneconomic justification of part or all of what is normally associated with CIM systems. Most works cited include brief annotations on subject inclusions and perhaps perceived strengths.

Parts 1 and 2,* on books and symposia, respectively, are mostly on CIM systems in general rather than on justification of CIM systems in particular. Most were not reviewed by this author/compiler (and hence not annotated), but they are included because many are thought likely to have some material on justification. Further, it is thought that the reader may well be interested in other aspects of CIM systems covered in the books and symposia.

Part 3 contains some 52 articles which are generally subjective overviews or guidelines on justification of part or all of CIM systems. Most are very good, and some are outstanding. A general gauge on reviewed quality and applicability can be obtained from the annotations. Of particular note are many of the articles in *Harvard Business Review*.

Part 4 contains some 35 articles which include quantitative techniques and analysis methodologies for justification of—usually of parts of—CIM systems. Of particular note are the works by Gustavson and Boothroyd. Probably the most comprehensive, yet oversimplified, economic analysis approaches somewhat paralleling parts of this chapter are by Meyer [Ref. 105] and General Dynamics Corporation [Ref. 95].

Part 5 lists selected books in the general area of engineering economics and decision analysis. These are not focused on CIM systems in particular; they are given as further backup to the analysis methodologies herein.

Perhaps no bibliography is complete without citing other bibliographies of references at least on the periphery of the subject. In addition to the very extensive 1983 IBM work cited above are the bibliographies in the table on page 4.436.

1 CIM-Related Books

1. Boehm, Barry W., *Software Engineering Economics*, Prentice-Hall, Englewood Cliffs, N.J., 1981, 767 pp.
(*First extensive book on this subject. Good for text and a basis for estimating software costs. Examples are mostly oriented toward general decision support systems rather than CIM.*)
2. Boothroyd, G., *Automatic Assembly*, Dekker, New York, 1982. TJ1317.B66.
3. Boothroyd, G., and P. Dewhurst, *Design for Assembly Handbook*, Automatic Assembly Program, M.E. Dept., University of Massachusetts, Amherst.
(*Not reviewed, but contains much of the material in other articles by Prof. Boothroyd. Provides a database for rapid estimation, gives a numerical rating for assembly efficiency, helps reduce manufacturing costs, etc.*)
4. Chasen, S. H., and J. W. Dow, *The Guide for the Evaluation and Implementation of CAD/CAM Systems*, CAD/CAM Decisions Co., Atlanta, Ga., 1980.
5. Dornfield, D. A., (ed.), *Automation in Manufacturing, Systems, Processes, and Computer Aids*, PED Series Vol. 4, ASME, New York, 1981.

*Most of the references in parts 1 and 2 were obtained from "CAD/CAM: A Bibliography" by A. H. Agajanian of the IBM Corp. General Technology Division (Technical Report 22.250, dated May 11, 1983). The report is outstandingly comprehensive with 549 references in some 30 subject categories.

Generic subject	Author	Number	Date	No. of works cited
Replacement decisions in: high-technology industries	Sullivan	111	1984	26
Economics of robots	Fleischer	52	1982	36
Productivity measurement and weighted evaluation	Sink	110	1984	21
Economics of robots	General Dynamics Corp. (for Air Force)	95	1980	83
Assembly automation economics	Boothroyd	40	1984	17
Software costs for manufacturing control systems	Hutchinson and Clementson	63	1984	33

6. Foundyller, C. M., *Turnkey CAD/CAM Computer Graphics: A Survey and Buyer's Guide*, Daratech Associates, Cambridge, Mass., 1980, 3 vols.

7. Grover, M. P., *Automation, Production Systems and Computer-Aided Manufacturing*, Prentice-Hall, Englewood Cliffs, N.J., 1980.

8. Grover, M. P., and E. W. Zimmers, *CAD/CAM: Computer-Aided Design and Manufacturing*, Prentice-Hall, Englewood Cliffs, N.J., 1984.

9. Gunn, T. G., Jr., *Computer Applications in Manufacturing*, Industrial Press, New York, 1981. TS155.G924.

10. Hatvany, J., M. E. Merchant, K. Rathmill, and H. Yoshikawa, *World Survey of CAM*, Butterworth, Kent, England, 1983.

11. Hayes, Robert H., and S. C. Wheelwright, *Restoring Our Competitive Edge: Competing through Manufacturing*, Wiley, New York, 1984.
(*Excellent definitive work based on years of research and writing.*)

12. Hunt, W. Daniel, *Industrial Robotics Handbook*, Industrial Press, New York, 1983.
(*Chapter 9, pp. 150–158, is on the economics of robotics. Example uses format and categories developed by General Dynamics Corporation for Air Force ICAM project. See 95.*)

13. Hunter, R. P., *Automated Process Control Systems: Concepts and Hardware*, Prentice-Hall, Englewood Cliffs, N.J., 1978. TS156.8.H86.

14. Hyer, N. L., (ed.), *Group Technology at Work*, SME, Dearborn, Mich., 1984, 254 pp.

15. Infotech Limited, *Factory Automation, State of the Art Report. 1. Analysis*, Infotech Limited, Maidenhead, England, 1980.

16. Kafrissen and M. Stephans, *Industrial Robots and Robotics*, Reston Publ. Co., Reston, Va., 1984.
(*Chapter 11, pp. 303–315, contains good statements about justification for robots.*)

17. Konen, Yoram, *Computer Control of Manufacturing Systems*, McGraw-Hill, New York, 1983, 287 pp.

18. Krouse, J. K., *What Every Engineer Should Know about Computer-Aided Design and Computer-Aided Manufacturing: The CAD/CAM Revolution*, Dekker, New York, 1982.

19. Pressman, R. S., and J. E. Williams, *Numerical Control and Computer-Aided Manufacturing*, Wiley, New York, 1977. TJ1189.P728.

20. SME, *CAD/CAM*, Dearborn, Mich., 1983, 587 pp.

21. Tanner, W., (ed.), *Industrial Robots, Fundamentals*, 2d ed., SME, Dearborn, Mich., 1981, 445 pp.
22. Tanner, W., (ed.), *Industrial Robots, Applications*, 2d ed., SME, Dearborn, Mich., 1981, 484 pp.
23. Taraman, Khalil, *CAD/CAM Integration and Innovation*, SME, Dearborn, Mich., 1984, 428 pp.
24. Wang, P., (ed.), *Automation Technology for Management and Productivity Advancements through CAD/CAM and Engineering Data Handling*, Prentice-Hall, Englewood Cliffs, N.J., 1983.

2 CIM-Related Symposia

25. Autofact West Proceedings, Anaheim, Calif., November 17–20, 1980; SME, Dearborn, Mich., 1980, 2 vols.
26. *Autofact 6 Proceedings*, 1984, SME, Dearborn, Mich., 1280 pp.
27. *CAD/CAM as a Basis for Development of Technology in Developing Nations, Proceedings of the IFIP WG5.2 Working Conference*, J. L. Encarnacao, O. F. F. Torres, and E. A. Warman, (eds.), Sao Paulo, Brazil, October 21–23, 1981; North-Holland, New York, 1982. CCA17-2820.
28. *Conference on Computer-Aided Manufacturing and Productivity, COMPRO 81*, London, October 21, 22, 1981; Institute of Production Engineering, London, 1981.
29. *Conference on Computer Graphics in CAD/CAM Systems, 1st, Proceedings*, Cambridge, Mass., April 9–11, 1979; MIT Press, Cambridge, Mass., 1979.
30. *Eleventh Conference on Production Research and Technology*, Carnegie-Mellon, 1984, SME, Dearborn, Mich., 432 pp.
31. *International Conference Cybernetics Society, Proceedings*, Cambridge, Mass., October 8–10, 1980; IEEE, New York, 1980.
32. *International Conference on Flexible Manufacturing Systems, 1st, Proceedings*, Brighton, England, October 20–22, 1982; IFS (Conferences), Ltd., Kempston, England, 1982. CCA18-4786.
33. *Man-Machine Communication in CAD/CAM, Proceedings of the IFIP WG5.2-5.3 Working Conference*, Tokyo, October 2–4, 1980; North-Holland, New York, 1981, CCA17-28129.
34. *NSF Grantees' Conference on Production Research and Technology, 8th*, Stanford, Calif., Jan. 27–29, 1981, T. Binford (ed.), NTIS Report PB81-183964.
35. *Numerical Control Society Annual Meeting and Technical Conference*, Pioneering in Technology . . . Building for the Future, 17th, 1980; Numerical Control Soc., Glenview, Ill., 1980.
36. *Robots 8 Conference Proceedings*, 1984, SME, Dearborn, Mich., 1800 pp.
37. Striving for Technological Excellence in Manufacturing through Communication, Planning, and Professionalism: *Proceedings of International Technical Conference, Numerical Control Society, 18th*, 1981; Numerical Control Soc., Glenview, Ill., 1981.

3 Articles on Justification of CIM; Generally Subjective Overview and Guidelines

38. Ayres, Robert U., and S. Miller, "Robotics, CAM, and Industrial Productivity," *National Productivity Review*, Winter 1981–1982, pp. 42–60.
(Overview on the impacts of "flexible" computer-aided technologies on the economics of batch production.)

39. Boothroyd, G., "Design for Productivity—The Road to Higher Productivity," *Assembly Engineering*, March 1982, 4 pp.
(*Early overview article on assembly savings possible through product redesign.*)

40. Boothroyd, G., "Use of Robots in Assembly Automation," apparently a keynote paper for (*Annals of ?*) *C.I.R.P. General Assembly*, 1984, 20 pp.
(*Seems to be a classic in describing various types of automated assembly systems with numerous illustrative diagrams and graphs on relative economics. Prof. Boothroyd is in the M.E. Dept. at the University of Massachusetts. Contains 17 references.*)

41. Boothroyd, G., and P. Dewhurst, "Computer-Aided Design for Assembly," *Assembly Engineering*, February 1983, 5+ pp.
(*Describes software programs available for Apple II Plus and IBM PC microcomputers. Six separate programs seem to be well developed.* Note: *The programs, used by 40+ firms, are available from Boothroyd and Dewhurst, 56 Sherman Lane, Amherst, MA 01002.*)

42. Boothroyd, G., and P. Dewhurst, "Design for Assembly," *Machine Design*, 1984, 22+ pp.
(*Reprints of four very well presented articles on the following topics: selecting the right method for manual assembly, automatic assembly, and Robot assembly and software systems.*)

43. Curtin, Frank T., "New Costing Methods Needed for Manufacturing Technology," *Management Review*, April 1984, pp. 29, 30.
(*Good, short overview. Mr. Curtin is General Manager of General Electric's Industrial Automation Systems Department.*)

44. Curtin, Frank T., "Planning and Justifying Automation Systems," *Production Engineering*, May 1984, 6 pp.
(*Excellently articulated philosophy of strategic considerations.*)

45. Devaney, C. William, "Building the Bridge between CAD/CAM/MIS," *Proceedings of Fall 1984 Conference*, Institute of Industrial Engineers, Atlanta.
(*Good general recognition of problems of CIM integrating including computer compatibility, database boundaries, and organizational relations.*)

46. Drozda, Thomas J., (ed.), "Robotized Multi-cells Best for Mid-value Output," *Robotics Today*, April 1983, pp. 53–56.
(*Gives good general example and philosophy.*)

47. DuPont-Gatelmand, Catherine, "A Survey of Flexible Manufacturing Systems," *Journal of Manufacturing Systems*, vol. 1, no. 1, 1982, pp. 1–15.
(*Rather definitive review and categorization of flexible manufacturing systems for machining and assembly used in the United States and 10 other major industrialized countries.*)

48. Eiler, R. G., W. T. Muir, and L. T. Michaels, "The Relationship between Technology and Cost Management," *Material Handling Engineering*, January 1984; "Improving Cost-Benefit Analyses," February 1984; "Cost Benefit Tracking Procedures," March 1984.
[*Suggests a series of steps that begins with factory modeling using the Air Force–developed integrated computer-assisted manufacturing definition (IDEF) modeling approach. IDEF is powerful for comprehensively describing functional activities as well as their complex interactions within engineering and manufacturing environments.*]

49. Eshleman, R. L., and F. S. Pagano, "What Managers Need to Know When Choosing Robots," *National Productivity Review*, Summer 1983, pp. 242–256.
(*General overview.*)

50. Evans, David, and P. C. Schwab, "Integrated Manufacturing-Financing: The Backing to Go Forward," *Production Engineering*, September 1984, pp. 122–124.
(*Recognizes that there are innovative financing vehicles to back modern technology investments.*)

51. Eversheim, W., and P. Hennman, "Recent Trends in Flexible Automated Manufacturing," *Journal of Manufacturing Systems*, vol. 1, no. 2, 1982, pp. 139–148.

(Good overview of degrees of manufacturing flexibility appropriate to volume and variability. Contains good graphical aids for analysis and planning.)

52. Fleischer, G. A., "A Generalized Methodology for Assessing the Economic Consequences of Acquiring Robots for Repetitive Operations," *Proceedings of May 1982 Annual Conference*, Institute of Industrial Engineers, Norcross, Ga., pp. 130–139. *(Good general overview. Contains 38 references and categorization of which advocate which method of analysis. Also contains an extensive breakdown of types of costs to be considered.)*
53. Foulkes, Fred. K., and J. L. Kirsch, "People Make Robots Work," *Harvard Business Review*, January–February 1984, p. 94–102. *("Most companies demonstrate concern for workers by assigning the first robots used to the 3 D's—dull, dirty, or dangerous—or the 3 H's—hot, heavy, and hazardous—jobs.")*
54. Gerwin, Donald, "Do's and Dont's of Computerized Manufacturing," *Harvard Business Review*, March–April 1982, pp. 107–116. *(One of the classic articles emphasizing importance of planning to take advantage of computers in process technology. Gerwin is Professor of Business Administration at University of Wisconsin-Milwaukee.)*
55. Gold, Bela, "CAM Sets New Rules for Production," *Harvard Business Review*, November–December 1982, pp. 88–94. *(Frequently cited classic that presents the case for serious consideration of computer-aided manufacturing process technology; gives particular recognition to benefits beyond those considered in usual financial evaluations.)*
56. Goldhar, Joel D., and M. Jelinek, "Plan for Economics of Scope," *Harvard Business Review*, November–December 1983, pp. 141–148. *(Substantial overview of importance of computers in manufacturing processing and good articulation of "scope" vs. "scale" economics. Goldhar is Dean of the School of Business Administration at IIT.)*
57. Grierson, Donald K., "Integrated Manufacturing—The Managers: Maestros or Magicians," *Production Engineering*, September 1984, pp. 38–43. *(Strong overview in re need to plan strategically about how to compete in the future and how to establish corporate architecture to implement. Grierson is a Senior Vice President of General Electric Co.)*
58. Grimm, James J., "Integrated Manufacturing—Technology's Tomorrow Just Arrived," *Production Engineering*, pp. 72–76. *(Overview with interesting "hierarchy of acronyms" and also forecasts in re availability of various types of manufacturing technologies.)*
59. Groover, Mikell P., "Meeting the Challenges of CAD/CAM," *National Productivity Review*, Winter 1982–1983, pp. 29–35.
60. Groover, Mikell P., and J. E. Hughes, "Job Shop Automation Strategy Can Add Efficiency to Small Operation Flexibility," *Industrial Engineering*, November 1981, pp. 67–74. *(General overview. Contains eight principles comprising a strategy for automating job shops.)*
61. Hayes, Robert H., and W. J. Abernathy, "Managing Our Way to Economic Decline," *Harvard Business Review*, July–August 1980, p. 67. *("Undue emphasis on analytic techniques bases decisions against the purchase of advanced manufacturing equipment that promotes long-term technical superiority.")*
62. Hayes, Robert H., and D. A. Garvin, "Managing as if Tomorrow Mattered," *Harvard Business Review*, May–June 1982, pp. 71–79. *(Frequently cited classic on the shortcomings of traditional investment analyses which fail to consider the effects of not investing in new, lower variable cost technology. Both authors are at Harvard Business School.)*
63. Hutchinson, George K., and A. T. Clementson, "Manufacturing Control Systems: An Approach to Reducing Software Costs," Presented at International Conference

on Manufacturing Science and Technology of the Future, M.I.T., Cambridge, Mass., October 9–12, 1984, 29 pp.
(*Shows that real-time control software costs and risk of failure to fulfill objectives are high. Suggests procedures to reduce problems. Contains 33 references.*)

64. Hyer, Nancy L., and V. Wemmerlov, "Group Technology and Productivity," *Harvard Business Review*, July–August 1984, pp. 140–149.
(*Light overview article on exploiting similarities in manufacturing problems or tasks.*)

65. Kaplan, Robert S., "Accounting and Control Systems for the New Industrial Competition," Working Paper 9-784-054, Harvard Business School, November 1983, 35 pp.
(*Scholarly work paralleling article of July–August 1984. See 66.*)

66. Kaplan, Robert S., "Yesterday's Accounting Undermines Production," *Harvard Business Review*, July–August 1984, pp. 95–101.
(*Excellent articulation of need to change outdated accounting and control systems to facilitate revitalization of manufacturing industries. Kaplan is at Harvard Business School.*)

67. Kester, W. Carl, "Today's Options for Tomorrow's Growth," *Harvard Business Review*, March–April 1984, pp. 153–159.
(*Presents framework for linking current capital budgeting decisions with strategic opportunities through recognizing different levels of competitive rivalry and options for the future.*)

68. Lambrix, Robert J., and S. S. Singhvi, "Preapproval Audits of Capital Projects," *Harvard Business Review*, March–April 1984, pp. 12–14.
(*Advocates and gives example of value of having multidisciplinary team independently study important investment proposals. Based on authors' work at Armco, Inc.*)

69. Levine, Susan J., and M. S. Yalowitz, "Managing Technology: The Key to Successful Business Growth," *Management Review*, September 1983, pp. 44–48.
(*Lists numerous benefits of technology management. Also outlines a process for technology management. Includes matrix of strategic options for various technologies and competitive positions.*)

70. Maller, Robert J., "Integrated Manufacturing—The Structure: Simplify and Focus, or Collapse," *Production Engineering*, September 1984, pp. 62–65.
(*Advocates focusing manufacturing operations with use of computer-aided technologies. Emphasizes need for discipline and attention to detail to implement.*)

71. Manchuck, Stanley, "Realistic Assessment of Prospects for Full CAD/CAM (CIMS) Integration," presented to 1984 Fall Conference, Institute of Industrial Engineers, October 1984, Atlanta, 12 pp.
[*Thoughtful discussion of difficulties of integration of engineering, manufacturing, and business systems. Reflects on needs for compatibility of data base(s) and communication system(s).*]

72. Michael, Gerald J., "Economic Justification of Modern Computer-Based Factory Automation Equipment," presented at first ORSA-TIMS Special Interest Conference on Flexible Manufacturing Systems, August 15–17, 1984, 14 pp.
(*Outstanding summary of inadequacies of traditional justification procedures. Good review of key literature.*)

73. McDonald, James L., and W. F. Hastings, "Selecting and Justifying CAD/CAM," *Assembly Engineering*, April 1983, pp. 24–27.
(*Describes a seven-step study process which begins with feasibility and needs analysis. Good articulation of practical considerations. The authors are with Productivity International, Inc.*)

74. National Research Council, *Computer Integration of Engineering Design and Production*, National Academy Press, November 1984, 62 pp.
(*Outstanding work of a blue chip committee on the CAD/CAM Interface of the Manufacturing Studies Board. Excellent expositions on benefits of CIM and issues*

that affect computer integration. Includes results of visits of five firms using CAD/CAM.)

75. Potter, Ronald D., "Analyze Indirect Savings in Justifying Robots," *Industrial Engineering*, November 1983, 3 pp.
(*Articulates sources and causes of various indirect savings.*)
76. Schmenner, Roger W., "Every Factory Has a Life Cycle," *Harvard Business Review*, March–April 1983, pp. 121–129.
(*Presents case for recognizing varying capabilities of plants over time and the need to plan for updating or otherwise taking advantage of capabilities.*)
77. Sepehni, Mehran, "Cost Justification before Factory Automation," *P & IM Review and APICS News*, April 1984, 3 pp.
(*Brief overview of costs and savings to be considered. Advocates breakeven analysis to study relative economics as function of volumes.*)
78. Shewchuk, John, "Justifying Flexible Automation," *American Machinist*, October 1984, pp. 93–96.
(*Recognizes need for nontraditional justification methods including risk simulation. Recognizes several pitfalls.*)
79. Shunk, Dan L., "Integrated Manufacturing-Factory Factors," *Production Engineering*, September 1984, pp. 50–53.
(*Advocates understanding ramifications of the factory of the future and quantifying parameters to compete.*)
80. Smith, Robert D., "Measuring the Intangible Benefits of Computer-Based Information Systems," *Journal of Systems Management*, September 1983, pp. 22–27.
(*Advocates use of value analysis approach for determining how extensive a computer-based information system is justified. Uses probabilities and expected values of benefits. Contains extensive checklists of intangible benefits which lead to improved business performance.*)
81. Stauffer, Robert N., (ed.), "Equipment Acquisition for the Automated Factory," *Robotics Today*, April 1983, pp. 37–40.
(*Good brief statements in re benefits of flexible automation.*)
82. Steele, Lowell, "Manager's Misconceptions about Technology," *Harvard Business Review*, November–December 1983, pp. 133–140.
(*Details seven common unrealistic assumptions about technology which make managing innovation unnecessarily difficult. The author was in corporate research and development with General Electric Co. for many years.*)
83. Stobaugh, Robert, and P. Telesio, "Match Manufacturing Policies and Product Strategy," *Harvard Business Review*, March–April 1983, pp. 113–120.
(*Presents strong case for recognizing importance of manufacturing—particularly in meshing production competence with business strategy.*)
84. Thompson, Arthur A., Jr., "Strategies for Staying Cost Competitive," *Harvard Business Review*, January–February 1984, pp. 110–117.
(*Presents case for strategic cost analysis to keep firms out of competitive pricing trap because of high variable operating costs characteristic of obsolete facilities.*)
85. Van Blois, John P., "Strategic Robot Justification: A Fresh Approach," *Robotics Today*, April 1983, pp. 44–48.
(*Good in calling for consideration of strategic and tactical attributes and factors in justification analyses for robots.*)
86. Warnecke, H. J., and G. Vettin, "Technical Investment Planning of Flexible Manufacturing Systems," *Journal of Manufacturing Systems*, vol. 1, no. 1, pp. 89–98.
(*Excellent ideas in re strategic planning procedure, including computer programs for the selection of machine tools and simulation of the manufacturing process sequence. Excellent visuals to show assessment of alternatives.*)
87. Waterbury, Robert, "The Total Approach to Automated Assembly," *Assembly Engineering*, March 1984, pp. 46–49.
(*Emphasis on General Electric Co.'s extensive automation programs.*)

88. Wheelwright, Steven C., "Japan—Where Operations Really are Strategic," *Harvard Business Review*, July–August 1981, pp. 67–74.
(*One of the early substantial articles on the discipline and dedication of Japanese manufacturers which improves the value of strategic operating policies.*)

89. Wheelwright, Steven C., "Reflecting Corporate Strategy in Manufacturing Decisions," *Business Horizons*, February 1978.
(*One of early significant articles on the topic. Suggests ways to mesh corporate and manufacturing decisions.*)

4 Articles on Justification for CIM, Including Quantitative Techniques and Analysis Methodologies

90. Boczany, William J., "Justifying Office Automation," *Journal of Systems Management*, July 1983, pp. 15–19.
(*Advocates simplified economic analysis technique.*)

91. Boothroyd, G., "Economics of Assembly Systems," *Journal of Manufacturing Systems*, (SME/CASA), vol. 1, no. 1, 1982, pp. 111–125.
(*Rather extensive derivation of simple cost formulas and graphical comparisons for manual vs. automatic assembly vs. programmable assembly. Describes different equipment types.*)

92. Boothroyd, G., "Economics of General-Purpose Assembly Robots," *Annals of C.I.R.P.*, vol. 33, no. 1, 1984, 8 pp.
(*Describes study of the comparative economics of various automatic assembly systems using general-purpose assembly robots. Good, clear descriptions with example quantification and excellent graphs.*)

93. Boucher, Thomas O., and J. A. Muckstadt, "Cost Estimating Methods for Evaluating the Conversion from a Functional Manufacturing Layout to Group Technology," Working Paper 84-125, Department of Industrial Engineering, Rutgers Univ., October 1984, 22 pp.
(*Develops useful formulas for estimating reduction in inventories and savings in some overhead costs due to automated systems with faster lead times, shorter manufacturing runs, etc.*)

94. Conn, Henry P., "Productivity: How to Define, Measure, and Improve It," *Production*, March 1984, pp. 50–63.
(*Not particularly on CIM. Contains eight extensive checklists with system for numerical evaluation of productivity. Two checklists concern manufacturing engineering and systems.*)

95. General Dynamics Corporation, *ICAM Robotics Application Guide*, Technical Report for Air Force Wright Aeronautical laboratories, TR-80-4042, vol. II, April 1980, 138 pp.
(*Contains listings of costs and savings and a recommended form for calculating present worths. Also contains 83 references.*)

96. Grieve, R. J., P. H. Lowe, and M. P. Kelly, "Robots: The Economic Justification," *Proceedings of Autofact Europe Conference* (sponsored by SME), Geneva, Switzerland, September 1983, 12 pp.
(*Good overview, recognition of cost components affected, and simple examples.*)

97. Gulati, Vince, "Presenting Project Justification in a Language Top Management Understands," *Proceedings of Fall 1984 Conference*, Institute of Industrial Engineers.
[*Illustrates use of financial statements and key financial* (ratio) *measures of performance as means of selling major project proposals, regardless of other attributes considered.*]

98. Gustavson, Richard E., "Engineering Economics Applied to Investments in Auto-

mation," *Proceedings of Second Assembly Automation Conference*, Brighton, England, 1981.
(*Shows derivation of formulas and has numerous excellent graphs for judging project desirability for combinations of three different variables. Mr. Gustavson is with Charles Stark Draper Laboratory.*)

99. Gustavson, Richard E., "Choosing Manufacturing Systems Based on Unit Cost," *13th ISIR/Robots 7 Conference*, Chicago, April 17–21, 1983. (SME Paper MS 83-322)
(*Develops and shows characteristic unit cost vs. annual production volume curves for manual vs. programmable automation vs. fixed automation.*)

100. Gustavson, Richard E., "Computer-Aided Synthesis of Least Cost Assembly Systems," *14th International Symposium on Industrial Robots*, Gothenburg, Sweden, October 1984, 12 pp.
(*Describes a top-down approach using cost and capability data for suitable technology alternatives to determine best choices. Algorithm contains the rudiments of dynamic programming. Reflects on-going research results which are very promising.*)

101. Gustavson, Richard E., J. L. Nevins, D. E. Whitney, and J. M. Rourke, "Assembly System Design Methodology and Case Study," *Robots 8 Conference*, June 1984, 29 pp.
(*Describes methodology for integrating assembly system design with product design. Shows application for an automobile manual transmission shaft assembly.*)

102. Hines, John R., "Predicting IC Costs," *Semiconductor International*, June 1983.
(*Shows simple equation for relating factory yields and costs to IC cost.*)

103. Hutchinson, George K., "Production Capacity: CAM vs. Transfer Line," *Industrial Engineering*, September 1976.
(*Contains models and graphs for use in comparing alternatives with differing fixed and variable costs. Professor Hutchinson is in School of Business Administration at the University of Wisconsin—Milwaukee and has done substantial publishing in flexible manufacturing systems, including a 300+ item bibliography.*)

104. Hutchinson, G. K., and J. R. Holland, "The Economic Value of Flexible Automation," *Journal of Manufacturing Systems*, vol. 1, no. 2, 1982, pp. 215–227.
[*Describes dynamic computer simulation studies to place an economic value on flexibility by comparing flexible manufacturing systems to transfer lines considering multiple products and marketplace (uncertainties).*]

105. Meyer, Ronald J., "A Cookbook Approach to Robotics and Automation Justification," *Proceedings of Robots VI Conference*, March 2–4, 1982, SME, Dearborn, Mich., 29 pp.
(*Excellent set of estimating guidelines for direct and indirect savings and suggested format for one-year justification analyses. Mr. Meyer is with Rockwell International.*)

106. Miller, Richard K., "The Bottom Line—Justifying a Robot Installation," *Robotics World*, April 1983.
(*Simple recognition of main costs involved and example calculation of yearly cash flows and payback.*)

107. Ogden, Harry, "Justifying Assembly Automation," *Proceedings of 2d Int'l Conf. on Assembly Automation*, Brighton, England, May 18–21, 1981.
(*Overview with examples of commonly used cash flow methods for economic analyses.*)

108. Robinson, Robbie, "CAD—the Financial Aspects," *Design Engineering*, November 1982, 125+ pp.
(*Contains simple example of quantifying economics of computer-aided drafting systems.*)

109. Seed, Allen H., "Cost Accounting in the Age of Robotics," *Management Accounting*, October 1984, pp. 39–43.
(*Good on topic. Contains example of differences in cost allocations for man-paced

vs. machine-paced environment. Mr. Seed is Senior Consultant with Arthur D. Little, Inc.)

110. Sink, D. Scott, and S. J. DeVries, "An In-Depth Study and Review of State-of-the-Art and Practice in Productivity Measurement Techniques," *Proceedings of May 1984 Annual Conference*, Institute of Industrial Engineers, pp. 333–344. (*Not on CIM justification, per se, but is very good overview of multiattribute and criteria evaluation methodology using utility or rating concepts. Contains handy graphical and tabular formats and 21 references.*)
111. Sullivan, William G., "Replacement Decisions in High Technology Industries—Where Are Those Models When You Need Them?," *Proceedings of 1984 Annual Conference, Institute of Industrial Engineers*, May 1984, pp. 119–128. (*Very good overview with suggested decision flow diagram and quantitative examples. Details costs and savings types for industrial robots; contains 26 references, most of which are included herein.*)
112. Sullivan, William G., "Towards an Understanding of Project Justification in High Technology Ventures," *Annual Conference of American Society for Engineering Education*, Salt Lake City, June 27, 1984, 12 pp. (*Good recognition of the issues. Suggests weighted evaluation model with attributes very specific to manufacturing technology alternatives.*)
113. Van Blois, John P., "Economic Models: The Future of Robotic Justification," *Proceedings of the 13th Industrial Symposium on Industrial Robots*, Chicago, April 17–21, 1983. (*Emphasizes standard economic analysis methodology.*)

5 Books on Engineering Economics and Decision Analysis

114. Bierman, H., and S. Smidt, *The Capital Budgeting Decision*, 4th ed., Macmillan, New York, 1975.
115. Bussey, Lynn E., *The Economic Analysis of Industrial Projects*, Prentice-Hall, Englewood Cliffs, N.J., 1978.
116. Canada, J. R., and J. A. White, *Capital Investment Decision Analysis for Management and Engineering*, Prentice-Hall, Englewood Cliffs, N.J., 1980.
117. DeGarmo, E. P., W. G. Sullivan, and J. R. Canada, *Engineering Economy*, 7th ed., Macmillan, New York, 1984.
118. Fabrycky, W. J., and G. J. Thuesen, *Economic Decision Analysis*, 2d ed., Prentice-Hall, Englewood Cliffs, N.J. 1980.
119. Fleischer, G. A., (ed.), *Risk and Uncertainty: Non-deterministic Decision Making in Engineering Economy*, Publication No. 2 in the monograph series of the Engineering Economy Division, American Institute of Industrial Engineers, Inc., Norcross, Ga., 1975.
120. Grant, E. L., W. G. Ireson, and R. S. Leavenworth, *Engineering Economy*, 7th ed., Wiley, New York, 1984.
121. Oakford, R. V., *Capital Budgeting: A Quantitative Evaluation of Investment Alternatives*, Wiley, New York, 1970.
122. Tarquin, A. J., and L. T. Blank, *Engineering Economy: A Behavioral Approach*, McGraw-Hill, New York, 1976.
123. Taylor, G. A., *Managerial and Engineering Economy*, 3d ed., Van Nostrand Reinhold, New York, 1980.
124. Thuesen, H. G., W. J. Fabrycky, and G. J. Thuesen, *Engineering Economy*, 5th ed., Prentice-Hall, Englewood Cliffs, N.J., 1977.
125. White, J. A., M. H. Agee, and K. E. Case, *Principles of Engineering Economic Analysis*, 2d ed., Wiley, New York, 1984.

Index

For simplicity, the double-numbering system has not been used in the index. The first part of the double number, which designates the part, has been dropped, and only the second part, which designates the actual page number, has been used.

About the Editors

Eric Teicholz is president of Graphic Systems, Inc., a firm specializing in CADD consulting, market research, and software training. The company has done extensive consulting in the areas of system evaluation, procurement, and management for designers, drafters, and facility managers. The author of numerous articles and books, Mr. Teicholz has over 17 years of experience with computers and design. Now teaching part-time at MIT, he was formerly an associate professor at the Harvard University Graduate School of Design.

Joel N. Orr is president of Orr Associates, a consulting firm working with some of the world's leading manufacturers in the area of industrial automation. With more than 19 years of experience, he is widely regarded as one of the pioneers of the computer-integrated manufacturing concept. Dr. Orr is a popular lecturer on the subject of computer-assisted manufacturing.